L. Luciani and F. Piscioli (Eds)

Aspiration Cytology in the Staging of Urological Cancer

Clinical, Pathological and Radiological Bases

Forewords by G. D. Chisholm and J. T. Grayhack

With 171 Figures

Springer-Verlag
London Berlin Heidelberg New York
Paris Tokyo

Lucio Luciani, MD
Head, Department of Urology, S. Chiara Hospital, I-38100 Trento, Italy

Francesco Piscioli, MD
Institute of Anatomic Pathology, Laboratory Cytology, S. Chiara Hospital, I-38100 Trento, Italy

Front cover illustration: Part 'c' taken from Fig. 38.14a–c. Aspirate from benign periureteral stenosis contains loose clusters of fibroblastic cells (a), sometimes with a corkscrew shape (b), and fine uniform nuclear texture (c). (H & E, x 61)

British Library Cataloguing in Publication Data
Aspiration cytology in the staging of urological cancer: clinical, pathological and radiological bases.
1. Man. Urogenital system. Cancer. Diagnosis. Applications of cytology I. Luciani, L. (Lucio), *1937–*
II. Piscioli, F. (Francesco), *1947–* . 616.99'4607582

Library of Congress Cataloging-in-Publication Data
Aspiration cytology in the staging of urological cancer.
Includes bibliographies and index.
1. Genitourinary organs—Cancer—Diagnosis. 2. Biopsy, Needle. 3. Diagnosis, Cytologic. 4. Tumors—Classification.
I. Luciani, L. (Lucio), 1937– . II. Piscioli, F. (Francesco), 1947– . [DNLM: 1. Biopsy, Needle. 2. Cytodiagnosis.
3. Genital Neoplasms, Male—pathology. 4. Neoplasm Staging—methods. 5. Urologic Neoplasms—pathology.
WJ 160 A841]
RC280.G4A77 1988 616.99'4607582 87-32199

ISBN-13: 978-1-4471-1454-3 e-ISBN-13: 978-1-4471-1452-9
DOI: 10.1007/978-1-4471-1452-9

This work is subject to copyright. All rights are reserved, whether the whole or part of the material is concerned, specifically the rights of translation, reprinting, re-use of illustrations, recitation, broadcasting, reproduction on microfilms or in other ways, and storage in data banks. Duplication of this publication or parts thereof is only permitted under the provisions of the German Copyright Law of September 9, 1965, in its version of June 24, 1985, and a copyright fee must always be paid. Violations fall under the prosecution act of the German Copyright Law.

© Springer-Verlag Berlin Heidelberg 1988
Softcover reprint of the hardcover 1st edition 1988

The use of registered names, trademarks etc. in this publication does not imply, even in the absence of a specific statement, that such names are exempt from the relevant laws and regulations and therefore free for general use.

Product liability: The publisher can give no guarantee for information about drug dosage and application thereof contained in this book. In every individual case the respective user must check its accuracy by consulting other pharmaceutical literature.

2128/3916–543210

Forewords

By Geoffrey D. Chisholm

The title of this book is a landmark in the emphasis that should now be placed on aspiration cytology in urological cancer. We all spend much time and money in trying to achieve the highest possible accuracy in tumour staging, yet often become frustrated by the subsequent course of events. We can all quote examples of patients who have had an apparently highly successful curative cystectomy, only to return within a year with distant metastases.

In our pursuit of better staging there has been intense interest in the development of imaging techniques. Progress continues to be made in this field and will surely bring us closer to the best possible delineation of the tumour. However, we all recognise that there is a fundamental weakness in both imaging and clinical staging: we cannot say for certain just how far the tumour has spread. Only the histopathologist and the cytopathologist can give us the exact answer, provided that there is an adequate specimen.

This book provides a most comprehensive account of the role of aspiration cytology. The opening two sections give a sound basis for all the current techniques, describing both the advantages and limitations. The Editors are to be congratulated on securing contributions from so many international leaders in urological oncology. This book has a good sense of authority, based on the personal experience of these contributors. I am confident that this will be the standard text on aspiration cytology for many years to come, and urologists are indebted to the Editors for this worthy addition to their literature.

Western General Hospital, University of Edinburgh, G.D.H.
Edinburgh, Scotland
1988

By J. T. Grayhack

This effort by Doctors Luciani and Piscioli to summarise the status of aspiration cytology as a staging tool in patients with urological cancer is an important contribution. Current medical practice is characterised by a proliferation of treatment options for a wide variety of diseases. Increased experience has usually identified specific circumstances in which the potential for a satisfactory therapeutic response to a given drug or surgical regimen is maximised. Selection and implementation of appropriate treatment of a particular disease characteristically requires increasingly exact diagnostic information as well as assessment of the phase of the disease that is present when the diagnosis is made. These general observations are clearly applicable to patients with a wide range of malignant diseases, including patients with urological malignancy. Cytology has played an increasingly important role in the diagnosis of urological cancer. The accuracy of needle sampling of suspicious regions in the kidney, adrenal and prostate has increased and the reliability of interpretation of cytological material has improved. Cytological assessment also plays a major role in the diagnosis of carcinoma of the bladder. Use of cytological studies to assess the biological potential of a tumour is being explored and documented with enthusiasm. The importance of histological evidence to establish a diagnosis of urological cancer before initiating specific therapies has been accepted for some time. The role of histological assessment in identifying tumour site or stage has increased appreciably over the past two decades. To a large extent the use of histology to identify extraorgan disease has concentrated on histological studies of excised lymph nodes or careful evaluation of excised organs and their surrounding tissues. When treatment options were (or are) confined to either surgical excision or palliative therapy, surgical exploration that employed operative staging as an initial part of the procedure was accepted, although not without recognition that this practice was less than optimal. As effective chemotherapy for testicular tumours and now probably bladder tumours has developed, and the limits of the role of surgical excision in the presence of extraorgan disease such as cancer of the prostate have been defined, the importance of identifying distant tumour sites in particular, but also periorgan disease, has increased. The sensitivity and specificity of imaging techniques in efforts to identify tumour in extraorgan sites have been disappointing. Histological confirmation is almost always desirable and usually essential for soft tissue staging. The experience with fine needle aspiration for tissue sampling has demonstrated that it has very limited immediate risk and, at least at this time, does not have recognised risk of appreciable long-term undesirable side-effects such as seeding. Use of imaging procedures has enhanced the accuracy of site sampling. Increasing experience has refined cytological criteria and their application for recognition of malignancy. The necessity for minimally invasive, highly accurate means of assessing extraorgan disease both prior to and following therapy will undoubtedly increase. Consequently, the balanced information presented in this book will help us as we attempt to assess and treat our patients with urological malignancy with the least risk and the greatest hope of benefit. The information presented will also serve a very useful purpose as a source of data for those who wish to explore new techniques to improve our current staging practices.

Northwestern University Medical School,
Chicago, Illinois, USA
1988

J. T. G.

Preface

The advances registered in the diagnosis and management of urological neoplasms in the last few years have far outpaced those of the past. Modern techniques in diagnostic uroradiology and uropathology have made available an ever increasing mass of knowledge, the synthesis and successful application of which, on a clinical basis, is becoming more and more difficult. Furthermore, improvement in the treatment of urological cancer has generated a pressing need to identify the "true" stage of the neoplasm being treated, mainly because of the unreliability of clinical staging.

Aspiration cytology has historical roots in the medical field and, as regards Urology, it has been widely used in the diagnosis of prostatic carcinoma in Europe and is now being accepted in the USA as well. Contributions to the literature on cytological diagnosis of prostatic cancer are therefore numerous. However, far fewer and more incidental studies have been dedicated to the role of aspiration cytology in establishing the real extension of prostatic carcinoma and other urological tumours, even though this knowledge should be considered essential, especially for planning adequate surgical therapy or justifying more conservative management.

The aim of this book is to focus on the value and applications of aspiration cytology in the staging of urological neoplasms. The volume contains a considerable number of important and original contributions on this topic from experts throughout the world, including our own experience over the past 10 years at Santa Chiara Hospital in Trento. It represents the most complete and up-to-date knowledge of the possibilities of this diagnostic procedure in the field of urological oncology, providing a sound pathological basis for the identification of the true stage of malignancies discovered by means of clinical and radiological investigation.

Although a limited number of publications have indeed dealt with cytological diagnosis in urological oncology, each one has been concerned only with individual organs and, almost exclusively, with diagnosis of the primary tumour. To the best of our knowledge, the present volume is the first to deal with aspiration cytology in the staging of all the urological neoplasms and it attempts to gather together the most relevant material on this topic.

As regards the general layout of the book, certain sections are more extensive than others, as the greater part of the studies carried out in the leading urological, pathological and radiological institutions have been mainly on very common neoplasms, such as carcinoma of the prostate, and less frequently on others. We also considered that it was appropriate to report the sometimes conflicting opinions on certain aspects. To support the usefulness of this diagnostic technique in the staging of the most common urological neoplasms, such as prostatic and bladder carcinoma, basic introductory clinical, pathological and radiological concepts have been included in the relative chapters, with emphasis on the therapeutic implications of aspiration cytological findings.

We would like to express our gratitude to all the participating authors for their outstanding contributions, thus making it possible to realise a wish we have had for many years, and for helping us in the exhausting task of compiling a reference book which we hope will be of use to all those involved in the diagnosis and treatment of urological cancers.

Acknowledgements

We wish to thank Dr F. Coccarelli for his help in the revision of the manuscript, and Marta Melchiori and Iole Pedrotti for their invaluable secretarial skills. We would also like to express our gratitude to Springer-Verlag for the excellent production of the book.

Trento
1988

Lucio Luciani
Francesco Piscioli

Contents

Contributors

L. Alasio, MD
Division of Anatomic Pathology and Cytology, Istituto Nazionale per lo Studio e la Cura dei Tumori, Milan, Italy

L. Balzarini, MD
Division of Diagnostic Radiology, Istituto Nazionale per lo Studio e la Cura dei Tumori, Milan, Italy

C. Bartolozzi, MD
Department of Clinical Physiopathology, Section of Diagnostic Radiology, University of Florence, Florence, Italy

L. Boccon-Gibod, MD
Clinique Urologique, Hospital Cochin, Paris, France

T. A. Bonfiglio, MD, FIAC
Department of Pathology and Laboratory Medicine, University of Rochester Medical Center, Rochester, New York, USA

S. Bosetti, MD
Division of Urology, S. Chiara Hospital, Trento, Italy

R. A. Castellino, MD
Division of Diagnostic Radiology, Stanford University, School of Medicine, Stanford, California, USA

E. Ceglia, MD
Division of Diagnostic Radiology, Istituto Nazionale per lo Studio e la Cura dei Tumori, Milan, Italy

G. D. Chisholm, ChM, FRCS, FRCSEd
University Department of Surgery ChM, FRCS, FRCSE Urology, Western General Hospital, Edinburgh, Scotland, UK

R. Correa, MD
Section of Urology, Department of Surgery, The Mason Clinic, Seattle, Washington, USA

G. Costantin, MD
Laboratory Cytology, Tumour Institute, Padua, Italy

P. Dalla Palma, MD
Institute of Anatomic Pathology, University of Padua, Padua, Italy

S. J. Dan, MD
Department of Radiology, Mount Sinai Hospital and School of Medicine, City University, New York, USA

J. B. deKernion, MD
Department of Surgery, Division of Urology, UCLA School of Medicine, Center for the Health Sciences, Los Angeles, California, USA

S. C. Efremidis, MD
Department of Radiology, Hippocration General Hospital, Thessaloniki, Greece

A. C. von Eschenbach, MD
Department of Urology, M. D. Anderson Hospital and Tumor Institute at Houston, Houston, Texas, USA

P. L. Esposti, MD
Cytology Laboratory, Radiumhemmet, Karolinska Hospital, Stockholm, Sweden

G. L. Failoni, MD
Division of Urology, S. Chiara Hospital, Trento, Italy

M. A. Fallon, MD
Department of Pathology and Laboratory Medicine, University of Rochester Medical Center, Rochester, New York, USA

W. J. Frable, MD
Section of Surgical and Cytopathology, Virginian Commonwealth University, Health Sciences Division, Medical College of Virginia, Richmond, Virginia, USA

G. Gadeholt, MD
Department of Radiology, University of Rochester School of Medicine and Dentistry, Rochester, New York, USA

R. Giardini, MD
Division of Anatomic Pathology and Cytology, Istituto Nazionale per lo Studio e la Cura dei Tumori, Milan, Italy

L. Giuliani, MD
Institute of Urology, University of Genoa, Genoa, Italy

D. F. Gleason, MD, PhD
Department of Laboratory Medicine and Pathology, University of Minnesota, Fairview Hospital, Minneapolis, Minnesota, USA

J. H. Göthlin, MD
Department of Diagnostic Radiology, University of Bergen, Haukeland University Hospital, Bergen, Norway

J. T. Grayhack, MD
Department of Urology, Northwestern University Medical School, Chicago, Illinois, USA

W. J. Highman, MB, ChB, DCH, MRCPath, MIAC
The Institute of Urology, London, England, UK

E. Hoppe, MD
Institut für Röntgendiagnostik, Charité Hospital, Humboldt University, Berlin, German Democratic Republic

P. B. Jajodia, MD
Departments of Urology and Pathology, University of California Medical School and Veterans Administration Medical Center, San Francisco, California, USA

B-S. Jing, MD
Department of Radiology, M. D. Anderson Hospital and Tumor Institute at Houston, Houston, Texas, USA

M. J. Kellett, DMRD, MIAC
The Institute of Urology, London, England, UK

R. Kidd, MD
Department of Radiology, The Mason Clinic, Seattle, Washington, USA

W. W. Koontz Jr., MD
Division of Urology, Virginia Commonwealth University, Health Sciences Division, Medical College of Virginia, Richmond, Virginia, USA

L. Luciani, MD
Division of Urology, S. Chiara Hospital, Trento, Italy

M. Lüning, MD
Institut für Röntgendiagnostik, Charité Hospital, Humboldt University, Berlin, German Democratic Republic

G. Masini, MD
Department of Clinical Physiopathology, Section of Diagnostic Radiology, University of Florence, Florence, Italy

E. Menichelli, MD
Division of Urology, S. Chiara Hospital, Trento, Italy

H. A. Mitty, MD
Department of Radiology, Mount Sinai Hospital and School of Medicine of the City University, New York, USA

E. Mukamel, MD
Sackler School of Medicine, Tel Aviv University, Beilinson Medical Center, Department of Urology, Petah Tikva, Israel

R. Musumeci, MD
Division of Diagnostic Radiology, Istituto Nazionale per lo Studio e la Cura dei Tumori, Milan, Italy

P. Narayan, MD
Departments of Urology and Pathology, University of California School of Medicine and Veteran Administration Medical Center, San Francisco, California, USA

R. T. D. Oliver, MD
Department of Radiotherapy and Oncology, The London Hospital Medical College, and The Institute of Urology, London, England, UK

S. R. Orell, MD, FIAC
Department of Pathology, Flinders Medical Center, Bedford Park, South Australia, Australia

F. Pagano, MD
Institute of Urology, University of Padua, Padua, Italy

D. F. Paulson, MD
Division of Urologic Surgery, Duke University Medical Center, Durham, North Carolina, USA

R. Petrillo, MD
Division of Diagnostic Radiology, Istituto Nazionale per lo Studio e la Cura dei Tumori, Milan, Italy

M. Piazza, MD
Institute of Anatomic Pathology, University of Padua, Padua, Italy

S. Pilotti, MD
Division of Anatomic Pathology and Cytology, Istituto Nazionale per lo Studio e la Cura dei Tumori, Milan, Italy

F. Piscioli, MD
Institute of Anatomic Pathology, Laboratory Cytology, S. Chiara Hospital, Trento, Italy

G. D. Pond, MD
Department of Diagnostic Radiology, University of Arizona Health Sciences Center, Tucson, Arizona, USA

T. Pusiol, CT
Institute of Anatomic Pathology, Laboratory Cytology, S. Chiara Hospital, Trento, Italy

F. Rilke, MD
Division of Anatomic Pathology and Cytology, Istituto Nazionale per lo Studio e la Cura dei Tumori, Milan, Italy

K. H. Rothenberger, MD
Department of Urology, Ludwig-Maximilians University Hospital, Landshut, Federal Republic of Germany

R. Scaletscky, MD
Formerly of The London Hospital Medical College, London, England, UK now of Porto Alegre, Brazil

P. Scappini, MD
Division of Urology, S. Chiara Hospital, Trento, Italy

C. C. Schulman, MD
Department of Urology, Erasme Hospital, University Clinics of Brussels, Brussels, Belgium

C. Selli, MD
Department of Clinical Physiopathology, University of Florence, Florence, Italy

A. Steg, MD
Clinique Urologique, Hospital Cochin, Paris, France

R. Stein, MD
Departments of Urology and Pathology, University of California School of Medicine and Veterans Administration Medical Center, San Francisco, California, USA

C. Stephenson, MB, BS, MRCP
The London Hospital Medical College, London, England, UK

J. D. Tesoro Tess, MD
Division of Diagnostic Radiology, Istituto Nazionale per lo Studio e la Cura dei Tumori, Milan, Italy

J. S. Train, MD
Department of Radiology, Mount Sinai Hospital and School of Medicine, City University, New York, USA

N. Villari, MD
Department of Clinical Physiopathology, Section of Diagnostic Radiology, University of Florence, Florence, Italy

S. Wallace, MD
Department of Radiology, Division of Diagnostic Imaging, M. D. Anderson Hospital and Tumor Institute at Houston, Houston, Texas, USA

Z. Wajsman, MD
University of Florida, College of Medicine, Department of Surgery, Division of Urology, Gainesville, Florida, USA

E. Wespes, MD
Department of Urology, Erasme Hospital, University Clinics of Brussels, Brussels, Belgium

G. Williams, MS, FRCS
Department of Surgery, Hammersmith Hospital and Royal Postgraduate Medical School, London, England, UK

F. Zattoni, MD
Institute of Urology, Tumour Centre, University of Padua, Padua, Italy

J. Zornoza, MD
Department of Diagnostic Radiology, Danbury Hospital, Danbury, Connecticut, USA

Historical Background

J. Zornoza

In 1883 Leyden [1] performed the first percutaneous aspiration biopsy in order to diagnose infectious diseases of the lung. Menetrier [2] in 1886 diagnosed the first carcinoma of the lung by aspirating tissue through a cannula inserted transthoracically. This method was widely used during the last part of the nineteenth and the beginning of the twentieth centuries; however, serious complications arose with this technique because of the use of large needles and the poor quality of the radiological equipment used for the biopsies. The use of this technique was reported in the literature in England in 1909 by Horder [3] and since then has received intermittent attention, ranging from enthusiasm to condemnation. The main objections to the procedure included the associated complications, such as pneumothorax, haemoptysis and tumour dissemination, and the inadequacy of the specimens obtained. In the early 1930s large-scale use of aspiration lung biopsy was made by Martin and Ellis [4], who attached a syringe to the biopsy needle and called the technique aspiration biopsy. At the same time, progress was being made with different pathological techniques, which in most cases provided more precise information and made aspiration biopsy a less desirable method of diagnosis.

Recently, with the improvement of tumour localising methods, the availability of thin needles and increased sophistication of cytological techniques, aspiration biopsy has been extended to many other organs in the body (e.g. liver, pancreas, lymph nodes, prostate, thyroid). The area of application of aspiration biopsy is the diagnosis of palpable or radiographically visible lesions. The recent development of radiological techniques, such as image intensification, fluoroscopy, ultrasound and computed tomography, allows the identification in organs of lesions not previously detected. These techniques permit precise positioning of the needle, which also increases the safety of the procedure. In the beginning only organs with a difficult surgical access route were referred for biopsy. However, recently aspiration biopsy has been extended to organs from which surgical specimens can readily be obtained, such as the breast, lymph nodes and thyroid.

The achievements of Dr G. N. Papanicolaou in the development of cytological techniques have been decisive in the success of aspiration biopsy. Although cytological methods will not eliminate the need for histological diagnosis, experience over the past 10 years demonstrates that aspiration biopsy has a definite role in the diagnosis of neoplastic disorders. The main reason is the clinicians' desire for confirmation of a presumptive diagnosis before the treatment is considered.

Another factor that has contributed to the success of this technique is the extremely high cost of medical care. Time and money can be saved by this simple technique. Further improvements can do much to increase the accuracy of the procedure further and decrease the limited number of complications.

References

1. Leyden DO (1883) Ueber infectiöse Pneumonie. Dtsch Med Wochenschr 9:52
2. Menetrier P (1886) Cancer primitif du poumon. Bull Soc Anat (Paris) 4:643
3. Horder TJ (1909) Lung puncture: a new application of clinical pathology. Lancet II:1345–1350
4. Martin HE, Ellis EB (1930) Biopsy by needle puncture and aspiration. Ann Surg 92:169–181

SECTION I

General

Chapter 1

Aspiration Cytology in Urological Oncology—the Urologist's Viewpoint

D. F. Paulson

Aspiration cytology is a diagnostic technique which should become a part of the armamentarium of every practising urologist. The contributors to this timely volume exhaustively address the topic of aspiration cytology in the diagnosis and staging of urological malignancy. Their chapters adequately document that aspiration cytology not only has a role in the diagnosis of the primary urological malignancies but also has a specific role in determining the extent of spread without resorting to open surgical exploration. The data presented herein and the data which have accumulated elsewhere indicate that aspiration cytology: (1) is highly reliable when there is an acceptably low incidence of false positive results, (2) has been proved as safe as open surgical biopsy and (3) has a reasonably high yield with few false negative results. It may be instructive to examine the three specific areas in which aspiration biopsy has proved most valuable to the urological surgeon: the confirmation of questionable renal mass lesions, the diagnosis of prostatic adenocarcinoma and the determination of the spread of metastatic disease.

Management of Renal Mass Lesions

The necessity for using aspiration cytology to diagnose solid renal mass lesions is dependent upon the methodology used to identify the unknown renal mass lesion. The published series would indicate that aspiration cytology provides an accuracy of diagnosis in confirmation of malignancy which varies between 50% and 100%, with an average of approximately 80% [1–6]. A false positive biopsy result occurred in 6 of 137 patients (a 4% false positive rate), with false negative cytological findings in 24 of 196 patients or 13%. It must be recognised that the range of accuracy in these various reports is a function of the prebiopsy studies. Current practice within the USA uses computed tomography (CT) to confirm or deny the presence of a solid space-occupying lesion within the kidney with this study alone dictating surgical intervention. Only in those cases in which the diagnosis is equivocal would aspiration cytology sway the decision for surgical intervention. Thus, the frequency with which aspiration cytology would provide information that would promote or deny surgical intervention is quite small. Patients whose disease is suspected to be other than of renal cell origin and who would presumably be managed preferentially by non-surgical means are appropriate candidates for aspiration cytological diagnosis. A typical instance would be the individual with a renal mass lesion, the characteristics of which cause concern that the disease is not renal cell malignancy. Such a lesion would be the hypovascular lesion which might be of metastatic or haematogenous origin.

The problem of tumour dissemination along the needle tract in percutaneous biopsies of renal mass lesions has proven negligible [7–9]. No adverse effect on the 5-year survival has been demonstrated in patients with carcinoma of the kidney who undergo percutaneous biopsy in comparison with controls [10]. The recorded data would indicate that seeding along the needle tract most commonly occurs when the needle is larger than a 20 to 21 gauge (outer diameter of 0.8 mm) [11–13].

Diagnosis of Prostatic Malignancy

The accumulated experience would indicate that fine needle aspiration will diagnose prostatic adenocarcinoma with a false negative rate as low as 2%, a rate two to five times lower than that identified when fine needle aspiration is compared with core biopsy for the diagnosis of prostatic malignancy [14–16].

Thus the accumulated data strongly supports the enhanced accuracy of aspiration biopsy in identifying malignancy of the prostate over that of a core biopsy. The contributions within this volume reflect the risk experience reported elsewhere, that aspiration biopsy is associated with a decreased risk of bleeding and infection over that seen with core biopsy. The incidence of febrile reactions with or without systemic sepsis is 1% or less, in contrast to the approximately 5%–7% incidence of a febrile episode with sepsis after transrectal or transperineal core biopsy [17–24]. Problems with significant bleeding, into the rectum, the perineum or the bladder, are unreported with aspiration biopsy as opposed to core biopsy.

One area that has provided considerable difficulty is the identification of atypia on cytopathological examination of aspirated cells. However, recent data would indicate that the diagnosis of atypia is associated with malignancy at other sites within the prostate in a significant number of patients. When this diagnosis is obtained following aspiration biopsy, this calls for more aggressive sampling of the gland by either multiple core biopsies or transurethral resection with histopathological review [25–27]. Traditionally, histopathological grading of the prostatic primary has been used as a prognostic indicator of the ultimate impact of therapy. However, the data would indicate that the histopathological grading of the primary tumour, while providing a visual indicator of the statistical probability of disease extent, does not function as an accurate predictor of the impact of therapy when compared with the anatomical distribution of disease in the patient at risk. Thus, with the progressive development of biochemical and imaging modalities which will allow the physician to determine the extent of disease prior to treatment selection, the necessity to determine, with accuracy, the histopathological grade of the primary malignancy is lessened. Thus, this argument against aspiration cytologies is reduced.

Identification of Metastatic Disease

Aspiration cytology has been promoted as a diagnostic technique to determine the presence or absence of metastatic disease. The clinician should

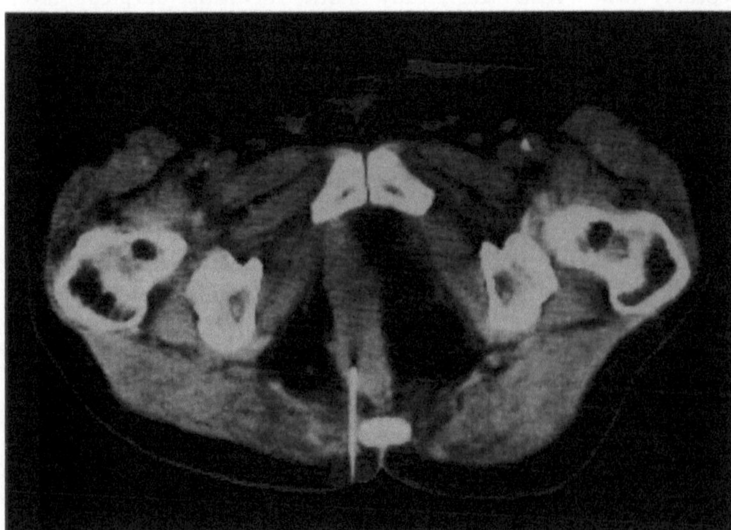

Fig. 1.1. CT scan showing placement of needle for aspiration biopsy in pelvic mass.

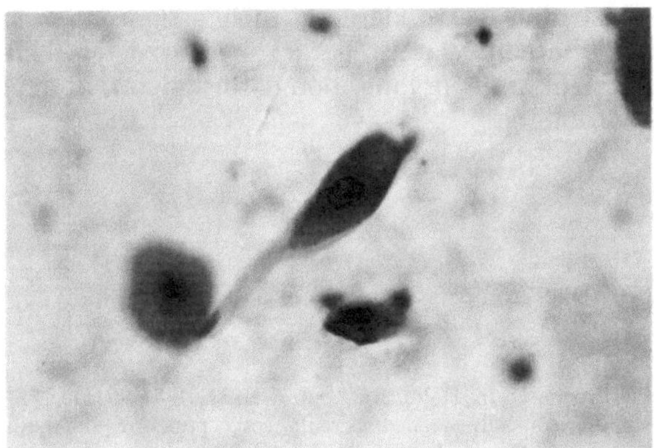

Fig. 1.2. Cytological preparation showing changes compatible with squamous cell carcinoma, similar to the bladder malignancy.

recognise that such diagnostic methodology must have the ability to diagnose the presence of metastatic disease with 100% accuracy if such diagnosis will influence the decision to proceed with a potentially curative therapy. Thus, aspiration cytology to determine metastatic nodal deposits can be safely used only when such biopsy can provide an accuracy rate of 100%. False positive biopsy results, in this situation, are unacceptable at any level. Aspiration biopsy has been promoted to diagnose the presence of nodal disease in prostatic malignancy. The accuracy rates are variable and they are a function of the imaging modality used to identify abnormal nodes. When lymphography or CT is used to identify abnormal nodes prior to proceeding with percutaneous aspiration biopsy, the clinician should recognise that the nodes are usually so abnormal as to permit their identification by these imaging modalities. Thus, the frequency with which a positive biopsy finding of an abnormal node occurs is greater than the frequency with which a positive aspiration finding occurs when the node is equivocal or normal. Thus, the use of aspiration biopsy to determine the presence or absence of disease in nodal structures which are normal by these same imaging criteria is improper. Use of this methodology to sample presumably normal nodes should be discouraged.

Percutaneous aspiration biopsy of abnormal structures to determine the presence or absence of malignant disease for treatment decisions provides the opportunity to confirm malignancy and select treatment without subjecting the patient to exploratory surgery. The following cases are illustrative of the impact of percutaneous biopsy.

Case 1: D.P. was a 61-year-old man who presented with haematuria and rectal bleeding. At the time of evaluation he was found to have a squamous cell carcinoma of the bladder and an adenocarcinoma of the rectum. He was subjected to a total pelvic exenteration. He reappeared 18 months later with a complaint of perineal pain. CT demonstrated a lesion low in the pelvis. The lesion could not be palpated through the perineum. A guided aspiration biopsy of this lesion returned the diagnosis of squamous cell carcinoma, thus permitting treatment to be applied on the basis of pathological diagnosis (Figs. 1.1, 1.2).

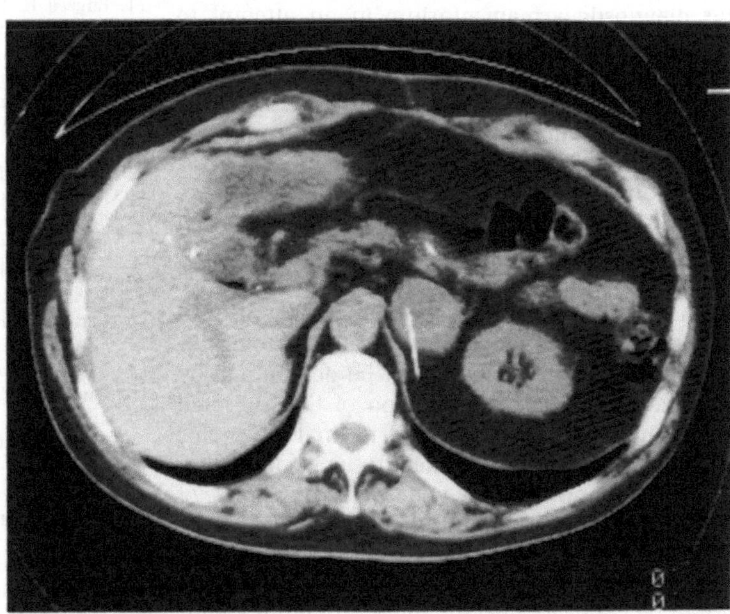

Fig. 1.3. CT scan showing placement of needle in left adrenal mass.

Case 2: T.B. was a 69-year-old man who 3 years earlier had undergone a right radical nephrectomy for renal cell carcinoma. CT follow-up 2 years later demonstrated a left suprarenal mass lesion. Diagnosis of this lesion was indeterminant and he underwent tomographically guided percutaneous aspiration biopsy (Figs. 1.3, 1.4). The diagnosis at this time was adrenal tissue without evidence of malignancy. He accordingly underwent exploration and was found to have an adrenal adenoma which was successfully removed. Eighteen months later he appeared with a lesion in his left chest. Percutaneous aspiration biopsy of the pulmonary nodule returned malignant cells compatible with renal adenocarcinoma.

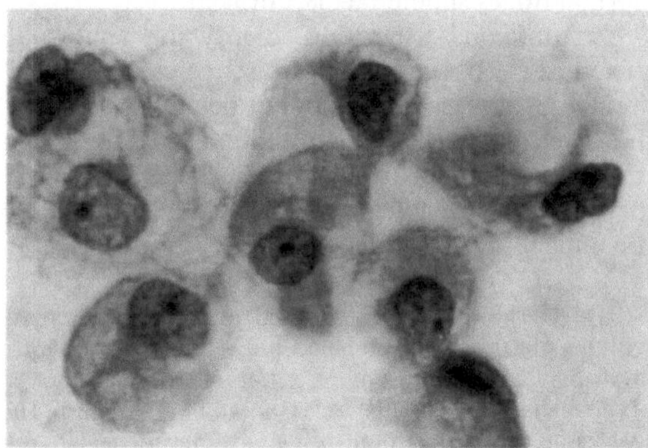

Fig. 1.4. Cytological preparation showing changes compatible with benign adrenal tissue.

These two cases illustrate the manner in which urological surgeons should use aspiration biopsy to diagnose recurrent or metastatic lesions in an attempt to make treatment decisions. The urological oncologist should incorporate aspiration biopsies in his diagnostic armamentarium in an attempt to provide information which will influence patient care decisions at minimal risk to the individual. Although current restrictions may result from the lack of a cytopathologist who can differentiate between inflammatory and malignant cells, this limitation is one which is demonstrably overcome with an extended experience and a determination to perfect these specialised skills. Furthermore, it may be that future improvements in flow cytometry or in the use of tumour-specific monoclonal antibodies will expand the role of aspiration cytology in detecting disease which cannot be determined by routine microscopic examination of the aspirated specimen. The identification of normal DNA patterns in the aspirated specimen or the identification of oncofetal antigens may widely expand the use of this staging methodology while enhancing accuracy and providing information of clinical benefit.

All the authors of chapters within this volume are to be congratulated for the thoroughness with which they have addressed their assigned topics, and the editors must be acknowledged for their foresight in the organisation of this important new contribution.

References

1. Kristensen JK, Holm HH, Rasmussen SN, Barlebo H (1972) Ultrasonically guided percutaneous puncture of renal masses. Scand J Urol Nephrol 6:49–56
2. Thornbury JR (1972) Needle aspirations of avascular renal lesions. Radiology 105:299–302
3. Thommesen P, Nielsen B (1975) The value of fine needle aspiration biopsy and intravenous pyelography in the diagnosis of renal masses. Fortschr Geb Rontgenstr Nuklearmed Erganzungsband 122:248–251
4. Stewart BH, Pasalis JK (1976) Aspiration and cytology in the evaluation of renal mass lesions. Cleve Clin Q 43:1–6
5. Karp W, Eklund L (1979) Ultrasound, angiography and fine needle aspiration biopsy in diagnosis of renal neoplasms. Acta Radiol 20:649–659
6. Helm CW, Burwood RJ, Harrison NW, Melcher DH (1983) Aspiration cytology of solid renal tumours. Br J Urol 55:249–253
7. Wehle MJ, Grabstald H (1986) Contraindications to needle aspiration of a solid renal mass: tumor dissemination by renal needle aspiration. J Urol 136:446–448
8. Gibbons RP, Bush WH, Burnett LL (1977) Needle tract seeding following aspiration of renal cell carcinoma. J Urol 118:865–867
9. McLoughlin MJ, Ho CS, Langer B, McHattie J, Tao LC (1978) Fine needle aspiration biopsy of malignant lesions in and around the pancreas. Cancer 41:2413–2419
10. von Schreeb T, Arner O, Skovsted G (1976) Renal adenocarcinoma. Is there a risk of spreading tumor cells in diagnostic puncture? Scand J Urol Nephrol 1:270–276
11. Engzell U, Jakobsson PA, Sigurdson A, Zajicek J (1971) Aspiration biopsy of metastatic carcinoma in lymph nodes of the neck. Acta Otolaryngol (Stockh) 72:138–147
12. Engzell U, Esposti PL, Rubio C, Sigurdson A, Zajicek J (1971) Investigation on tumor spread in connection with aspiration biopsy. Acta Radiol (Stockh) 10:385–398
13. Sinner WN, Zajicek J (1976) Implantation metastases after percutaneous transthoracic needle aspiration biopsy. Acta Radiol [Diagn] (Stockh) 17:473–480
14. Chodak GW, Steinberg GD, Bibbo M, Wied G, Straus FS II, Vogelzang NJ, Schoenberg HW (1986) The role of transrectal aspiration biopsy in the diagnosis of prostatic cancer. J Urol 135:299–302
15. Carter HB, Riehle RA, Koizumi JH, Amberson J, Vaughan ED (1986) Fine needle aspiration of the abnormal prostate: a cytohistological correlation. J Urol 135:294–298
16. Ljung BM, Cherrie R, Kaufman JJ (1986) Fine needle aspiration biopsy of the prostate gland: a study of 103 cases with histological followup. J Urol 135:955–958
17. Bissada NK, Rountree GA, Sulieman JS (1977) Factors affecting accuracy and morbidity in transrectal biopsy of the prostate. Surg Gynecol Obstet 145:869–872
18. Esposti PL (1974) Aspiration biopsy cytology in the diagnosis and management of prostatic carcinoma. Thesis, Stockholm: Stahl & Accidens Tryck, p 46
19. Thompson PM, Talbot RW, Packham DA, Dulake C (1980)

Transrectal biopsy of the prostate and bacteraemia. Br J Surg 67:127–128
20. Thompson PM, Pryor JP, Williams JP, Eyers DE, Dulake C, Scully MF, Kakkar WW (1982) The problem of infection after prostatic biopsy: the case for the transperineal approach. Br J Urol 54:736–740
21. Ruebush TK, McConville JH, Calia FM (1979) A double blind study of trimethoprim-sulfamethoxazole prophylaxis in patients having transrectal biopsy of the prostate. J Urol 122:492–494
22. Davison P, Malament M (1971) Urinary contamination as a result of transrectal biopsy of the prostate. J Urol 105:545
23. Wendel RG, Evans AT (1967) Complications of punch biopsy of the prostate gland. J Urol 97:122–126
24. Esposti PL, Elman A, Norlen H (1975) Complications of transrectal aspiration biopsy of the prostate. Scand J Urol Nephrol 9:208
25. McNeal JE (1965) Morphogenesis of prostatic carcinoma. Cancer 18:1659
26. Tannenbaum M (1975) Differential diagnosis in uropathology, carcinoma in situ of prostate gland. Urology 5:143–146
27. Oyasu R, Bahnson RR, Nowels K, Garnett JE (1986) Cytological atypia in the prostate gland: frequency, distribution and possible relevance to carcinoma. J Urol 135:959–962

Chapter 2

Percutaneous Needle Aspiration in Urological Oncology—the Radiologist's Viewpoint

G. D. Pond and R. A. Castellino

Introduction

The use of needles as a biopsy instrument can be traced to the mid nineteenth century, when needle aspiration was first used to obtain specimens under direct vision [1]. Since then, needles have been used successfully to perform biopsies on patients with suspected inflammatory or neoplastic disease in a wide variety of anatomical sites.

Radiological guidance for biopsy was initially done using fluoroscopy after intravenous injection of contrast media. Needle placement into the kidney for evaluation of renal cysts and solid tumours was facilitated. For diagnosing processes involving lymph nodes, either lymphoproliferative or metastatic in type, fluoroscopic guidance made it possible to obtain histological material from nodes opacified with lymphographic contrast. The routine use of these techniques has accelerated with the introduction of small gauge needles. Since biopsy sites may be adjacent to large vessels and since abdominal viscera are often traversed during needle biopsy, flexible needles with small diameters such as the Chiba needle were developed to reduce the risk of major complications.

Advances in imaging technology have further facilitated performance of needle aspiration biopsy and expanded the number of applications. Cross-sectional imaging using either ultrasound (US) or computed tomography (CT) has vastly improved our ability to perform needle localisations in three dimensions. The ease, speed, and safety of these procedures has now made fine needle aspiration biopsy (FNAB) commonplace. Today it is not unusual for needle aspirate to be obtained, stained, and interpreted in a matter of minutes, often on an outpatient basis.

Perhaps most gratifying to the radiologist performing needle aspiration is sparing patients the greater risk and discomfort of major surgery. Surgical procedures are often required in oncological staging not for any therapeutic purpose, but only for diagnosis and treatment planning.

Technique

After injection of local anaesthetic, a small skin nick is made. Introduction of the needle into the organ or mass to be biopsied is done in small increments under fluoroscopic control. With US and CT, the needle is placed in one step to a predetermined depth. A needle stop may be used. Position is confirmed fluoroscopically by obtaining two different views. "Shadowing" at the needle tip is used to confirm correct needle placement with US. On CT the needle tip can be identified and its position ascertained by proper slice selection. A syringe is then used to apply strong suction while the needle is manoeuvred up and down along its long axis through the lesion. Rotation of the needle is also used to break tissue fragments free from the biopsy "target".

Although cores of tissue (which are fixed immediately with formalin for *histological* examination) are occasionally obtained with small needles, even 23 gauge, more commonly the aspirate is extremely small in volume and contains small fragments of tissue which may even be too small to resolve with the naked eye. These aspirates are therefore best termed aspiration *cytologies* as opposed to core *biopsies* since they should not be fixed in formalin, but rather placed in ethanol, to preserve the cytological characteristics of individual cells and small clumps of cells.

Ideally, a portion of the specimen is immediately placed on a slide, smeared, and "quick stained". This specimen can be reviewed by a pathologist or experienced cytology technologist and an assessment of specimen adequacy made within minutes. If the specimen is inadequate, another needle pass can be made immediately, thereby avoiding a return visit by the patient. It is possible to obtain adequate material with no more than two needle passes in 90% of patients [2]. However, five or even more passes may be necessary to approach this yield, depending on the biopsy site [3]. Infrequently, a histospecific diagnosis is made from a single slide obtained on the first needle pass. More commonly, the material which was not immediately stained is spun down into a cell block, fixed, sectioned and permanently stained for further pathological review. The final pathological interpretation usually awaits review of this cell block.

There are a variety of needles available for FNAB. They vary in size from 23 gauge (Chiba needle) to 16 gauge (Turner needle). The most commonly used needles tend to be 20 gauge or smaller, since fewer complications are thought to occur with them [4]. However, larger needles may improve diagnostic yield, and therefore decrease the number of needle passes necessary. Fewer passes decrease the incidence of complications. A true "cutting" needle, designed to produce a core of tissue, can occasionally be used safely when larger specimens are required. However, some small-gauge needles are now available which improve the chances of obtaining a core specimen through modifications of the needle tip. Flat bevel needles, some with longitudinal notches, trephine needles, and needles with side slots have been developed and are available in 20 and 22 gauge [4]. This compromise of improved specimen size while maintaining a small needle size has proved quite popular.

A major consideration in selecting an appropriate needle is the type of pathological process suspected. Relatively small numbers of cells may be adequate to diagnose a carcinoma. However, larger tissue specimens are usually needed to diagnose lymphoma, since architectural detail is required to determine subtype, which may profoundly affect treatment. A core specimen might be necessary. If immunopathological techniques are contemplated, even larger specimens are required. In such instances, the radiologist should confer with the referring clinician and the pathologist to determine whether a needle biopsy is even appropriate, since a sufficient volume of tissue for definitive diagnosis may not be obtainable by FNAB.

Another consideration when choosing the appropriate needle is the approach to a suspected abnormality. The risk of complications during abdominal biopsy is minimal when using "skinny" needles with good flexibility. Even when transperitoneal and transhepatic approaches are used to perform FNAB the procedure is still extremely safe [2, 5–9]. Nonetheless, complications can occur and it therefore seems prudent to use approaches that minimise the risk of puncturing gut, solid abdominal viscera or great vessels, when access to the same area is possible by other routes. For example, a posterior, retroperitoneal approach to a lymph node is usually preferable to an anterior, transperitoneal route.

The risks of FNAB are small. Occasional cases of haemorrhage are reported [4, 9, 10]. Transpleural biopsy may result in pneumothorax [2, 3]. Another reported risk is "seeding" of the needle tract with malignant cells, causing spread of tumour or at least a potential site of recurrence [3, 11–14]. Fortunately, this complication is extremely uncommon.

The choice of an imaging technique for FNAB is usually based upon which modality provides best visualisation. In patients with renal masses, fluoroscopy after intravenous contrast (preferably with biplane or C-arm fluoroscopy) has historically been the commonest choice. The lesions are easily seen, and needle position can quickly be ascertained. Fluoroscopy is the usual choice for lymph node biopsy in patients who have undergone lymph(angi)ography (LAG), since biplane fluoroscopy allows meticulous needle positioning and confirmation of intranodal placement. This is the best approach to minimally enlarged or normal size lymph nodes which are suspicious in appearance or clearly have abnormal internal architecture.

Ultrasonography and CT can also be used for adrenal, renal and lymph node biopsy [2–5]. Modern ultrasound equipment usually provides excellent visualisation, does not require intravenous contrast or ionising radiation, and is less expensive than CT. In patients with little body fat, US is often superior to CT for identifying masses and may be the modality of choice in athletic, asthenic or cachectic patients. CT is preferable in obese patients or in patients with considerable abdominal gas, who are difficult to study with US. US remains an extremely operator-dependent cross-sectional imaging technique. The skill and experience of the ultrasonographer should be taken into account when choosing this imaging method.

Computed tomography is perhaps the most versatile technique for guidance during needle biopsy. Lymph nodes can be identified in locations not opacified by LAG, such as renal and splenic hili, mesentery, porta hepatis, and retrocrural space. Adrenal and renal lesions are seen with ease. Internal characteristics of masses, such as necrotic centres, can be identified and avoided so that the biopsies are obtained from a viable area in a given lesion. Larger needles can sometimes be used since the aorta, inferior vena cava, and iliac vessels are visualised after intravenous contrast and can therefore be avoided. Needle location is confirmed by obtaining the appropriate CT slices before a specimen is taken. Limitations of CT include (1) difficulty in imaging individuals with little body fat, (2) longer time periods required for biopsy compared with fluoroscopy and US and (3) lack of "real-time" imaging during the procedure.

Magnetic resonance imaging (MRI) provides cross-sectional images similar to CT, but can display anatomy in sagittal and coronal, as well as axial orientation. This may prove to be an important advantage in the future, but for now the logistical problems of MRI, including lengthy scan times, limited access to scanners, and the ferromagnetic materials used in biopsy needles, prevent routine clinical use.

Results

The objective of FNAB is recovery of a specimen adequate to provide a specific histological diagnosis. This is dependent on the technical skill of the radiologist and the diagnostic skill of the pathologist. The diagnostic yield varies depending on the biopsy site, as access to some anatomical locations is more technically difficult. Similarly, some pathological processes are challenging to the pathologist, even with "adequate" specimens. "Accuracy" of the technique is the percentage of patients undergoing the procedure in whom the correct diagnosis is made. To make the correct diagnosis, "adequate" specimens must be obtained by the radiologist. Once this is accomplished, the pathologist must correctly interpret the specimen.

Partly because of this shared responsibility in guaranteeing that FNAB is highly accurate, the terms "diagnostic material" and "adequate specimen" have created some confusion in the literature. Usually it can be assumed that these terms indicate that a known and previously visualised pathological process has been biopsied and that the samples are "diagnostic" for a specific histological diagnosis. In that situation an "adequate" specimen would be synonymous with the term "diagnostic". This would reflect "accuracy" in the statistical sense. Unfortunately, it is not always clear whether "inadequate" specimens (i.e. "acellular", "blood only", or "normal liver" on a transhepatic adrenal biopsy) have been included or excluded in the various calculation of accuracy. Most data suggest that the yield of "adequate" material by FNAB of kidney, adrenal gland, and lymph nodes averages about 90% [2, 3, 8, 9, 15]. When "adequate" specimens are provided for pathological review, a similar percentage of specific diagnoses are made. Sometimes, no pathological diagnosis results, even from an "adequate" (i.e. cellular) specimen. The normal parenchyma of an organ adjacent to a tumour mass may be inadvertently sampled. One area of the organ may prove negative for metastases while all the adjacent parenchyma is extensively involved. Pathological evaluation of "adequate" specimens can also be compromised by gut or other viscera traversed during a biopsy. Normal cells from gut admixed with nodal cytological material could create confusion. Keeping the stylet in place until the target is entered diminishes this problem, as does terminating the aspiration prior to needle withdrawal. Finally, sometimes the small size of FNAB specimens makes it impossible to reach a definitive pathological diagnosis, even though the specimen is clearly abnormal. Taking into account all cases, even in the unsuccessful procedures, the "accuracy" of FNAB is excellent, between 80% and 90%. The alternative for these patients is, after all, major surgery.

Adrenal Gland

Most larger series of adrenal FNAB have used CT guidance [2, 8, 16], although smaller numbers of cases have been done using US and even fluoroscopy [17, 18]. Success in obtaining diagnostic material correlates closely with the size of the adrenal gland or adrenal mass. Pagani obtained "adequate" material in only 29 of 43 adrenals (67%), but all were of normal size by CT criteria. These patients had known bronchogenic tumours, and he was searching for metastases [16]. Bernardino et al. biopsied adrenal masses demonstrable on CT, and FNAB led to correct diagnosis in 44 of 53 (83%). The nine non-diagnostic cases were rebiopsied, increasing the diagnostic yield to 48 of 53 (91%) [2]. Heaston et al. were successful in 13 of 14 patients (93%) with CT-proven adrenal masses [8].

Minor risks may accompany adrenal FNAB, although no mortality is reported. Pneumothorax or haemorrhage may occur, especially when a transpleural or transhepatic route is used [2, 3, 10]. Pain, pancreatitis and sepsis have also been reported [3].

Kidney

The diagnostic yield of renal FNAB by fluoroscopy, US, or CT is high, averaging over 90% in obtaining "adequate" specimens, with a similar high percentage of diagnostic interpretations made [5, 9, 17, 18]. US is probably the commonest imaging modality now used for these biopsies because of its advantages of excellent visualisation of masses, speed, "real-time" imaging, lower cost and the lack of ionising radiation.

Since smaller needles are most commonly used for renal aspiration biopsy, fewer complications are encountered compared with the large-core biopsies normally obtained to diagnose non-oncologic medical disorders, such as glomerulonephritis (1% major morbidity) [10]. Renal biopsies are safe because of the kidney's retroperitoneal location. Most often, FNAB is done without the need of traversing the peritoneal cavity. As previously mentioned, one rarely documented but potential complication of FNAB is "seeding" of the needle tract with malignant cells [11–14]. Considering that thousands of renal FNABs have been performed, the relative paucity of documented "seeding" cases suggests that this complication is distinctly rare.

Retroperitoneal and Pelvic Lymph Nodes

Documentation of retroperitoneal or pelvic lymph node metastases which are due to the spread of a variety of genitourinary neoplasms (bladder, prostatic, testicular, penile) is often critical for treatment planning and determining prognosis. Since serious understaging of such tumours will result if nodal metastases are overlooked, regional lymphadenectomy is often performed. This procedure diminishes understaging, but the morbidity of lymphadenectomy is high, between 20% and 30% [19].

In order to identify nodal sites of metastases from prostatic, bladder and testicular carcinomas, bipedal LAG and CT have both been employed extensively. Unfortunately, these techniques have high false negative rates, and a smaller but significant incidence of false positive studies [20–22]. False negative results occur with CT since metastases are common in normal size nodes or only slightly enlarged nodes. Micrometastases are inapparent even on the LAG. Reactive and post-inflammatory changes can lead to false positive LAG interpretation. The average accuracy of LAG and CT are comparable, between 60% and 80%, for detecting metastases from urological malignancies [23].

In order to improve staging and still avoid the need for lymphadenectomy, FNAB has been advocated [7, 10, 15, 23]. Performing multiple FNABs, including the sampling of nodes which appear normal on the LAG, can spare patients lymphadenectomy if a positive result is found. A negative result is of much less value, since *all* nodes are not being sampled. A sampling error even within an involved node may occur. For those with negative FNAB findings, lymphadenectomy is sometimes still necessary.

Luciani and his co-workers compared the accuracy of LAG alone with fluoroscopically guided FNAB of nodes (average 3.3 nodes biopsied/patient) in 30 patients with bladder cancer who later underwent lymphadenectomy [24]. The accuracy of LAG alone was 70%, while the LAG-guided FNAB accuracy was 96%. There were 24 true negative results and no false positives among the LAG/FNAB (specificity 100%). One false negative result was confirmed at surgery. Five of a total of six positive cases were detected by FNAB (sensitivity 83%). Macintosh et al. found FNAB accurate in 45 of 47 patients (96%) [15]. Kidd et al. reported LAG accurate in 74% overall, but with a 15% false negative incidence (4 of 27 patients) [25]. A subgroup of 11 of these patients also had FNAB. The accuracy was improved to 91% after FNAB. Gibod et al. reported a much higher incidence of false positive LAGs than Luciani. Of 54 patients, 32 had LAGs interpreted as "positive". These patients underwent FNAB of the suspicious nodes. The results showed neoplasm in only 11 of the 32 patients (34%) [10]. Negative FNABs were confirmed by adenectomy in 21 patients (from the total of 54). Similarly, four positive FNABs were confirmed by adenectomy. Gibod's series seems to illustrate the frequency with which hilar defects, reactive change and fibrosis can mimic nodal metastases. This problem of low specificity was addressed by Spellman et al., who recommended maintaining very strict criteria for interpreting metastatic disease on LAGs to assure high specificity [20]. In this report, relaxing criteria for interpretation of LAGs did not improve sensitivity significantly but decreased both specificity and accuracy.

It has been suggested that false negative LAGs may sometimes result from failure to opacify the obturator nodes, the primary lymphatic drainage from bladder and prostate [20, 26, 27]. This opinion persists even in the face of compelling evidence that these nodes are consistently opacified by lymphographic contrast media. Merrin et al. documented the presence of contrast in obturator nodes and confirmed the cases surgically [26]. Zoretic et al. also confirmed obturator node opacification in 100% of 25 cases and performed FNAB in several. Proof in his 25 cases was surgical [27].

Conclusions

The development and implementation of fluoroscopic, CT, and US guided FNAB has enhanced the contribution of diagnostic radiology to patient care. When LAG, CT or US demonstrate bulky lymphadenopathy, a diagnosis of metastasis can confidently be made. However, at times these imaging studies demonstrate findings which can lead only to a tentative interpretation, especially in patients with suspected lymph node metastases. Benign conditions affecting nodes, such as fatty replacement or scarring can be extremely difficult to differentiate from tumour. In these instances, FNAB is a useful tool for making the specific diagnosis by providing material for histological evaluation. If the results are positive for tumour, lymphadenectomy is avoided. However, a "negative" lymph node FNAB does not necessarily exclude the presence of metastases, and in such cases more invasive staging procedures such as surgical lymph node biopsy may be necessary. Nevertheless, for most patients FNAB is fast, easy and safe. Pathological results are quickly obtained with an extremely low level of morbidity. The low cost compared with surgery is also a substantial advantage of this diagnostic approach. A fundamental for success remains close cooperation between the diagnostic radiologist and pathologist, with a mutual understanding of the technical needs and limitations encountered by each.

References

1. Kunn MA (1847) A new instrument for the diagnosis of tumours. Mth J Med Sci 7:853–857
2. Bernardino ME, Walther MM, Phillips VM, Graham CW, Sewell K, Gedaudus-McClees BR, Baumgartner I, Torres WE, Erwin BC (1985) CT-guided adrenal biopsy: accuracy, safety, and indications. AJR 144:67–69
3. Ferrucci JT, Wittenberg J, Mueller PR, Simeone JF, Harbin WP, Kirkpatrick RH, Taft PD (1980) Diagnosis of abdominal malignancy by radiologic fine-needle aspiration biopsy. AJR 134:323–330
4. Lieberman RP, Hafez GR, Crummy AB (1982) Histology from aspiration biopsy: Turner needle experience. AJR 138:561–564
5. Holm HH, Pederson JF, Kristensen JK, Rasmussen SN, Hanke S, Jensen F (1975) Ultrasonically guided percutaneous puncture. Radiol Clin North Am 13:493–503
6. Goldstein H, Zornoza J, Wallace S (1977) Percutaneous fine needle aspiration biopsy of pancreatic and other abdominal masses. Radiology 123:319–322
7. Göthlin JH, Höiem L (1981) Percutaneous fine-needle biopsy of radiographically normal lymph nodes in the staging of prostatic carcinoma. Radiology 141:351–354
8. Heaston DK, Handel DB, Ashton PR, Korobkin M (1982) Narrow gauge needle aspiration of solid adrenal masses. AJR 138:1143–1148
9. Sundaram M, Wolverson MK, Heiberg E (1982) Utility of CT-guided abdominal aspiration procedures. AJR 139:1111–1115
10. Gibod LB, Katz M, Cochand B, Le Portz B, Steg A (1984) Lymphography and percutaneous fine needle node aspiration biopsy in the staging of bladder carcinoma. J Urol 132:24–26
11. Clark BG, Leadbetter WF, Campbell JS (1953) Implantation of cancer of the prostate in site of perineal needle biopsy: report of a case. J Urol 70:937–939
12. Engzell U, Esposti PL, Rubio C, Sigurdson A, Zajicek J (1971) Investigation in tumors spread in connection with aspiration biopsy. Acta Radiol Oncol Radiat Phys Biol 10:385–388
13. Sinner WN, Zajicek J (1976) Implantation metastasis after percutaneous transthoracic needle aspiration biopsy. Acta Radiol [Diagn] (Stockh) 17:473–480
14. Ferrucci JT, Wittenberg J, Margolies MN, Carey RW (1979) Malignant seeding of needle tract after thin needle aspiration biopsy; a previously unrecorded complication. Radiology 130:345–346
15. Macintosh PK, Thomson KR, Barbaric ZL (1979) Percutaneous transperitoneal lymph node biopsy as a means of improving lymphographic diagnosis. Radiology 131:647–649
16. Pagani JJ (1983) Normal adrenal glands in small cell lung carcinoma: CT-guided biopsy. AJR 140:949–951
17. Pepeiras RV, Meiers W, Kunhardt B, Troner M, Hutson D, Barkin JM, Viamonte M (1978) Fluoroscopically guided thin needle aspiration biopsy of the abdomen and retroperitoneum. AJR 131:197–202
18. Buonocore E, Skipper GJ (1981) Steerable real-time sonographically guided needle biopsy. AJR 136:387–392
19. Babcock JR, Grayhack JT (1979) Morbidity of pelvic lymphadenectomy. Urology 13:483–486
20. Spellman MC, Castellino RA, Ray GR, Pistenma DA, Bagshaw M (1977) An evaluation of lymphography in localized carcinoma of the prostate. Radiology 125:637–644
21. Benson KH, Watson RA, Spring DB, Agee RE (1981) The value of computed tomography in evaluation of pelvic lymph nodes. J Urol 126:63–69

22. Sherwood T, O'Donohue EPN (1981) Lymphograms in prostatic cancer: false-positive and false-negative assessment in radiology. Br J Radiol 54:15–17
23. Castellino RA, Marglin SI (1982) Imaging of abdominal and pelvic lymph nodes: lymphography or computed tomography. Invest Radiol 17:433–443
24. Luciani L, Piscioli F, Pusiol T, Scappini P (1986) The value of aspiration cytology in the definitive staging of bladder cancer. Br J Urol 58:26–30
25. Kidd R, Crane RD, Dail DH (1984) Lymphangiography and fine-needle aspiration biopsy: ineffective for staging early prostate cancer. AJR 141:1007–1012
26. Merrin C, Wajsman Z, Baumgartner G, Jennings E (1977) The clinical value of lymphangiography: are the nodes surrounding the obturator nerve visualized. J Urol 117:762–764
27. Zoretic SN, Wajsman Z, Beckley SA, Pontes JE (1983) Filling of the obturator nodes in pedal lymphangiography: fact or fiction. J Urol 129:533–535

Chapter 3

Fine Needle Aspiration Biopsy of Urological Neoplasms—an American Experience

W. J. Frable and W. W. Koontz Jr.

Applications of fine needle aspiration biopsy (FNAB) to the urogenital system have not as yet enjoyed wide popularity within the USA. This biopsy method may be applied directly to tumours within the urogenital system, or imaging by computed tomography (CT), ultrasonography and arteriography can be used. Specific areas where aspiration biopsy has been employed include the kidney [1–4], testis [2] and prostate [2, 4–8].

Prostate

The prostate has been the most frequent target of FNAB, via the transrectal approach, using the Franzen needle guide to reach the target precisely [2, 9–11]. Applications of FNAB include not only diagnosis of prostatic cancer, but prognosis and hormonal management. Most of these reports are still currently from Scandinavian and other European countries. One monograph by an American author has recently been published on the subject of FNAB of the prostate [12]. Walsh wonders why it has taken so long to popularise the fine needle biopsy method for the prostate, particularly in a cost-conscious environment, since it is an outpatient procedure. He points out that two recent series from American institutions have reported excellent results [5, 13, 14]. Furthermore, the urologist should be more inclined to aspirate a minimally suspicious nodule since the biopsy can be performed at the initial outpatient examination.

It therefore seems ironic that more than 50 years ago the diagnosis of prostatic carcinoma by aspiration biopsy was first described by Ferguson, a urological surgeon at the Memorial Center for Cancer [15]. The approach was transperineal using an 18-gauge, 15-cm needle. Ferguson injected local anaesthesia into the perineal area and guided the biopsy needle down to the prostatic capsule with a finger placed in the rectum. The puncture site was to one side of the midline and the rectal wall and urethra were avoided. Although Ferguson was trying for a small core of tissue, smears were prepared and stained with haematoxylin and eosin [15].

No results were included in Ferguson's paper and the method lay dormant until the report of Franzen et al. in 1960, describing preliminary results with the transrectal approach [16]. A needle guide designed by Franzen and his colleagues allowed the biopsy needle to be placed accurately and directly into a suspicious prostatic nodule through the rectum. Use of the fine needle ensured a non-traumatic aspiration biopsy requiring no anaesthesia. The procedure could be repeated many times. These favourable and efficient features remain true today.

Fine needle aspiration biopsy was begun at The Medical College of Virginia in the late 1970s through the interest and cooperation of the urology service. Only a few cases were investigated initially, but from 1982 onwards the series has grown slowly but steadily. One hundred and fifty-seven separate biopsies have been performed on 150 patients. Table 3.1 documents the results. There have been 48 diagnoses of cancer, corroborated clinically in follow-up or by a surgical procedure that included obtaining tissue demonstrating prostatic carcinoma. Two cases are currently carried as false positives and there are four false negative cases documented by follow-up tissue studies. Seven aspirates were considered unsatisfactory, while seven additional cases were reported as satisfactory but with smears containing limited numbers of cells.

Table 3.1. Fine needle aspiration biopsy of the prostate, Medical College of Virginia experience

Prostatic cancer: true positive	Negative for cancer: true negative	False positive	False negative
48	95	2	4
Unsatisfactory aspirates			7
Satisfactory but limited cells			7
Sensitivity of positive aspirate			92%
Specificity of negative aspirate			97%

With regard to unsatisfactory aspirates of the prostate, Table 3.2 supports the contention that there is a learning curve with aspiration biopsy of the prostate. From 1982 until the time of writing, a period when most of the aspirates were performed, physician A, who has generally excellent technical skill and great interest in this procedure, had only an 11% failure rate. Physician B, who took less time

Table 3.2. Fine needle aspiration biopsy of the prostate: comparison of technical skill in specimen procurement

Year biopsy taken	Number of unsatisfactory aspirates		
	Physician A	Physician B	Physician C
1982	1		
1983		2	3
1984	1	3	1
1985		5	
1986	4	5	3
1987	1		2
Total unsatisfactory	7	15	9
Total biopsies taken	59	58	52
Percentage unsatisfactory	11%	25%	17%

with the details of the aspiration procedure, had twice the biopsy failure rate of Physician A. Physician C, representing a composite group of physicians on the urology service, did somewhat worse than Physician A, but not significantly so. The actual number of aspirates, however, is lower, with only a few for any given physician in group C. The data support the conclusion that this particular type of aspiration biopsy requires skill and experience.

The overall sensitivity and specificity reported for the present authors' series are well within the range of others. Chodak et al. have recently compared aspiration biopsy and core biopsy of the prostate in the same group of patients. They achieved 98% sensitivity for aspiration biopsy but only 81% for core biopsy [14]. Carter et al. made a similar study, also using transurethral resection to obtain tissue. They confirmed 49 of 57 FNAB diagnoses of prostatic carcinoma by core needle sample or transurethral resection. Seven core biopsies were considered benign and one was inadequate [5]. Similar studies in the past have shown a lower performance of core biopsy, when carcinoma was actually present, in comparison with aspiration biopsy. The present authors are aware of one case of aspiration biopsy of the prostate seen in consultation where the smears were considered quite diagnostic of prostatic carcinoma. Core biopsy and subsequent transurethral resection were both negative for carcinoma. Two years later the patient had a second core biopsy because the prostate still felt abnormal. Carcinoma was detected.

Three other areas of interest have been pursued with respect to FNAB of the prostate: diagnosis using DNA flow cytometric measurements; grading of prostatic carcinoma by both flow cytometric measurements of DNA and subjective light microscopic criteria; and staging of prostatic and other urological cancers by a combination of lymph(angi)ography and needle aspiration of enlarged and suspicious lymph nodes, confirming or denying the presence of metastatic disease. A number of contributions in this monograph address this particular issue.

Esposti had previously reported some success in the grading of prostatic carcinoma on the basis of the cellular morphology of the aspirate. There was good correlation in his series of 469 patients, separating well-differentiated from poorly differentiated carcinoma with respect to survival. Carcinomas reported as moderately well differentiated fell somewhere in between the survival figures for well- and poorly differentiated carcinomas of the prostate, with a tendency towards the better survival of the well-differentiated tumours. The most typical feature of the well-differentiated carcinoma was the microglandular complex. This is also the most significant initial diagnostic features of FNAB of the prostate [7, 17].

The grading system involving three grades employed by Esposti and outlined in the monograph by Linsk and Franzen has been quite easy to follow in the present authors' laboratory [18]. It provides a good correlation with any subsequent tissue confirmation. Three grades also appear to be quite adequate for clinical management and treatment planning, in contrast to the elaborate system of Gleason.

The results of DNA measurements of prostatic carcinoma cells by flow cytometry and a correlation with survival have shown variable results. The first application by Sprenger et al. was used for diagnosis. There were many false positives (29.7%) and false negatives (11.4%) in comparison with the diagnosis by light microscopic features [19]. In a more recent paper, Seppelt and his colleagues were able to reduce the false positive results to 4.3% with DNA flow cytometry measurements, but the false negative rate remained the same. These same authors reported only a 39.5% accuracy with malignancy grading [20]. A much larger series reported by Ronstrom et al. noted that 90% of patients with benign prostatic disease had a diploid or tetraploid pattern. The prostatic carcinoma patients had cells of increasing aneuploidy in correlation with an ascending grade of anaplasia as observed by light microscopy [21]. Koss concluded that these studies showed promise and undertook a similar investigation in his own laboratory [22].

In the past the reported results had been too incomplete to assess accurately the utility of FNAB of the prostate, ranging from 63% [23] to 91% [24]. Recently, however, 90% accuracy has been reported for the diagnosis of carcinoma [5, 14]. These same authors have not documented false positives, but two prior papers have reported false positive rates of 2% [24] and 28% [25].

Eight prior studies have compared simultaneous fine needle aspiration and tissue needle biopsy in diagnosing prostatic carcinoma [8, 22, 26, 27–30]. Accuracy has ranged between 85% and 95%, with a fair degree of concordance between core and aspiration biopsy. Chodak and co-workers reported their recent comparative study of core and aspiration biopsy. Core biopsy had a 14% false negative rate, while no cancer was missed by aspiration. Also, 14% of the core biopsies had insufficient material for diagnosis while this was true for only 3% of the aspiration biopsies [31]. Ekman's group made an interesting observation in their comparative series. Of seven cases originally diagnosed as carcinoma by aspiration but not by needle biopsy, repeat core biopsy demonstrated prostatic carcinoma in six [28].

The largest comparative series also resulted in the highest false positive rate of 6.5%. There were 1700 biopsies performed with a cytological diagnostic accuracy for prostatic carcinoma of 93.3%. There were 2.39% of the cases of carcinoma not detected by aspiration biopsy while 3.7% were diagnosed cytologically but not confirmed by histology [32]. It could be concluded that attempts to diagnose very well differentiated tumours, while raising the overall accuracy of FNAB, does lead to an increase in false positive cases. Some of these may actually be occult carcinomas which cannot be detected histologically with only core biopsy.

Fine needle aspiration biopsy of prostatic carcinoma has also been used as an indication of response to therapy, both radiation and orchiectomy. It lends itself readily to that evaluation because of its relative atraumatic nature and utility as an outpatient procedure. Tomic et al. in evaluation of patients with prostatic cancer following orchiectomy, found two cell patterns on aspiration biopsy at 6 and 12 months. The first pattern was unmodified carcinoma cells and the second was carcinoma cells with regressive changes. Clinical regression was much more frequent at 36 months of follow-up in those patients whose aspiration biopsy showed regressive changes in cancer cells at 6 and 12 months [33]. The present authors have noted the same prognostic significance following radiation therapy with follow-up fine needle aspiration of the prostate, if unmodified carcinoma cells remain, in a small number of cases studied to date.

Evidence of acute prostatitis should be considered a contraindication for transrectal aspiration biopsy. The Karolinska Institute reported among 1400 transrectal needle aspiration biopsies of the prostate four cases of Gram-negative sepsis. One of these patients had a fatal septicaemia. The authors also found three additional case reports of this complication [34]. Other complications reported in a 10-year period during which there were 3002 transrectal aspiration biopsies of the prostate included two cases of slight transient haematuria, five cases of transient pyrexia, one of acute epididymitis and three of slight haemospermia [2].

It should also be noted that there have been no reported cases of implantation of carcinoma following FNAB of the prostate. Recently, the seventh case of this type of complication using a core needle biopsy through the perineum was reported by Brausi et al. [35]. All previous cases had also been the result of core needle biopsy for prostatic cancer.

Pelvic Lymph Nodes

As a staging procedure for urological cancer, fine needle aspiration of pelvic lymph nodes following lymph(angi)ography with or without fluoroscopy has been advocated. There is an apparent dichotomy of results, however, between European and American authors. Piscioli and his colleagues have reported two series of patients using lymph(angi)ography and fluoroscopic guidance for fine needle aspiration of pelvic lymph nodes. From a study of 31 patients with clinically localised prostatic carcinoma the overall accuracy was 93.5% with a false negative rate of 6.4% [36]. In another series of 71 patients with bladder, prostatic or penile cancer they studied 257 nodal chains with no false positive diagnoses and 6.0% false negative biopsy result [37]. From another series by Luciani et al., aspiration biopsy increased the sensitivity of staging results in 35 patients with prostatic cancer over lymph(angi)ography alone from 57% to 91% and the specificity from 47% to 100%. Positive cytological findings in pelvic lymph nodes were conclusive for stage D disease, but negative cytological findings in pelvic nodes were only definitive if the prostatic cancer was well differentiated and/or the Gleason grading sum was 2, 3 or 4 [38].

Kidd et al. and Kidd and Correa reported a very negative view of lymph(angi)ography and needle aspiration as an enhanced staging method for prostatic and bladder cancer. In one paper they studied 436 patients with carcinoma of the prostate by lymph(angi)ography and fine needle aspiration of abnormal opacified lymph nodes. They found that this procedure was of very limited value unless the patient had at least clinical stage C disease. It did not have any value in patients with a Gleason grade of less than 6. The authors conclude that a positive result in clinical stage C disease will save the morbidity of a staging lymph node dissection [39]. In a smaller series of both prostatic and bladder cancer patients, 49 cases, FNAB performed on lymph(angi)ographically normal nodes failed to detect any cases of metastatic disease [40]. Likewise, Flanigan et al. demonstrated low sensitivity, 50% and 53.8% for CT scan with aspiration biopsy versus lymph(angi)ography with aspiration of pelvic nodes in patients considered surgically acceptable for treatment of prostatic cancers. Seven of 14 patients ultimately found to have stage D_1 disease were confirmed by FNAB of lymph(angi)ographically normal nodes. The authors concluded that the method was cost effective since it identified preoperatively 50% of the patients with lymph node metastases, thus saving a major surgical procedure [41].

Kidney

Although renal masses are easily biopsied by fine needle aspiration, the major experience is again being reported from Europe. Soderstrom uses a direct approach to aspiration without imaging while others rely on pyelography, angiography and fluoroscopic control [1, 2, 42]. A recent series by Juul et al. reported 301 ultrasonically guided FNABs of renal masses. A cytological diagnosis was established in 82% of cases. Suitable specimens were obtained in 95% of the biopsies. The number of false positives, 14, seems unacceptably high. These cases were all overinterpretation of well-differentiated renal cell carcinoma [43]. Orell and his colleagues reported thin-needle aspiration biopsies of 83 solid renal and adrenal masses over an 8-year period. His group used fluoroscopy and ultrasound, relying on the latter more often as the series accumulated. Seventy-seven of the cases were renal masses. Diagnostic sensitivity was 90%. There was one false positive, a renal angiomyolipoma interpreted as low-grade renal cell carcinoma [44].

Two other cases of renal angiomyolipoma have been reported as diagnosed by needle aspiration. One was actually reported as renal angiomyolipoma [45] while the other was considered a benign mesenchymal tumour of unspecified type [46]. Another expectedly difficult interpretation when encountered with aspiration biopsy of a renal mass is oncocytoma. Four cases were found from a review of the literature. In one case the diagnosis was suggested when coupled with a characteristic angiographic pattern [47]. In another report of three cases, two were diagnosed as renal oncocytoma but a grading system was employed. Two of the cases were considered grade I while the last case was regarded as granular cell renal cell carcinoma, the histology being diagnosed as renal oncocytoma grade II. All three patients were treated by radical nephrectomy. It is difficult to determine from the photomicrographs what these tumours actually represent. The follow-up is also short [48].

Linsk and Franzèn have detailed the cytological features of renal cell carcinoma in both primary and metastatic sites. They found the most useful feature identifying renal cell carcinoma in aspiration smears to be the partially extruded nucleus. The appearance of these nuclei probably results from the delicate nature of the cell cytoplasm in many renal cell cancers, particularly those with a clear cell pattern. They were able to identify 7 of 15 primary renal cell carcinomas from the aspiration biopsy of a metastatic focus [49].

Prior to the routine use of CT scan and ultrasonography most of this author's experience has been with intraoperative aspiration of fluid from renal cysts. This has been an unrewarding procedure cytologically as no cases of malignancy have been found in 15 years. These fluids are usually acellular or contain only a few mononuclear cells probably of renal tubular origin. Since 1981 33 fine needle aspirations of renal masses using imaging have been performed in 30 patients in the radiology department at the Medical College of Virginia. Twelve of these cases have been for both diagnosis and drainage of large renal cysts. No malignancy has been found in this group. The remaining aspiration biopsies have been performed on solid renal masses. Seven cases were reported as renal cell carcinoma and six have been confirmed either by subsequent resection or clinical features of the follow-up. One case was found to be a large adenocarcinoma of the caecum with extensive retroperitoneal growth involving the lower pole of one kidney. Two additional diagnoses of malignant tumour were considered metastatic. Subsequent clinical features or prior tissue diagnoses of the primary lesion confirmed the aspiration biopsy findings. The nine remaining cases were reported as having no malignant cells or evidence of necrosis and/or inflammation. One of these nine cases is a false negative report. The patient at surgical exploration was found to have a small renal cell carcinoma of the left kidney as well as an incidental phaeochromocytoma of the right adrenal. Thus, though there have been few cases of renal fine needle aspiration in the present authors' laboratory, the reports of malignant neoplasms have been accurate.

The present authors have not dealt with enough aspiration biopsies of renal cell carcinoma to appreciate the differences with respect to tumour grade as described in detail by Zajicek. He divides the renal cell carcinomas into well-differentiated, moderately well differentiated, and poorly differentiated. Zajicek and his colleagues have advocated preoperative radiotherapy in those aspirations of renal cell carcinoma classified as poorly differentiated [2]. Correlation with treatment and survival has been good using this evaluation of a poorly differentiated renal cell neoplasm [2, 50].

Nurmi et al. recently expanded this type of analysis by aspirating excised renal cell carcinoma and correlating the cytological and histological grade with prognosis, independently, in a blind study. Their assessment was good based on a division of the tumours into three grades: well, moderately well and poorly differentiated. The cytological evaluation undergraded the tumour in relation to histological findings in 19.1% of cases and overgraded the renal cell carcinoma in 5.0%. The authors concluded that the cytological evaluation required several fine needle aspiration samples from different areas of the tumour and was most difficult with well-differentiated renal cell carcinoma [51]. There are some practical considerations if accurate grading requires multiple aspiration biopsies of the same tumour in vivo.

Results from several studies have provided reasonably accurate diagnoses of both cysts and solid tumours of the kidney. Edgren et al. reported a series combining aspiration with angiography, studying 55 patients of whom 42 had malignant tumours. Angiography was 94% accurate in diagnosis, while fine needle aspiration was accurate in only 71% of the cases. Combined accuracy for both techniques was 97%. Fine needle aspiration was positive in all of the four cases of renal pelvis carcinoma, whereas angiography was negative [42]. Holm et al. diagnosed 43 of 49 solid renal tumours by FNAB. This group was also successful in aspirating 57 of 61 renal cysts [52].

Few, if any, complications have been described by those reporting major series of FNABs of the kidney. Soderstrom described some cases of microhaematuria, whereas Zajicek reported no complications [1, 2]. Two cases of needle tract seeding following needle biopsy have been reported [50, 52, 53]. Riches cites a case reported by Hanley of dissemination of tumour cells in the needle tract. The case is poorly documented and a thick needle was used. This type of case bears no relationship to FNAB of renal tumours [50, 54].

The case of Gibbons et al. appears to be an authentic example of needle tract implantation of renal cell carcinoma following 20 months after FNAB. Two separate attempts at biopsy were made through a posterior approach. No cells were obtained using an 18-gauge needle. Despite the unsatisfactory aspiration, tumour later appeared in the site of the needle tract. Interpreting this case is difficult. It is stated that tumour was removed intact via an anterior subcostal transperitoneal approach. The exact relation of the surgical incision to the aspiration is not described. The fact that the aspiration did not yield any diagnostic cells and yet tumour cells appeared in the needle tract is bizarre [55].

von Schreeb's group found no evidence of spread of tumour cells following renal puncture. Their series encompassed a minimum follow-up period of 5 years and a control group that did not have renal puncture. They used needles varying from 0.75 to 1.5 mm external diameter in a series of 150 cases of which one-half had puncture [56]. Orell et al. reported post-biopsy haemorrhage within a large necrotic renal cell carcinoma following fine needle aspiration. This resulted in severe pain [44]. In a small number of cases the present authors have experience of, there have been no complications or evidence of needle tract seeding.

Testis

Only the Scandinavian literature yields any information concerning FNAB of testicular masses [2, 57, 58]. The American literature does not contain any comprehensive reports of needle aspiration biopsy of the testicle. Koss notes one report in his monograph, an aspiration of a suspected hydrocele, reported by Japko et al. as mesothelioma of the tunica vaginalis [22, 59]. He also illustrates the recognition of seminoma when large tumour cells were found within the aspirate from a clinically suspected hydrocele [22].

The present authors have seen several examples of metastatic seminoma and have aspirated for diagnosis several pelvic teratomas in youngsters. The seminomas contain not only large tumour cells with particularly prominent nucleoli but also a characteristic striated or tigroid background of stroma, probably cytoplasm from fragile tumour cells, that is quite striking morphologically. This stromal or cytoplasmic pattern is seen only with air-dried Romanovsky-stained smears. The teratomas contained multiple tissue elements. While they are difficult to recognise specifically, their heterogeneous patterns point immediately to the possibility of teratoma, usually malignant.

References

1. Soderstrom N (1966) Fine needle aspiration biopsy. Almquist and Wiksell, Stockholm, pp 137–147
2. Zajicek J (1979) Aspiration biopsy cytology. Part 2. Cytology of infradiaphragmatic organs. Monogr Clin Cytol 7:1–37, 104–128, 129–166
3. Meier WL, Willscher MK, Novicki DE, Pischinoer RJ (1979) Evaluation of perihilar and central renal masses using the Chiba needle. J Urol 121:414–416
4. Orell SR, Langlois SL, Marshall VR (1985) Fine needle aspiration cytology in the diagnosis of solid renal and adrenal masses. Scand J Urol Nephrol 19:211–216
5. Carter HB, Reihle RAJ, Koizumi JH, Amberson J, Vaughan ED Jr (1986) Fine needle aspiration of the abnormal prostate: a cytohistological correlation. J Urol 135:294–298
6. Maier U, Czerwenka K, Neuhold N (1984) The accuracy of transrectal aspiration biopsy of the prostate: an analysis of 452 cases. Prostate 5:147–151
7. Esposti P (1966) Cytologic diagnosis of prostatic tumors with the aid of transrectal aspiration biopsy. A critical review of 1,110 cases and a report of morphologic and cytochemical studies. Acta Cytol 10:182–186
8. Epstein NA (1976) Prostatic biopsy: a morphological correlation of aspiration cytology with needle biopsy histology. Cancer 38:2078–2087
9. Koss IG, Woyke S, Schreiber K, Kohlberg W, Freed SZ (1984) Thin-needle aspiration biopsy of the prostate. Urol Clin North Am 11:237–251
10. Tchakarov S (1984) Results of 317 transrectal puncture biopsies of the prostate. Int Urol Nephrol 16:133–139
11. Ljung BM (1985) Fine-needle aspiration biopsy of the prostate gland: technique and review of the literature. Semin Urol 3:18–26
12. Kline TS (1985) Guides to clinical aspiration biopsy. Vol 1. Prostate. Igaku-Shoin, New York
13. Walsh PC (1986) Fine needle aspiration biopsy of the prostate—why has it taken so long to accept? J Urol 135:334–335
14. Chodak GW, Steinberg GD, Bibbo M, Straus FH II, Wied GL (1986) The role of transrectal aspiration biopsy in the diagnosis of prostatic cancer. J Urol 135:299–302
15. Ferguson RS (1930) Prostatic neoplasms, their diagnosis by needle puncture and aspiration. Am J Surg 9:507–511
16. Franzen S, Giertz G, Zajicek J (1960) Cytological diagnosis of prostatic tumours by transrectal aspiration biopsy: a preliminary report. Br J Urol 32:193–196
17. Esposti PL (1971) Cytologic malignancy grading of prostatic carcinoma by transrectal aspiration biopsy. Scand J Urol Nephrol 5:199–209
18. Linsk JA, Franzèn S (1983) Clinical aspiration cytology. Lippincott, Philadelphia, pp 243–266
19. Sprenger E, Michaelis WE, Vogt-Schaden M et al. (1976) The significance of DNA flow-through fluorescence cytophotometry for the diagnosis of prostate carcinoma. Beitr Pathol 159:292–298
20. Seppelt U, Sprenger E, Hedderich J (1985) DNA oriented automated malignancy diagnosis and malignancy grading in prostatic cancer (Ger). Urol Int 40:76–81
21. Ronstrom L, Tribukait B, Esposti P-L (1981) DNA pattern and cytologic findings in fine-needle aspirates of untreated prostatic tumors. A flow-cytofluorometric study. Prostate 2:79–88
22. Koss LG, Woyke S, Olszewski W (1984) Aspiration biopsy cytology. Cytologic interpretation and histologic basis. Igaku-Shoin, New York, p 246
23. Sunderland H, Lederer H (1971) Prostatic aspiration biopsy. Br J Urol 43:603–607
24. Rheinfrank RE, Nulf ThH (1969) Fine needle aspiration biopsy of the prostate. Endoscopy 1:27–32
25. Ventura M, Barasolo E, Morano E et al. (1977) Franzen needle transrectal prostatic biopsy in the cytologic diagnosis of prostatic cancer. J Urol Nephrol (Paris) 83:858–862
26. Kline TS, Kelsey DM, Kohler FP (1977) Prostatic carcinoma and needle aspiration biopsy. Am J Clin Pathol 67:131–133
27. Alfthan O, Klintrup HE, Koivuniemi A et al. (1970) Cytological aspiration biopsy and Vim-Silverman biopsy in the diagnosis of prostatic carcinoma. Ann Chir Gynaecol 59:226–229
28. Ekman H, Hedberg K, Persson PS (1967) Cytological versus histological examination of needle biopsy specimens in the diagnosis of prostatic cancer. Br J Urol 39:544–548
29. Schnurer LB, Fritjofsson A, Lindgren A et al. (1969) Fine needle versus coarse needle in puncture diagnosis of prostatic carcinoma. Acta Pathol Microbiol Scand 76:150–160
30. Alfthan O, Klintrup HE, Koivuniemi A, Taskinen E (1968) Comparison of thin-needle and Vim-Silverman needle biopsy

in the diagnosis of prostatic cancer. Duodecim 84:506–511

31. Chodak GW, Bibbo M, Straus FH, Wied GL (1984) Transrectal aspiration biopsy versus transperineal core biopsy for the diagnosis of carcinoma of the prostate. J Urol 132:480–482
32. Sonnenscheim R (1975) The effectiveness of transrectal aspiration cytology in the diagnosis of prostatic cancer. Eur Urol 1:189–192
33. Tomic R, Bergman B, Hietala SO, Angstrom T (1985) Prognostic significance of transrectal fine-needle aspiration biopsy findings after orchiectomy for carcinoma of the prostate. Eur Urol 11:378–381
34. Esposti PL, Elman A, Norlen H (1975) Complications of transrectal aspiration biopsy of the prostate. Scand J Urol Nephrol 9:208–213
35. Brausi M, Latini A, Palladini PD (1986) Local seeding of anaplastic carcinoma of the prostate after needle biopsy. Urology 27:63–64
36. Piscioli F, Leonardi E, Reich A, Luciani L (1984) Percutaneous lymph node aspiration biopsy and tumour grade in staging of prostatic carcinoma. Prostate 5:459–468
37. Piscioli F, Scappini P, Luciani L (1985) Aspiration cytology in the staging of urologic cancer. Cancer 56:1173–1180
38. Luciani L, Scappini P, Pusiol T, Piscioli F (1985) Comparative study of lymphography and aspiration cytology in the staging of prostatic carcinoma. Report of 35 cases with histological control and review of the literature. Urol Int 40:181–189
39. Kidd R, Crane RD, Dail DW (1984) Lymphangiography and fine-needle aspiration biopsy: ineffective for staging early prostate cancer. AJR 142:1007–1012
40. Kidd R, Correa R Jr (1984) Fine-needle aspiration biopsy of lymphangiographically normal lymph nodes: a negative view. AJR 142:1005–1006
41. Flanigan RC, Mohler JL, King CT, Atwell JR, Umer MA, Loh F-K, McRoberts JW (1985) Preoperative lymph node evaluation in prostatic cancer patients who are surgical candidates: the role of lymphangiography and computerised tomography scanning with directed fine needle aspiration. J Urol 134:84–87
42. Edgren J, Taskinen E, Alfthan O, Mäkinen J, Juusela H (1975) Radiology and fine needle aspiration biopsy in the diagnosis of tumours of the kidney. Ann Chir Gynaecol Fenn 64:209–216
43. Juul N, Torp-Pedersen S, Gronvall S, Holm HH, Kock F, Larsen S (1985) Ultrasonically guided fine needle aspiration biopsy of renal masses. J Urol 133:579–581
44. Orell SR, Langlois SL, Marshall VR (1985) Fine needle aspiration cytology in the diagnosis of solid renal and adrenal masses. Scand J Urol Nephrol 19:211–216
45. Glenthoj A, Partoft S (1984) Ultrasound-guided percutaneous aspiration of renal angiomyolipoma. Report of two cases diagnosed by cytology. Acta Cytol 28:265–268
46. Nguyen GK (1984) Aspiration biopsy cytology of renal angiomyolipoma. Acta Cytol 28:261–264
47. Alanen KA, Tyrkko JE, Nurmi MJ (1985) Aspiration biopsy cytology of renal oncocytoma. Acta Cytol 29:859–862
48. Nguyen GK, Amy RW, Tsang S (1985) Fine needle aspiration biopsy cytology of renal cell oncocytoma. Acta Cytol 29:33–36
49. Linsk JA, Franzèn S (1984) Aspiration cytology of metastatic hypernephroma. Acta Cytol 28:251–260
50. Riches E (1964) Endoscopic and laboratory investigations. In: Riches ES (ed) Tumours of the kidney and ureter. Livingstone, Edinburgh, p 157
51. Nurmi M, Tyrkko J, Puntala P, Sotarauta M, Antila L (1984) Reliability of aspiration biopsy cytology in the grading of renal adenocarcinoma. Scand J Urol Nephrol 18:151–156
52. Holm HH, Pederson JF, Kristensen JK, Rasmussen SN, Hanke S, Jensen F (1975) Ultrasonically guided percutaneous puncture. Radiol Clin North Am 13:493–503
53. von Schreeb T, Franzèn S, Ljungqvist A (1967) Renal adenocarcinoma. Scand J Urol Nephrol 1:265–269
54. Hanley HG (1963) Discussion on cyst and tumor occurring in the same kidney. Trans Am Assoc Gen Urin Surg 55:126–128
55. Gibbons RP, Bush WH, Burnett LL (1977) Needle tract seeding following aspiration of renal cell carcinoma. J Urol 118:854–867
56. von Schreeb T, Abner O, Skovsted G, Wikstad G (1967) Renal carcinoma. Is there a risk of spreading tumor cells in diagnostic puncture? Scand J Urol Nephrol 1:270–276
57. Persson PS, Ahren C, Obrant KS (1971) Aspiration biopsy smear of testis in azoospermia. Scand J Urol Nephrol 5:22–26
58. Ekelund L, Gothlin J (1976) Fine needle biopsy of metastases at retrograde pyelography, directed by fluoroscopy. Report of a case with malignant teratoma of the testis. Scand J Urol Nephrol 10:261–262
59. Japko L, Alamada Horta A, Schreiber K, Mitsudo S, Karwa GL, Singh G, Koss LG (1982) Malignant mesothelioma of the tunica vaginalis testis: report of first case with preoperative diagnosis. Cancer 48:119–127

Section II

Diagnostic Imaging, Aspiration Techniques and Cytological Diagnosis

Chapter 4

Anatomy and Metastatic Invasion of the Ilio-Pelvic-Aortic Lymphatic System

Z. Wajsman

Many patients with clinically localised prostatic and bladder cancer have lymph node metastases. Unfortunately, even now, with the availability of many new imaging techniques, there is a significant incidence of understaging of bladder and prostate neoplasms. One of the more important reasons for this is the lack of an accurate method to detect microscopic or even gross involvement of regional lymph nodes.

The primary lymphatic drainage of pelvic organs is to the regional lymph nodes which are located around the major blood vessels, namely the external and internal iliac arteries and veins. In addition, presacral lymph nodes and, in the case of bladder and prostate, perivesical lymph nodes are part of the direct, regional lymphatic system of these organs. In rare cases, a direct communication may occur between the bladder and a common iliac lymph node [1]. In general, however, it is accepted that lymph nodes located above the bifurcation of the common iliac artery and the inguinal nodes are part of the juxtaregional system and only in extremely exceptional conditions may they be involved primarily, bypassing the regional lymph nodes' drainage.

The external iliac group that drains the leg is divided into three interconnecting chains: (1) external iliac chain, (2) middle chain and (3) internal iliac or hypogastric chain (Fig. 4.1). The nodes around the prostate and bladder primarily drain to the internal chain and the obturator lymph nodes, located along the obturator nerve, are part of this internal chain. It is not surprising, then, that the obturator lymph node is most frequently involved in metastasis from bladder or prostate cancer.

Surgeons who have performed many pelvic lymph node dissections have for a long time observed the intimate lymphatic connections between the external iliac and the hypogastric lymphatic network, and it has been difficult for us to understand the frequently quoted lack of filling of the obturator lymph nodes at the time of pedal lymphography. Indeed, this was a major reason why pedal lymphography was not used in the staging of prostate or bladder cancer. It was first shown [2] that, following pedal lymphography, the obturator nodes always contain the contrast material. This was proved by dissecting the lymph node located along the obturator nerve and X-raying it. Following this study, pelvic lymph node dissection was performed in 25 consecutive patients and pelvic X-rays were taken once the external iliac nodes were completely excised, leaving the obturator nodes intact (Fig. 4.2). The X-rays clearly documented obturator lymph nodes filled with dye intact in the pelvis (Fig. 4.3). Following the removal of the lymph node located along the obturator nerve, no more opacified lymph nodes could be seen on X-rays. It was proved, then, that the obturator lymph node not only contains the contrast material, but also that it can be recognised on pedal lymphography. A fine needle can be easily directed into

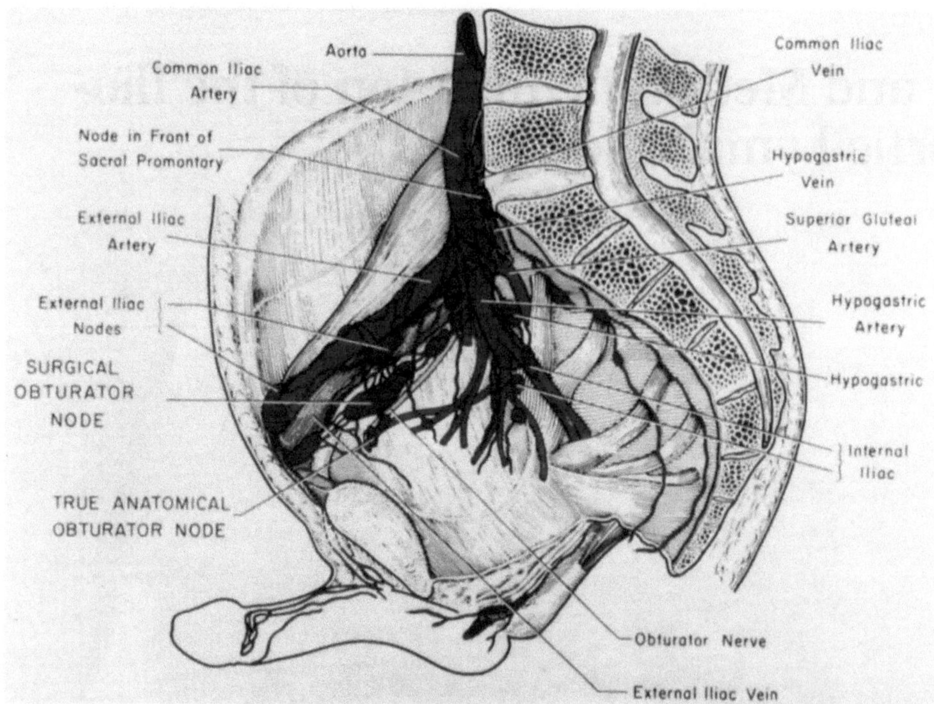

Fig. 4.1. Major lymphatic drainage of pelvic organs. Sagittal view of pelvic lymphatic system. The three lymphatic channels—the external iliac, the middle chain (with the surgical obturator nodes) and the hypogastric chain—are clearly visualised. Note that the anatomical obturator nodes are located near the exit of the obturator nerve into the obturator foramen. The hypogastric and presacral nodes are rarely visualised on lymphography, unlike contrast visualisation of the surgical obturator lymph gland.

the obturator lymph node after some experience has been acquired [3] (Figs. 4.4–4.7).

A careful review of the anatomy reveals that the so-called anatomical obturator lymph node is located near the obturator foramen. It is a rare entity and is not involved in the lymphatic drainage of the bladder or prostate. This "true" obturator node is present in only 6%–8% of all individuals [4, 5]. Its clinical value is unknown, and it usually is not seen at the time of lymphography.

In order to prevent confusion, we have suggested that lymph nodes surrounding the obturator nerve should be designated "surgical" obturator nodes in contrast to "true" anatomical obturator nodes occasionally found in the obturator foramen. None of our 25 patients who have documented obturator lymph node filling had gross lymphatic involvement at the time of lymph node dissection. Only five of them had microscopic invasion and all of them had normal lymphography. It is safe to assume, then, that the constant visualisation of these nodes occurs naturally and not as a result of lymphatic obstruction and subsequent retrograde filling of the obturator nodes.

In recent years, lymphography has been extensively used in clinical staging of prostatic, bladder and cervical cancer. False positive lymphograms are relatively uncommon and should not exceed 10% if strict criteria of interpretation are used and based on experience. Small filling defects, less than 1 cm, or delayed filling or displacement of lymphatic channels should not be used as criteria for positive lymphography. Inflammatory changes caused by previous surgery or lymphography itself, and benign reactive hyperplasia may cause significant changes in lymphographic appearance. Although most reported studies have only around 20% false negative results [6], lymphography was found to reveal the presence of nodal metastases in only 50% of patients [7–9]. It is clear now that failure to visualise hypogastric and presacral nodes may be responsible for this lack of sensitivity of lymphography in detection of lymph node metastasis. These nodes are quite frequently involved in pelvic malignancies [10, B. Hughes 1985, personal communication]. An additional and most significant cause of this failure is that lymphography is not able to demonstrate microscopic lymph node metastases.

In summary, the major advantage of lymphography is its ability to demonstrate clearly the majority of lymphatic drainage systems of pelvic organs. Small filling defects of 1 cm or larger are easily demonstrated and may be aspirated under

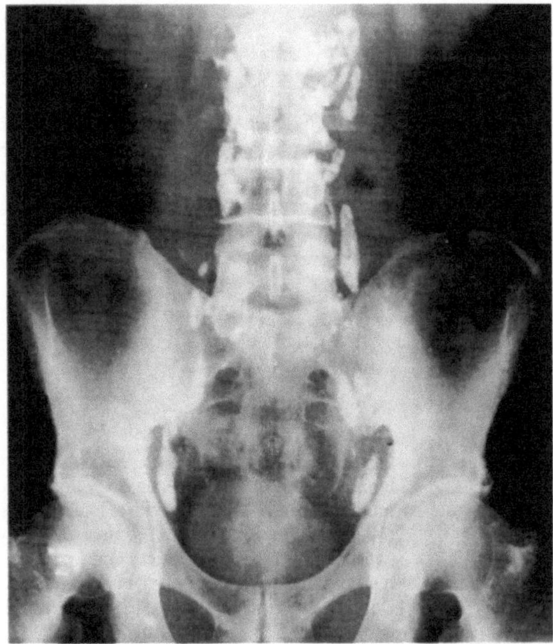

Fig. 4.2. Lymphography prior to lymph node dissection (right and left are reversed). Observe the lack of presentation of hypogastric nodes. The glands located along the common iliac artery and external iliac artery are demonstrated. Note two lymph glands located at the most distal part of the picture. The more lateral lymph node is located around the external iliac artery, the medially located one is the surgical obturator node (see Fig. 4.3).

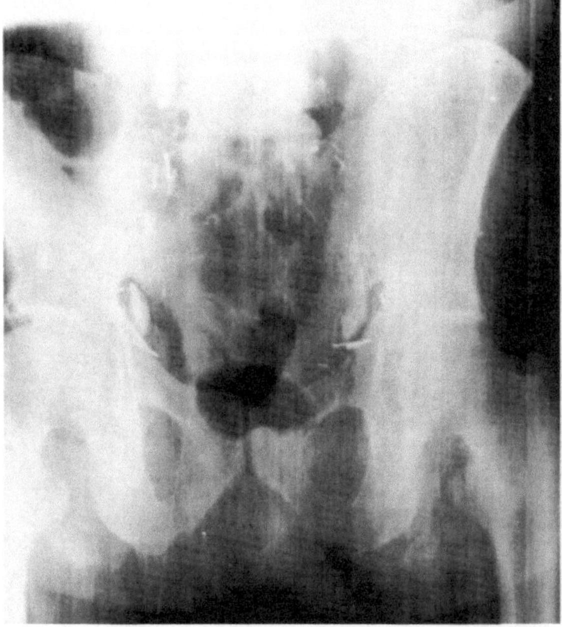

Fig. 4.3. Separate visualisation of surgical obturator node. X-ray taken on completion of dissection of all nodes located along the external iliac artery and vein. The lymph gland located along the obturator nerve was left intact on both sides and is clearly seen filled with dye.

fluoroscopy. Lymphography is able to demonstrate the most frequently involved lymph node in the metastatic process, which is the obturator lymph node located along the obturator nerve. However, it is rarely able to visualise the hypogastric or presacral nodes which are also involved in regional metastatic spread. It cannot reveal microscopic metastases and does require a very experienced radiologist. It is also time consuming and not an inexpensive procedure. Therefore, lymphography alone or combined with lymph node aspiration is indicated in a relatively limited number of clinical situations and can be effectively performed only in major medical centres.

Computerised Tomographical Scanning of Pelvic Lymph Nodes

Much hope was placed in computed tomography (CT) for detection of lymph node metastases. Although this technique is an extremely important clinical tool, it should be interpreted with care and,

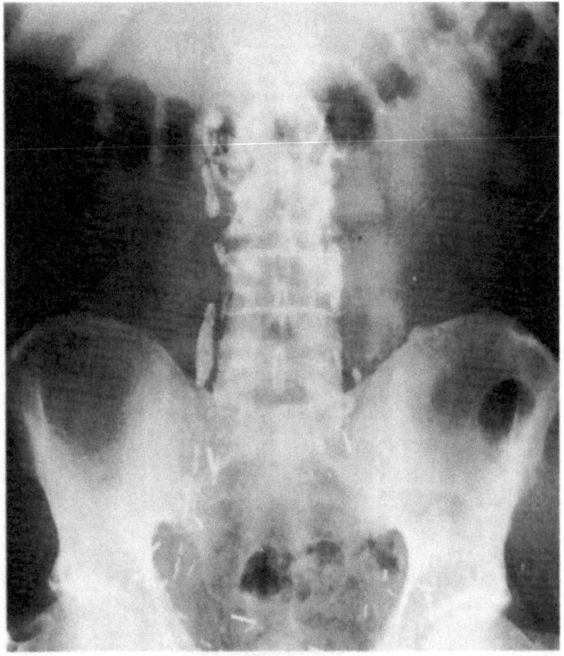

Fig. 4.4. X-ray view following completion of pelvic lymph node dissection. X-ray taken following dissection of the surgical obturator node. No more lymph node glands are seen in the pelvis.

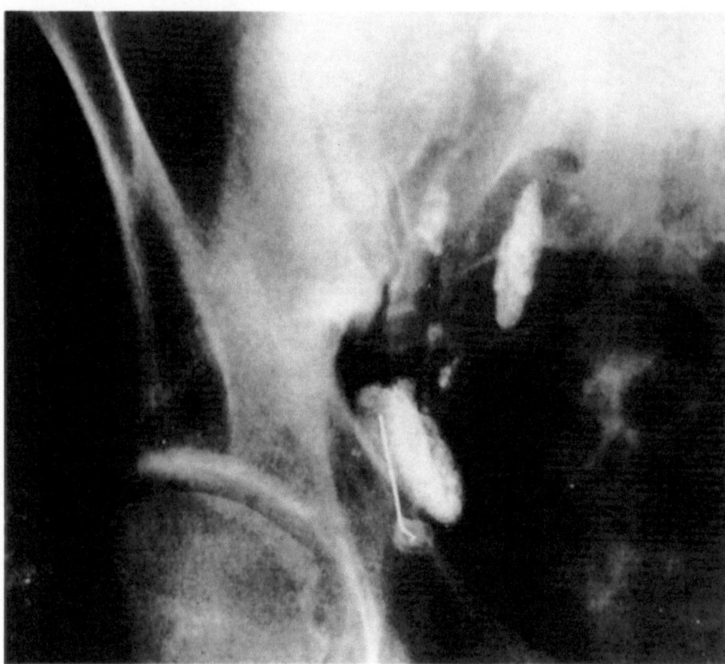

Fig. 4.5. Fine needle aspiration of pelvic lymph node, AP exposure. Same patient as in Figs. 4.2–4.4. A fine needle is located in the pelvic lymph node, but, because of overlapping, it is impossible to define the lymph gland where the needle is located. Only the fact that a short distance was passed to enter the node suggests the location of the needle in the external iliac gland.

again, it requires a great deal of expertise. Recent reports reveal a high incidence of false negative and, more important, a significantly high incidence of false positive interpretations of CT scanning. The sensitivity of this method was reported recently by Weinerman [11] as 66% with a 10% false positive rate. The presence of enlarged benign lymph nodes or a poorly opacified or non-opacified loop of bowel may be interpreted as enlarged positive nodes. Pelvic blood vessels are also frequently confused

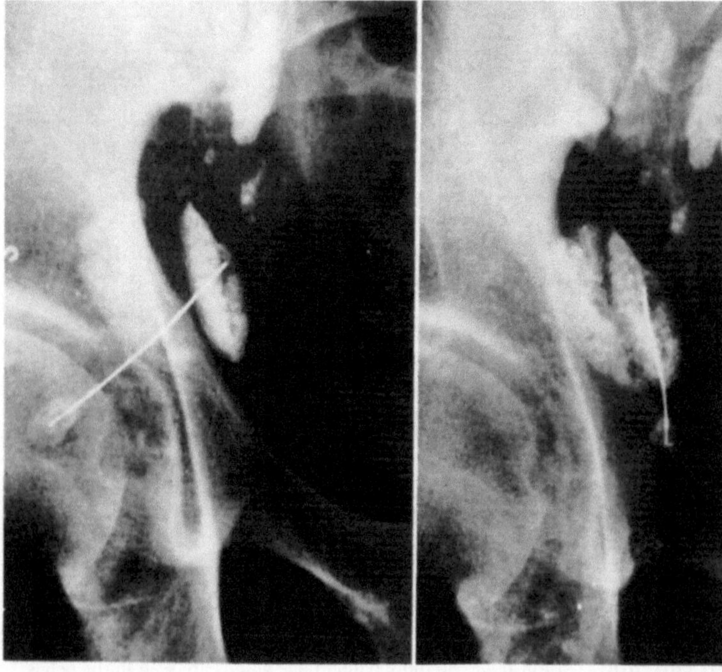

Fig. 4.6. Fine needle aspiration of obturator node, oblique view. Same patient as in Fig. 4.5. The fine needle is located in the surgical lymph node (compare with Fig. 4.3).

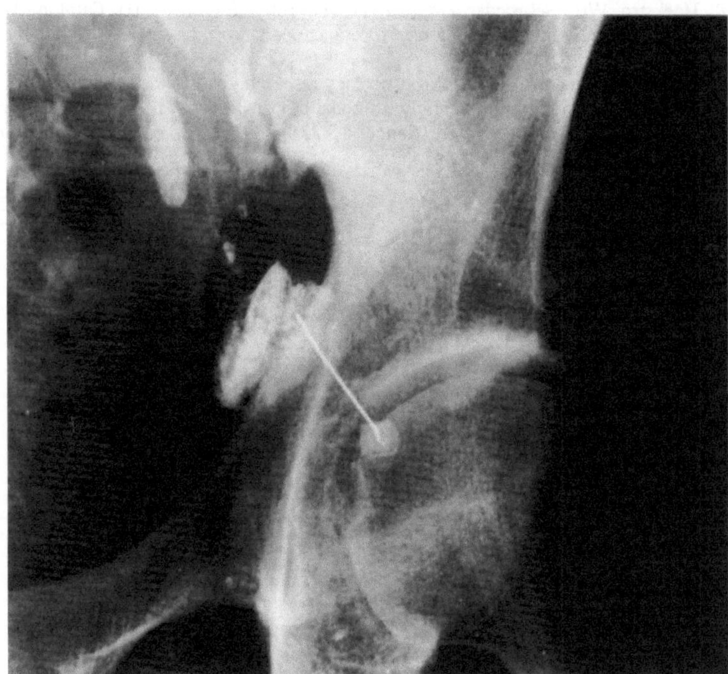

Fig. 4.7. Fine needle aspiration of external iliac node, same patient, same site (but mirror effect). The fine needle is located in the external lymph node. Observe the much shorter distance needed to enter this node.

with pelvic lymph nodes. At present, CT can demonstrate nodes of 2 cm, but it will rarely demonstrate a filling defect in such nodes. It cannot be used to detect microscopic lymph node involvement. On the other hand, a CT scan may reveal enlarged lymph nodes in the areas not demonstrated by lymphography, such as lymph nodes in the presacral or hypogastric area.

Computed tomography remains a powerful tool as a relatively accurate staging method, providing a great deal of information on anatomy and pathology of the retroperitoneum and abdominal organs. When positive it can easily be used to direct a fine needle towards a point of interest. At present, it remains the most frequently used method of staging of prostate and bladder cancer, while lymphography is reserved for patients who have a negative CT scan but who are at high risk for lymphatic metastases (see subsequent chapters).

Magnetic Resonance Imaging

The new modality of magnetic resonance imaging (MRI), with its ability to perform direct sagittal, coronal and oblique imaging, is currently under evaluation for its clinical usefulness. MRI has the potential to show the motion of blood within the blood vessels and to help distinguish pelvic vessels from adjacent lymph nodes. We can not yet detect a significant advantage of MRI over CT in evaluation of lymph nodes, but this method is a very accurate one in the determination of local tumour extension and its relation to other organs.

References

1. Herman PG (1974) The normal lymphogram. In: Gooneratne BWM (ed) Lymphography—clinical and experimental. Clowes, London
2. Merrin C, Wajsman Z, Baumgartner GC, Jennings E (1977) The clinical value of lymphography: are the nodes surrounding the obturator nerve visualised? J Urol 117:762–764
3. Zoretic SN, Wajsman Z, Badeley SA, Pontes JE (1983) Filling of the obturator nodes in pedal lymphangiography: fact or fiction? J Urol 129:533–535
4. Cuneo B, Marcille M (1901) Topographie des ganglions ilio pelviens. Bull Mem Soc Anat (Paris) pp 653–663
5. Rouviere H (1932) Anatomie des lymphatiques de l'homme. Systeme lymphatique de l'abdomen et du bassin. Masson, Paris, pp 283–293
6. O'Donoghue EPN, Shridhar P, Sherwood T, Williams JP, Chisholm GD (1976) Lymphography and pelvic lymphadenectomy in carcinoma of the prostate. Br J Urol 48:689–696
7. Grossman IC, Carpiniello V, Greenberg SH, Malloy TR, Wein AJ (1980) Staging pelvic lymphadenectomy for carcinoma of the prostate: Review of 91 cases. J Urol 124:632–634

8. Hoekstra WG, Schroeder FM (1981) The role of lymphangiography in the staging of prostatic cancer. Prostate 2:433–440
9. Liebner EG, Stefani S (1980) Uro-oncology research group. An evaluation of lymphography with nodal biopsy in localized carcinoma of the prostate. Cancer 45:728–734
10. Catalona WJ (1984) Prostate cancer. Grune and Stratton, Orlando
11. Weinerman PM, Arges PH, Coleman BG, Pollack HM, Banner MP, Wein AJ (1983) Pelvic and bladder adenopathy from bladder and prostate carcinoma: detection by rapid sequence computed tomography. AJR 140:95–99

Chapter 5

Lymphangiography in Cancer of the Genitourinary Tract

B-S. Jing and S. Wallace

The diagnosis and management of the patient with genitourinary malignancy is a multidisciplinary approach by the urologist, diagnostic radiologist, pathologist, radiotherapist and medical oncologist. Management depends, to a great extent, upon a thorough assessment of the extent of the neoplastic disease. Lymphangiography (LAG) is employed in patients with cancer of the testes, bladder, prostate and penis in the evaluation of the pelvic and para-aortic lymph nodes. Using rigid criteria, when the lymphangiogram is considered positive it has an accuracy of 90%–95%; when the lymphangiogram is considered negative, 15%–20% of patients prove to have metastatic disease.

Testis

Lymphatic Drainage

Testicular lymphatics accompany the internal spermatic artery and vein and drain into the lumbar nodes. The right trunks terminate in the right lumbar glands, which lie between the level of the aortic bifurcation and the renal vein. In 10% these trunks end in a node in the angle between the renal vein and the inferior vena cava. The left testicular lymphatics drain into the para-aortic nodes near the left renal vein in approximately two-thirds of the patients. They may also end in glands at the level of the bifurcation of the aorta or even into the common iliac nodes (Fig. 5.1) [1].

Lymphatics from the epididymis may accompany the testicular lymphatics to the lumbar nodes or terminate in the external iliac nodes. Crossover of right lymphatics to the left may be seen in the para-aortic area. Crossover from left to right is unusual.

LAG in Cancer of the Testis

Nodal metastases from testicular malignant disease show several different architectural patterns. The single most reliable criterion for the diagnosis of metastatic disease is a defect in a node not traversed by lymphatics (Fig. 5.2) [2]. This defect is usually at the periphery; the remaining functioning portion of the node is frequently crescent shaped. The lymphatics leading to the defect are interrupted by the destructive process in the node. A similar configuration may be produced by most carcinomas as well as certain lymphomas, metastatic sarcoma, abscess formation, caseation necrosis and fibrosis. A node totally replaced by neoplasm is not opacified, although the lymphatics may be distorted or obstructed. Obstruction of these channels may result in visualisation of collateral pathways, that is, lymphatic to lymphatic, lymphatic to prelymphatic, and lymphatic to venous.

Nodal metastases from testicular neoplasms occasionally may have an abnormal internal archi-

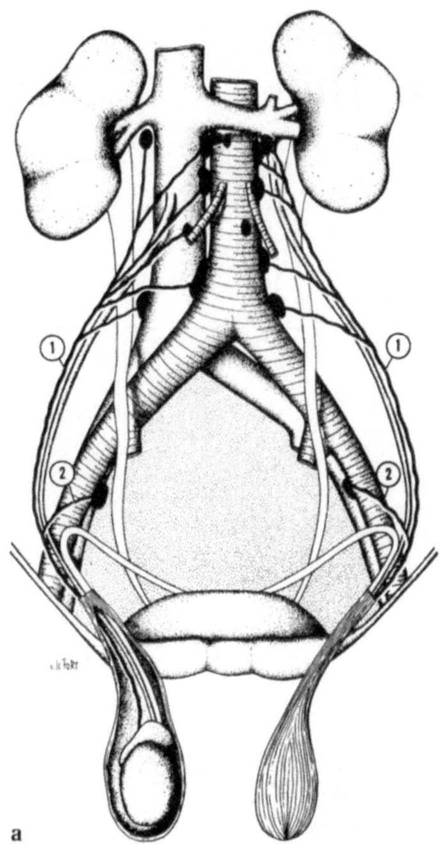

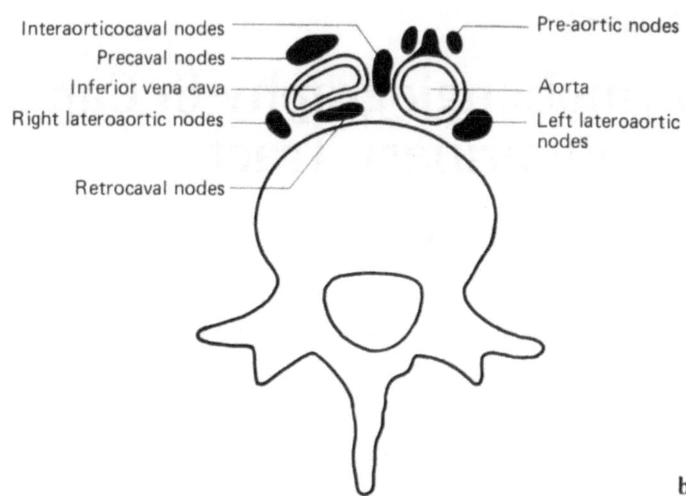

Fig. 5.1. a Lymphatics of the testes: *1*, para-aortic (lumbar); *2*, external iliac. **b** Various groups of abdominoaortic lymph nodes. (Courtesy of The University of Texas System Cancer Center, M.D. Anderson Hospital and Tumor Institute, Houston, Texas)

tecture with a relatively intact marginal sinus, most frequently seen in lymphomas. This picture is seen in some seminomas and rhabdomyosarcomas as well as in lymphomas of the testicle (Fig. 5.3) [3]. Therefore, metastases from testicular neoplasms may show most commonly a "carcinoma" pattern and less frequently a "lymphoma" pattern or even a mixed pattern.

Our experience with 291 cases of testicular malignancies studied by pedal LAG is described in Table 5.1. Patients with seminoma, of whom 24% had abnormal findings by LAG, were treated by radiotherapy alone, making nodal correlation impossible. All other testicular malignancies with lymph node metastases were managed by a combination of surgery, chemotherapy and radiotherapy for the primary neoplasm, permitting closer scrutiny of lymphangiographic findings. Surgical exploration was not uniformly undertaken, especially in patients with very advanced disease. More recently, metastatic carcinoma from the testicle to the para-aortic nodes is treated by systemic chemotherapy. Surgery is only resorted to for residual disease including cystic teratomatous transformation of choriocarcinoma.

Table 5.1. Lymphangiograms of carcinoma of the testicle (291 cases)

	Positive	Negative
Seminoma	28	76
Carcinoma	76	105
Rhabdomyosarcoma	2	4
Total	106	185

Surgical or autopsy findings or both in 121 cases were correlated with lymphangiographic findings (Table 5.2). Retroperitoneal node dissections were performed at intervals following radiotherapy. Of the 50 dissections in which positive nodes were

Table 5.2. Lymphangiographic pathological correlation in 121 surgical and/or autopsy cases

Positive	37	Carcinoma	117
False negative	13	Rhabdomyosarcoma	4
Negative	70[a]		
False positive	1		
Total	121		121
Sensitivity	37 of 50 = 74%		
Specificity	70 of 71 = 98.6%		
Accuracy	107 of 121 = 88.4%		

[a] For the purpose of correlation and patient management, all equivocal or suspicious readings were considered negative.

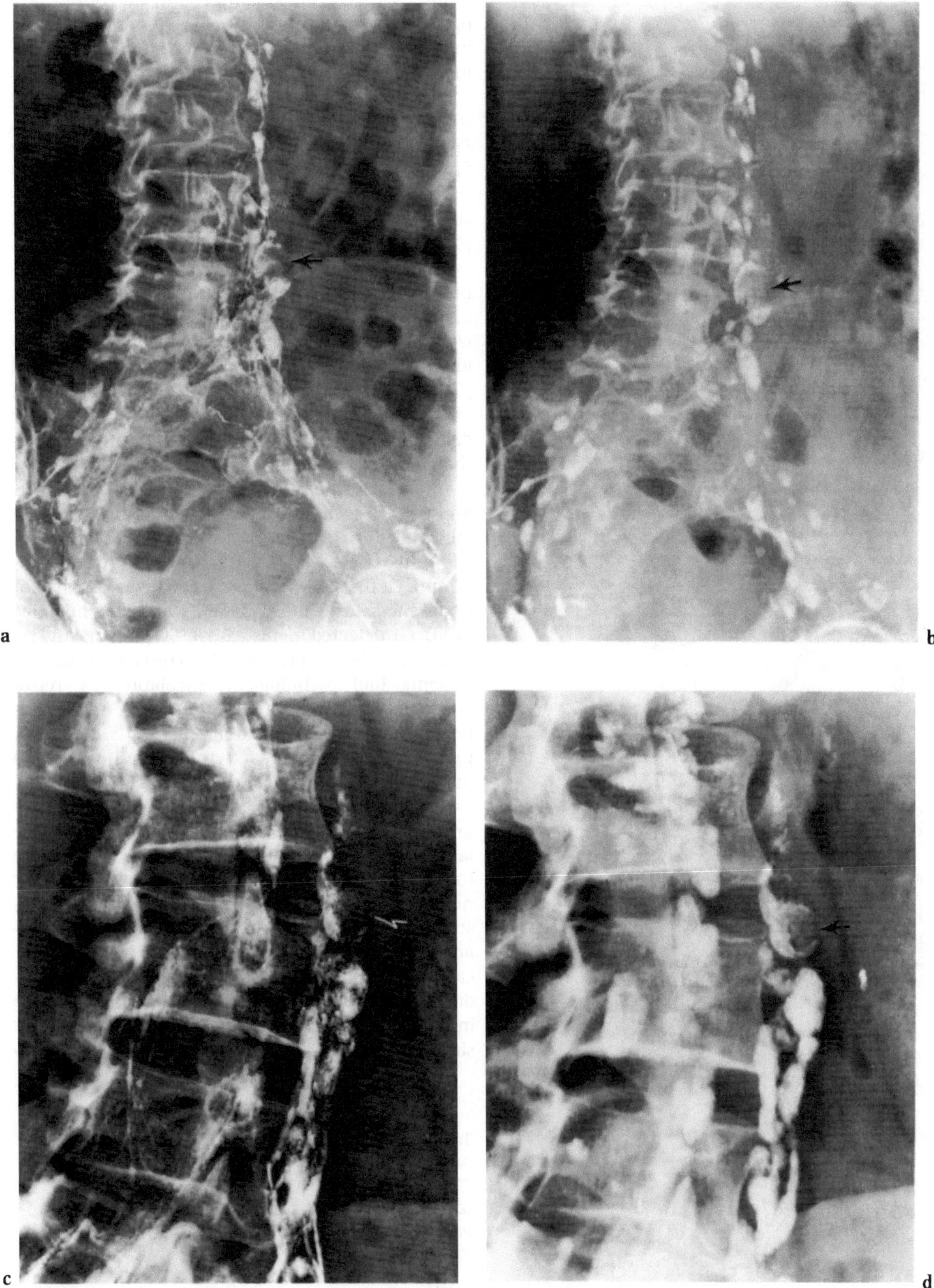

Fig. 5.2a–d. Carcinoma of the testis: nodal metastases with "carcinoma" pattern. **a, b** Ipsilateral metastases from carcinoma of the left testis. **a** Lymphatic phase; **b** nodal phase. **c, d** Cross-over metastases from carcinoma of the right testis. **c** Lymphatic phase; **d** nodal phase.

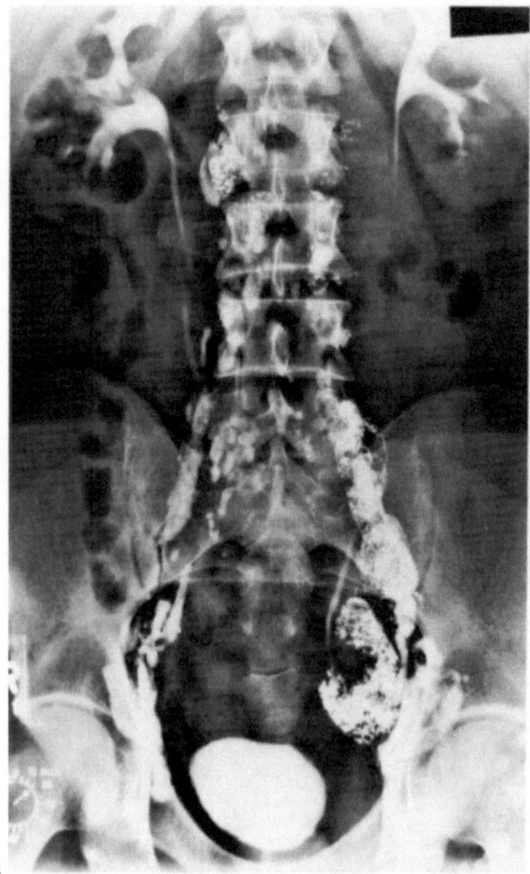
a

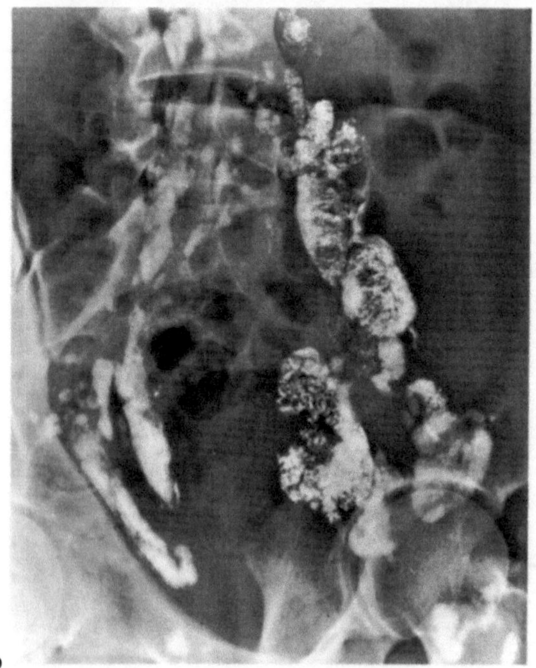
b

Fig. 5.3a,b. Seminoma of the testis: nodal metastasis with "lymphoma" pattern. **a** Lymphatic phase; **b** nodal phase.

found, 37 were in patients (74%) diagnosed preoperatively by pedal LAG. Positive radiographic interpretation had excellent correlation with pathological findings (97%). Of a total of 83 patients whose LAG findings were considered negative, 70 (84%) had negative findings at surgical exploration. Eight of the 13 patients who exhibited false negative findings were found to have metastases lateral to those usually opacified by pedal LAG; these metastases may have been diagnosed by testicular LAG [4].

When extensive disease is present in the retroperitoneum, computed tomography (CT) becomes a useful adjunct to the lymphangiogram because CT is most accurate at defining nodal enlargement and identifying the presence of tumour masses (Fig. 5.4) [5, 6]. Thus the two studies are complementary, one defining intranodal architecture and the other delineating extranodal tumour extension as well as the enlarged unopacified lymph nodes.

In our institution, in a comparative study of LAG and CT, CT scans were performed within a month following LAG in 103 patients with carcinoma of the testis (82 with carcinoma and 21 with seminoma). Of these 103 patients, LAG was positive for nodal metastases in 53 patients and was negative in 50. CT detected metastases in 50 patients and no metastases in 53 (Table 5.3). Thirty-nine of 103 patients had pathological correlation by retroperitoneal lymph node dissection or percutaneous needle biopsy (34 with carcinoma and 5 with seminoma). LAG was proved to be positive in 29 patients, with an overall accuracy of 92.3%. Twenty-six patients had positive CT findings, with an overall accuracy of 84.6% (Table 5.4). The five false negatives in the CT group (12.8%) were due to small lesions less than 1.5 to 2.0 cm. Two false negatives in LAG (5.1%) resulted from a microscopic lesion in one case, and a lesion in the preaortic and the interaorticocaval regions in another case. One false positive (2.5%) in the CT group was due to lymphoid hyperplasia, and one false positive in the LAG group was caused by a benign lesion, sinus histiocytosis. In advanced lesions of car-

Table 5.3. Findings of LAG and CT in 103 patients with carcinoma of the testis[a]

	LAG	CT
Positive	53	50
Negative	50	53
Total	103	103

[a] From Jing and Wallace [19]; reproduced by kind permission of the publisher, Williams and Wilkins Co., Baltimore, Maryland.

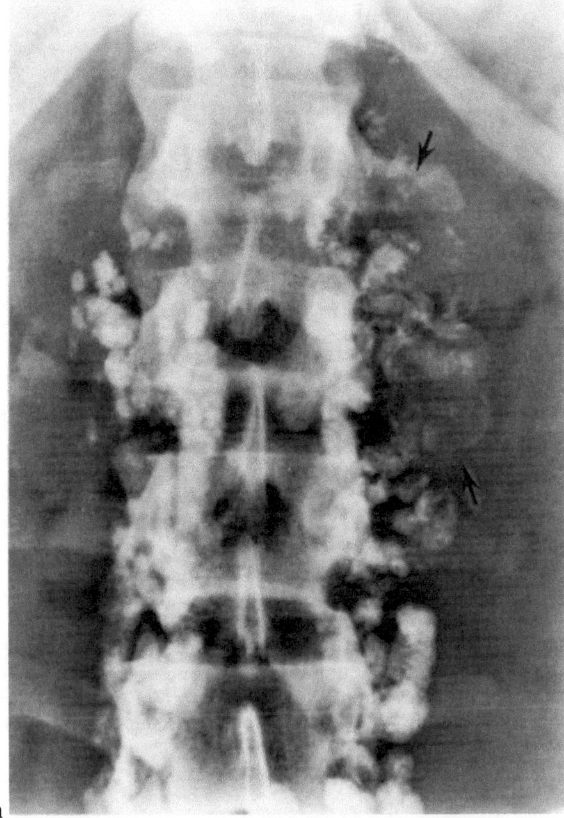

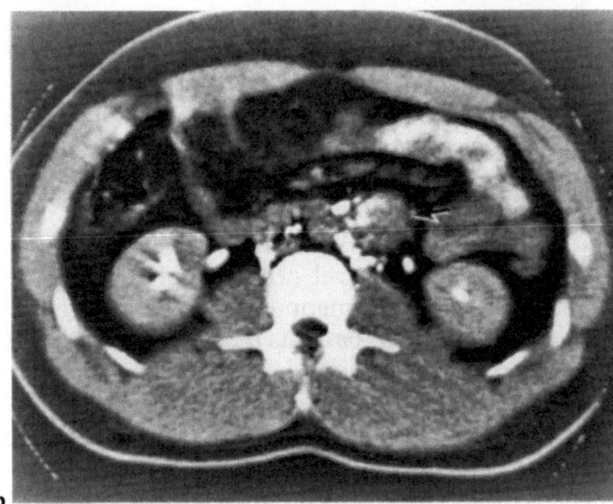

Fig. 5.4a,b. Carcinoma of the left testis. **a** Lymphangiogram showing nodal metastases with irregular deposition of contrast medium in left para-aortic area (*arrow*). **b** CT scan showing nodal metastases, partially opacified and non-opacified (*arrow*).

cinoma of the testis, LAG often failed to reveal the upper limits of the lesion; however, CT is of definite value in demonstrating the extent of nodal metastases and the involvement of the adjacent organs. Sometimes when the nodal lesion is high in the renal hilar region, it is often partially visualised by LAG, and CT offers more information to delineate the extent of the lesion (Fig. 5.5) [6–8].

Table 5.4. Pathological correlation in 39 patients with carcinoma of the testis[a]

	LAG	CT
Positive	29	26
False negative	2	5
Negative	7	7
False positive	1	1
Total	39	39
Sensitivity	29 of 31 = 93.5%	26 of 31 = 83.8%
Specificity	7 of 8 = 87.5%	7 of 8 = 87.5%
Accuracy	36 of 39 = 92.3%	33 of 39 = 84.6%

[a] From Jing and Wallace [19]; reproduced by kind permission of the publisher, Williams and Wilkins Co., Baltimore, Maryland.

Urinary Bladder

Lymphatic Drainage

The lymph drainage of the bladder is by three routes: the collecting trunks of the trigone, the collecting trunks of the posterior wall, and the collecting trunks of the anterior wall. The first echelon of drainage for the lymphatics of the urinary bladder is represented by the external iliac nodes, particularly the middle and medial groups and occasionally by hypogastric nodes and lateral sacral nodes of the common iliac group (Fig. 5.6) [1].

LAG in Cancer of the Urinary Bladder

LAG opacifies the external and common iliac lymph nodes. On occasion, hypogastric nodes close to the origin of the internal iliac artery may be demonstrated. Negative findings may result even though the paravesical, hypogastric or the occasional node present in the obturator fossa may be affected by metastases, for these nodes escape visualisation.

The diagnostic criteria for nodal metastases in carcinoma of the bladder are the same as those described earlier. The single most reliable criterion is a filling defect in a node not traversed by lymphatics (Fig. 5.7) [9]. When a node is completely replaced, the lymphatics may be displaced or obstructed (Fig. 5.8). Lymphatic obstruction with or without col-

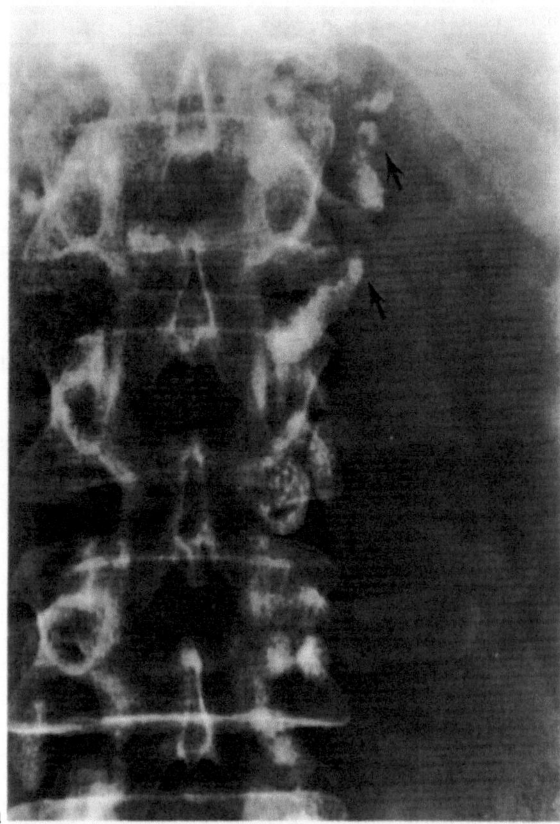

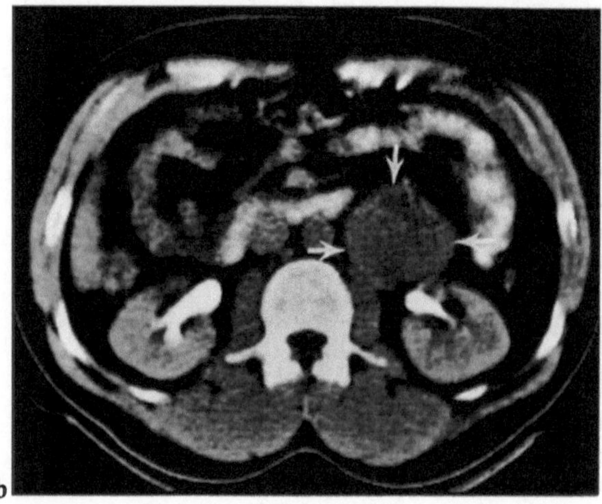

Fig. 5.5a,b. Carcinoma of the left testis. **a** Lymphangiogram showing irregular nodes in left upper para-aortic area (*arrow*). **b** CT scan showing nodal metastases of considerable size in left upper para-aortic area high in the renal pelvis (*arrows*).

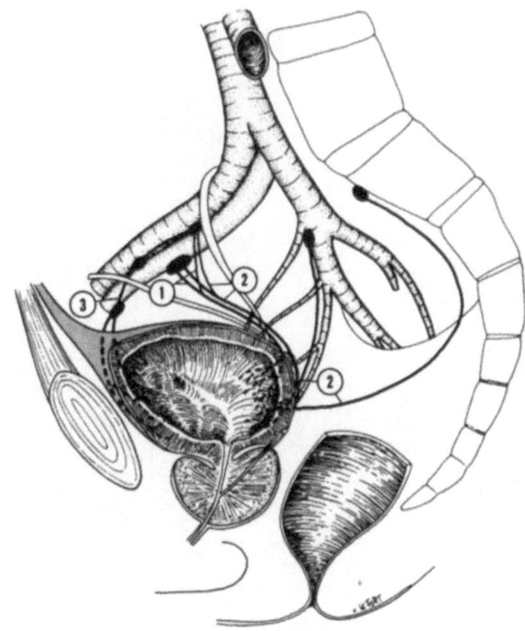

Fig. 5.6. Lymphatics of the bladder. *1*, The collecting trunks of the trigone; *2*, the collecting trunks of the posterior wall; *3*; the collecting trunks of the anterior wall. (Courtesy of The University of Texas System Cancer Center, M.D. Anderson Hospital and Tumor Institute, Houston, Texas).

lateral channels may be the sole secondary evidence of nodal replacement. Confirmation by complementary procedures, notably CT, ultrasonography, as well as magnetic resonance imaging (MRI) and at times venography [10, 11], is of value to demonstrate the presence of enlarged nodes (Figs. 5.8b, 5.9). Percutaneous lymph node biopsy has been of considerable assistance using any of the imaging modalities as a guide. A negative biopsy finding does not exclude disease.

In our institution, 91 patients with carcinoma of the bladder had bilateral pedal lymphangiograms as part of the assessment of metastatic disease. However, 60 of these patients received preoperative irradiation (50 Gy tumour dose in 5 weeks through 10 × 10 cm fields) 6 weeks prior to surgical exploration. In all cases surgical exploration included careful palpatory scrutiny of the pelvic and para-aortic lymph nodes. Realising the shortcomings of palpatory evaluation, only enlarged or suspicious lymph nodes were removed for histological examination. Three patients underwent bilateral pelvic lymphadenectomy. The results of lymphangiographic and surgical correlation are shown in Table 5.5. No false positive lymphangiographic readings were encountered. In the nine patients with confirmed nodal metastases diagnosed by LAG, an accuracy of 100%, the specific sites of the nodal involvement were: (1) the common iliac chain in four patients, with a concomitant involvement of a

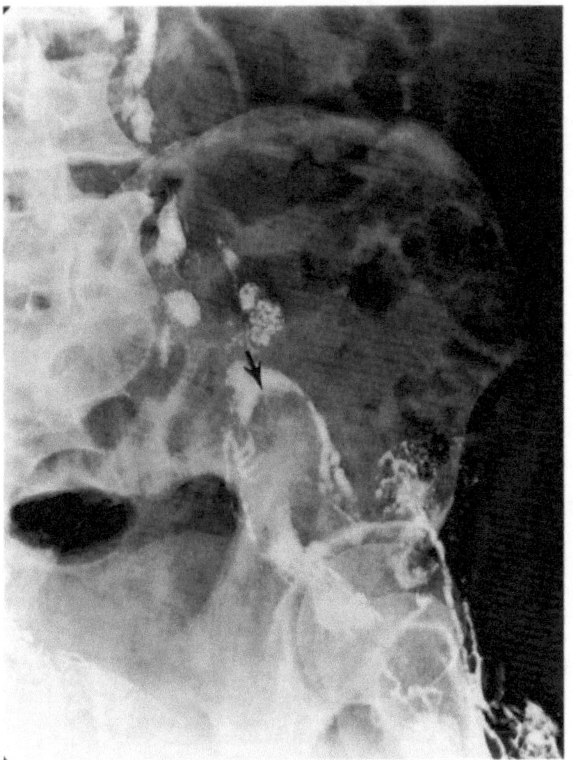
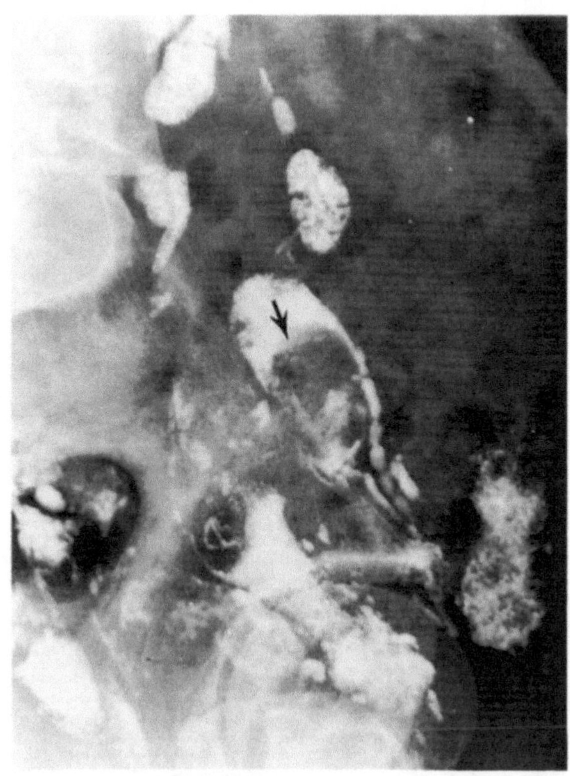

Fig. 5.7a,b Nodal metastasis from carcinoma of the bladder. **a** Lymphatic phase. Lymphatics do not permeate the areas of replacement by metastatic carcinoma of the left external iliac node (*arrow*). **b** Nodal phase. Filling defects in the node are areas of tumour deposition. The remaining functioning portion of the node is opacified, producing a crescent configuration (*arrow*).

para-aortic node in one; (2) the external iliac chain, exclusively involved in two patients; and (3) both the common and external iliac chains, involved in two patients. The remaining one patient had metastatic carcinoma of the bladder to a node involved with histiocytic lymphoma. Lymphangiograms which gave positive readings for metastatic disease were encountered more frequently with the more advanced clinical stages, as shown in Table 5.6. Our positive lymphangiogram findings in 7% of patients with stages B_1 and B_2, 14% with stage C and 40% of patients with stage D underscore the value of this procedure in reducing the number of patients subjected to unnecessary and futile surgical exploration.

Table 5.5. Carcinoma of the bladder: lymphangiographic/surgical correlation (60 surgical cases)[a]

Positive	9
False negative	5
Negative	46
False positive	0
Total	60
Sensitivity	9 of 14 = 64%
Specificity	46 of 46 = 100%
Accuracy	55 of 60 = 91.6%

[a] From Jing and Wallace [19]; reproduced by kind permission of the publisher, Williams and Wilkins Co., Baltimore, Maryland.

In our series, 67 patients with carcinoma of the bladder had lymphangiograms, followed within a month by CT scans (Table 5.7). Thirty-seven had pathological correlation with cystectomy or lymph node biopsy (Table 5.8). LAG was proved to be positive in 10, with an overall accuracy of 97.2%. On CT scanning, there were eight patients found to have nodal metastasis, with an overall accuracy of 91.8%. One false negative (2.7%) in the LAG group was due to a microscopic metastatic lesion. There were three false negatives (8.1%) on CT scans; two were due to small lesions less than 1.5–2.0 cm in diameter, and one was missed in the common iliac region because of obliquity of the projection.

The CT scan is less sensitive because the criterion—presence of a tumour-bearing node greater than 1.5 cm in size—is difficult to apply to pelvic

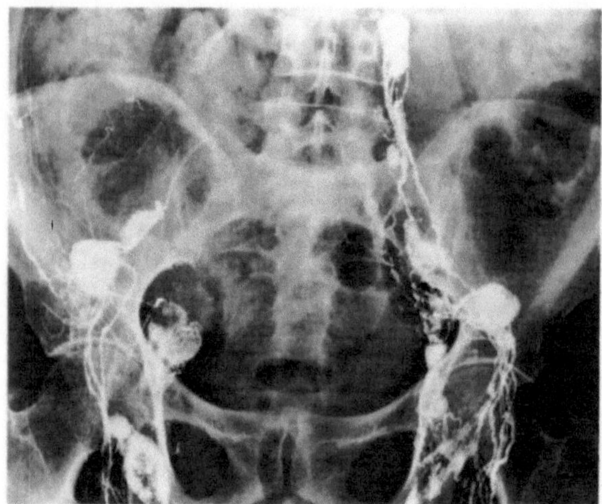

a

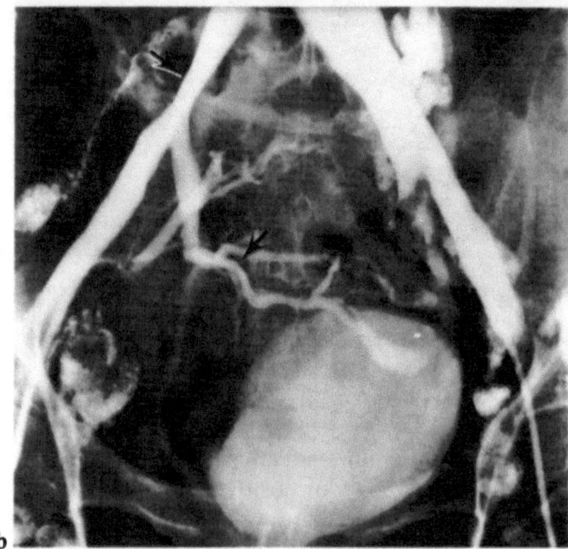

b

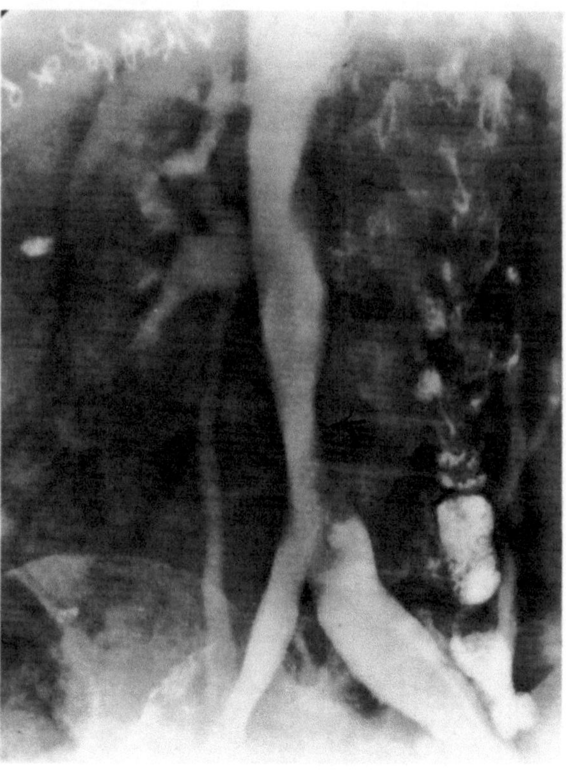

c

Fig. 5.8a–c. Carcinoma of the bladder. **a** Total replacement of nodes by metastases with lymphatic obstruction and collateral circulation. **b,c** Pelvic venography (**b**) and inferior venacavography (**c**) demonstrating extrinsic compression by the tumour masses (*arrows*).

Table 5.6. Carcinoma of the bladder: lymphangiographic/surgical correlation according to clinical stage of disease[b]

Interpretation	Clinical stage B_1	B_2	C	D	No. of patients
Positive	1	1	3	4	9
False positive	0	0	0	0	0
Negative[a]	14	12	16	4	46
False negative	1	0	2	2	5

[a] For the purpose of correlation and patient management, all equivocal or suspicious readings were considered negative.
[b] From Jing and Wallace [19]; reproduced by kind permission of the publisher, Williams and Wilkins Co., Baltimore, Maryland.

Table 5.7. Findings of LAG and CT in 67 patients with carcinoma of the bladder[a]

	LAG	CT
Positive	20	15
Negative	47	52
Total	67	67

[a] From Jing and Wallace [19]; reproduced by kind permission of the publisher, Williams and Wilkins Co., Baltimore, Maryland.

Table 5.8. Pathological correlation in 37 patients with carcinoma of the bladder[a]

	LAG	CT
Positive	10	8
False negative	1	3
Negative	26	26
False positive	0	0
Total	37	37
Sensitivity	10 of 11 = 91%	8 of 11 = 72.7%
Specificity	26 of 26 = 100%	26 of 26 = 100%
Accuracy	36 of 37 = 97.3%	34 of 37 = 91.9%

[a] From Jing and Wallace [19]; reproduced by kind permission of the publisher, Williams and Wilkins Co., Baltimore, Maryland.

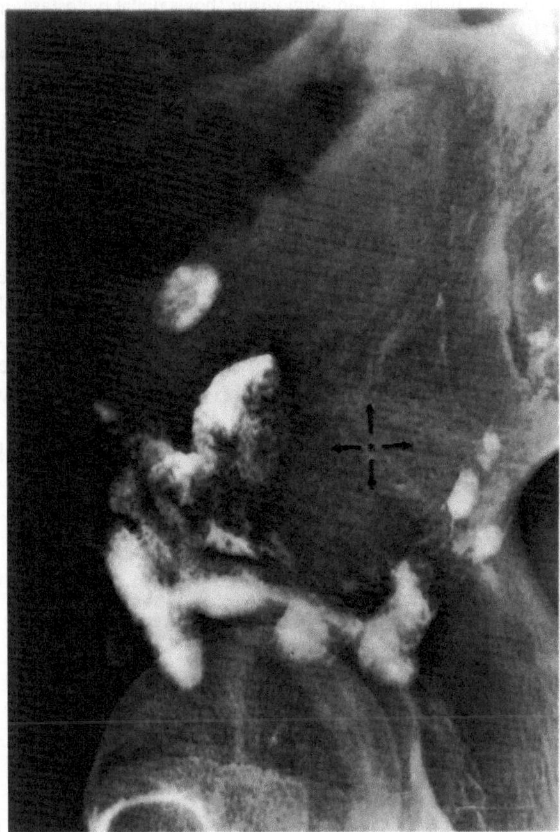

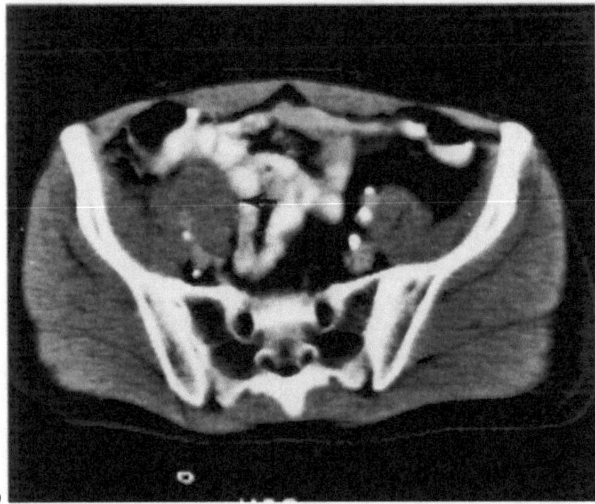

Fig. 5.9a,b. Carcinoma of the bladder. **a** Lymphangiogram shows massive nodal metastasis, mostly non-opacified, in the right external iliac region with upper margin undetermined (*asterisk*). **b** CT scan of the pelvis at the level of S-1 demonstrates non-opacified nodal metastases in the right distal common iliac region (*arrow*). The right para-aortic area is free of nodal metastases (not shown here).

nodes. Because the CT scan evaluates nodal size and the lymphangiogram evaluates nodal architecture, these studies should be complementary and, used together, provide more information than either alone [7, 12].

Prostate

Lymphatic Drainage

The first echelon of drainage from the prostate is by means of the following four collecting trunks: (1) the external iliac pedicle terminating in nodes of the middle group of the external iliac chain, (2) the hypogastric pedicle draining to the hypogastric nodes, (3) the posterior pedicle terminating in presacral lymph nodes and (4) the inferior pedicle draining to the hypogastric nodes near the origin of the internal iliac artery (Fig. 5.10). The lymphatics of the prostate communicate with those of the bladder, the seminal vesicles, and the rectum [1].

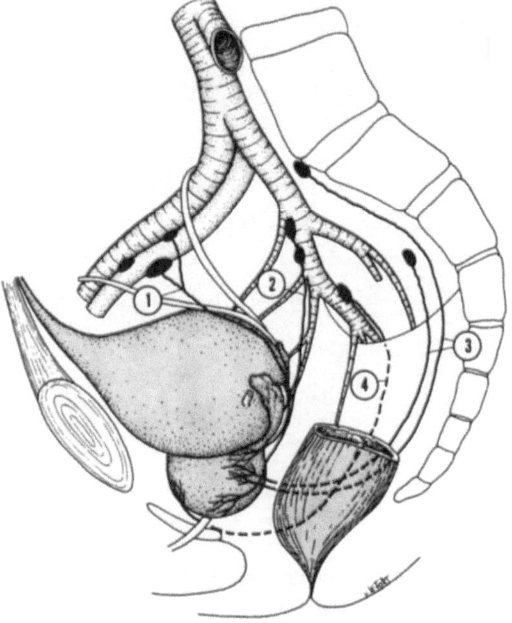

Fig. 5.10. Lymphatics of the prostate. *1*, External iliac pedicle; *2*, hypogastric pedicle; *3*, posterior pedicle; *4*, inferior pedicle. (Courtesy of The University of Texas System Cancer Center, M.D. Anderson Hospital and Tumor Institute, Houston, Texas)

Fig. 5.12a,b. Nodal metastasis from carcinoma of the prostate: "carcinoma" pattern. Lymphangiogram shows nodal metastases in the left external iliac and common iliac regions. a Lymphatic phase; b nodal phase.

→

LAG in Cancer of the Prostate

The lymphangiographic findings in nodal metastases from carcinoma of the prostate are varied and show several architectural patterns. The abnormal lymph nodes may be normal in size or enlarged, most often with marginal filling defects, that is, partial replacement, having a crescent appearance with lymphatics failing to traverse the defect, the "carcinoma" pattern (Figs. 5.11, 5.12). The involved nodes may occasionally appear moderately enlarged, with an irregular internal architecture, the "lymphoma" pattern (Fig. 5.13) [3]. Fragmented nodes with multiple filling defects may also be seen [13–15]. Lymphatic obstruction with or without collateral channels is often associated with these abnormal nodes. When the lymph node is completely replaced by tumour, the presence of mass may be confirmed by venography, ultrasound, CT, or MRI [10, 16, 17]. Percutaneous lymph node biopsy is performed when there is any doubt. Only positive biopsy findings are of significance.

Of the regional lymph nodes, only the external iliac nodes can be demonstrated on the lym-

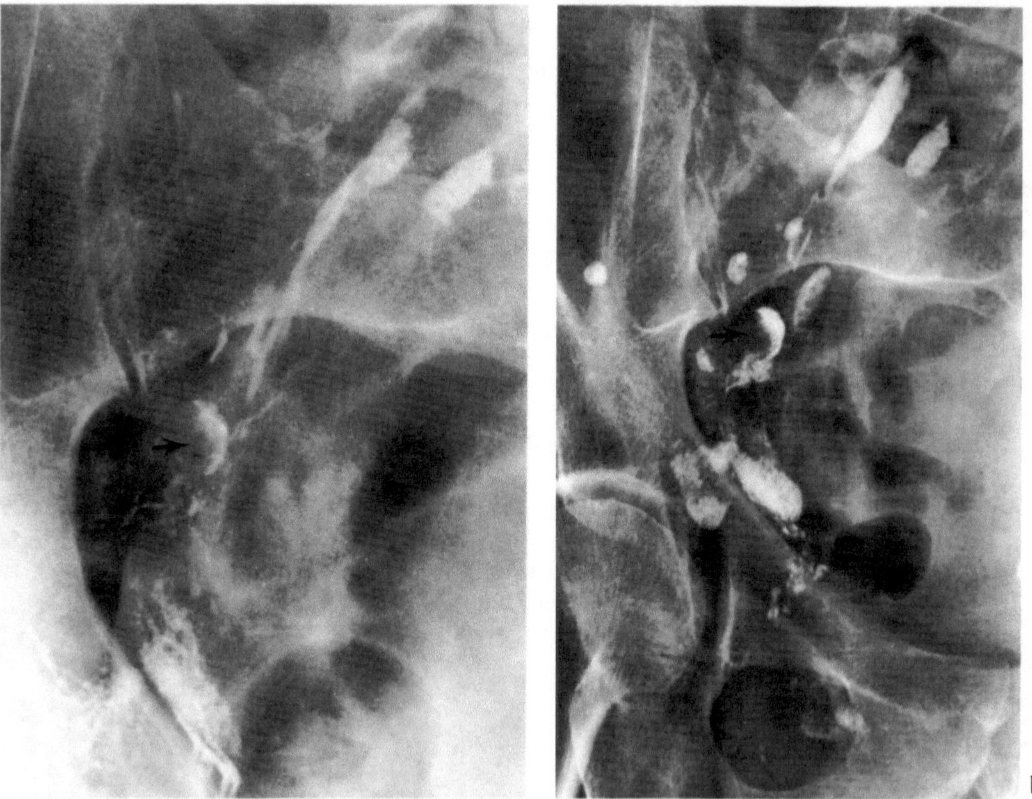

Fig. 5.11a,b. Nodal metastasis from carcinoma of the prostate: "carcinoma" pattern. Lymphangiogram shows nodal metastases in the right external iliac region. It is a typical carcinoma pattern. a Lymphatic phase; b nodal phase.

→

Fig. 5.13a,b. Nodal metastases from carcinoma of the prostate: "lymphoma" pattern. Lymphangiogram shows extensive nodal metastasis; its pattern simulates that seen in malignant lymphomas. a Lymphatic phase; b nodal phase.

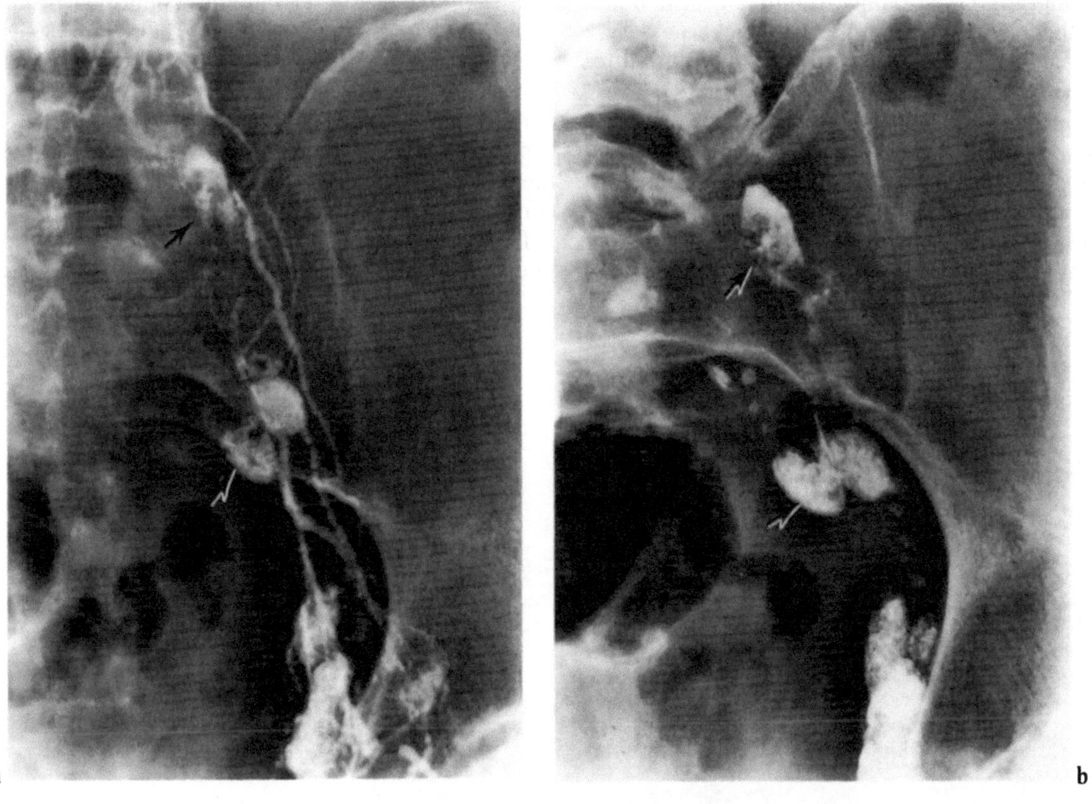

Fig. 5.12 ▲

Fig. 5.13 ▼

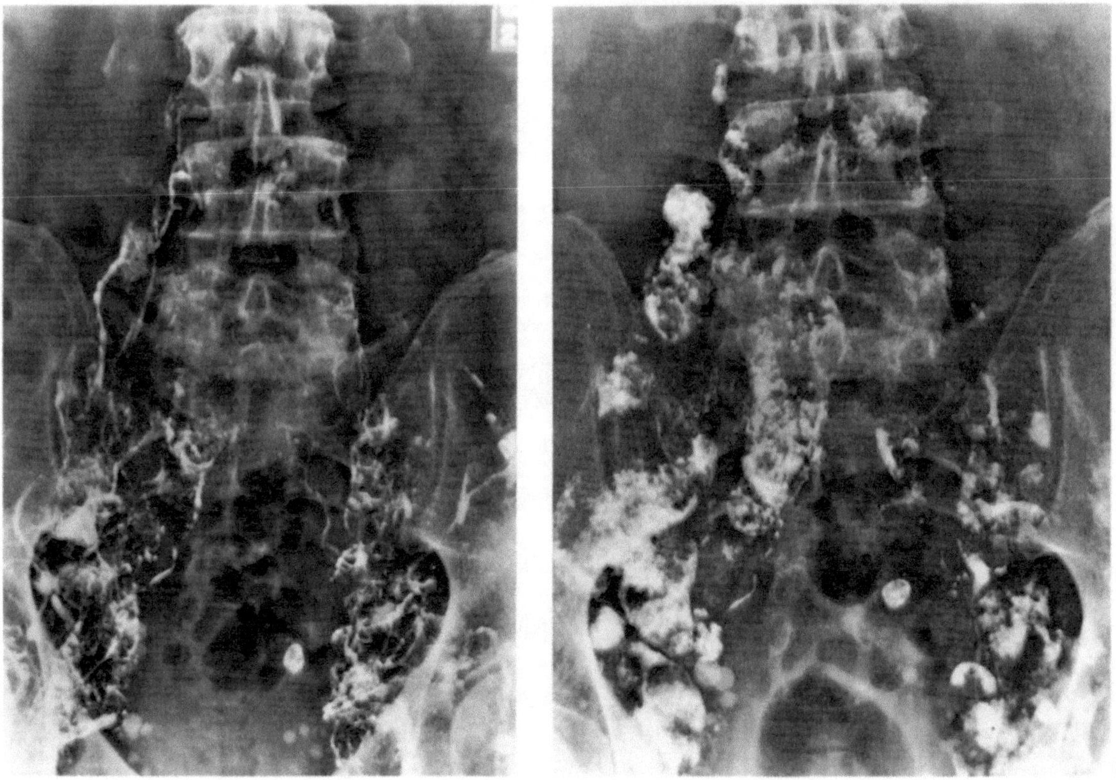

phangiogram. Occasionally, the hypogastric nodes near the origin of the internal iliac artery may be visualised. The opacification of the promontory nodes is usually inconsistent.

In our series, 208 patients with prostatic carcinoma had pedal LAG. Of these patients, 47 (23%) had nodal metastases with abnormal results on lymphangiogram, 40 in stage C and 7 in stage D. Four of these 47 had nodal architectural changes most compatible with configuration seen in lymphoma caused by metastatic carcinoma, confirmed histologically by percutaneous needle biopsy in 3. Of the 40 patients with stage C disease and abnormal results on lymphangiogram, lymph node biopsy confirmed the findings in 20 [18]. Percutaneous aspiration biopsy of the lymph nodes established metastatic disease in 11; pelvic lymphadenectomy confirmed the presence of metastases in 9.

Lymphangiographic visualisation of para-aortic nodal metastases in the absence of iliac movement is

a

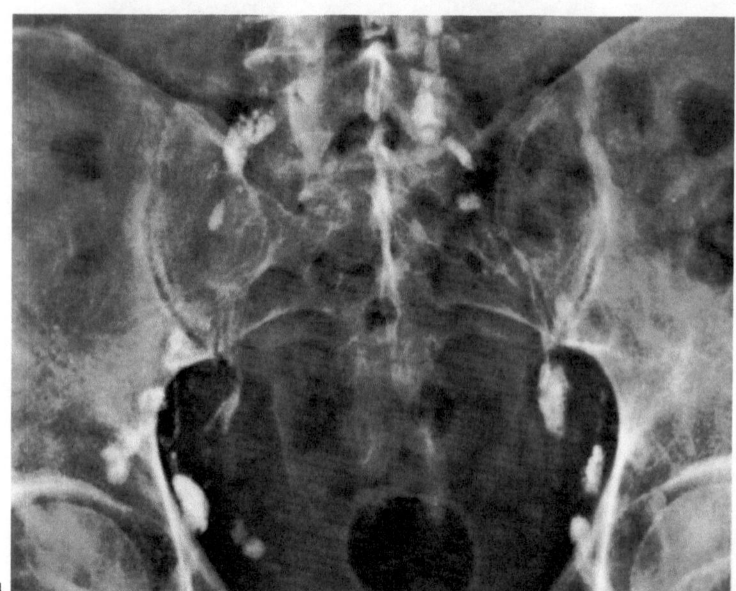

b

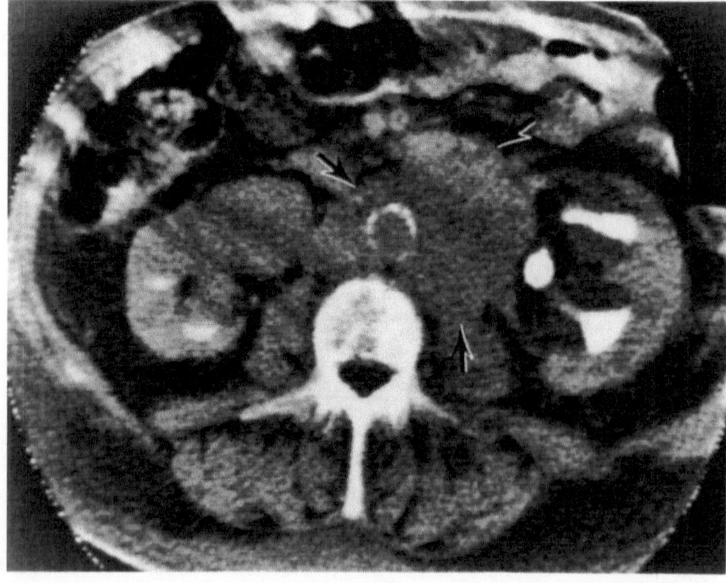

Fig. 5.14a,b. Carcinoma of the prostate with para-aortic nodal metastases. **a** Lymphangiogram shows no evidence of nodal metastases in the pelvic region. **b** CT scan shows extensive nodal metastases in para-aortic areas, mainly on the left side (*arrows*) with bilateral hydronephroses.

unusual but does occur (Fig. 5.14). This distribution could still be in continuity if the metastases in the pelvis are not opacified, that is, the hypogastric or presacral areas, or in nodes completely replaced by tumour [12]. The advantage of CT is its ability to detect enlarged nodes (greater than 1.5 cm in diameter) in hypogastric and presacral areas. Totally replaced nodes suggested by lymphatic abnormalities of obstruction with or without collateral circulation can be well delineated by CT. In advanced lesions with lymphatic obstruction in the pelvic region and non-opacification of the node in the para-aortic area, CT is of definite value in revealing the size and extent of the lesion (Fig. 5.15).

a

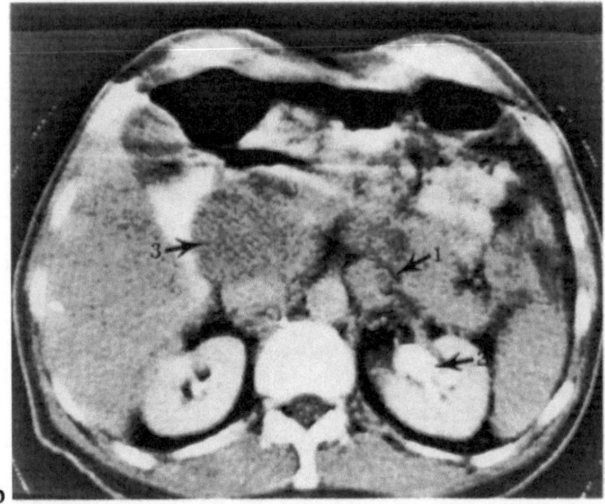

b

Fig. 5.15a,b. Advanced carcinoma of the prostate. **a** Lymphangiogram shows extensive metastases in the pelvic region with poor opacification of the lymph nodes above the level of S-1. **b** CT scan of the abdomen demonstrates nodal metastases in left para-aortic area (*1*) and with left hydronephrosis (*2*). There is a peripancreatic mass, probably caused by metastases (*3*).

Penis

Lymphatic Drainage

The lymphatics of the prepuce drain to the superior and medial groups of the superficial inguinal nodes. The lymphatics of the glans penis join those of the urethra and prepuce and proceed in the following two groups: (1) those that follow the femoral canal, terminating in the deep inguinal nodes, including the node of Cloquet and the medial retrocrural lymph nodes; and (2) those that follow the inguinal canal and drain into the lateral retrocrural lymph nodes. The lymphatics of the corpora cavernosa penis end in the superior and medial groups of the superficial inguinal nodes or the retrocrural iliac nodes (Fig. 5.16) [1]. The right and left inguinal lymph nodes have a rich communication with each other through subcutaneous lymphatics.

LAG in Cancer of the Penis

LAG by both the penile and pedal routes has been used to define lymph node involvement from carcinoma of the penis. For the most part the pedal approach is more universally performed. Only a portion of the inguinal nodes are opacified by this technique. Although the inguinal lymph nodes are frequently involved by chronic inflammatory disease, the information obtained by pedal LAG is still valid and useful. Again, the single most reliable criterion representing metastasis is a defect in a node not traversed by lymphatics (Fig. 5.17). The other criteria enumerated previously are operative here.

The imaging modality of CT is not as useful in detecting early nodal metastases as is LAG. However, in advanced lesions, CT is of definite value in detecting pelvic extension of the lesion and defining distant metastases.

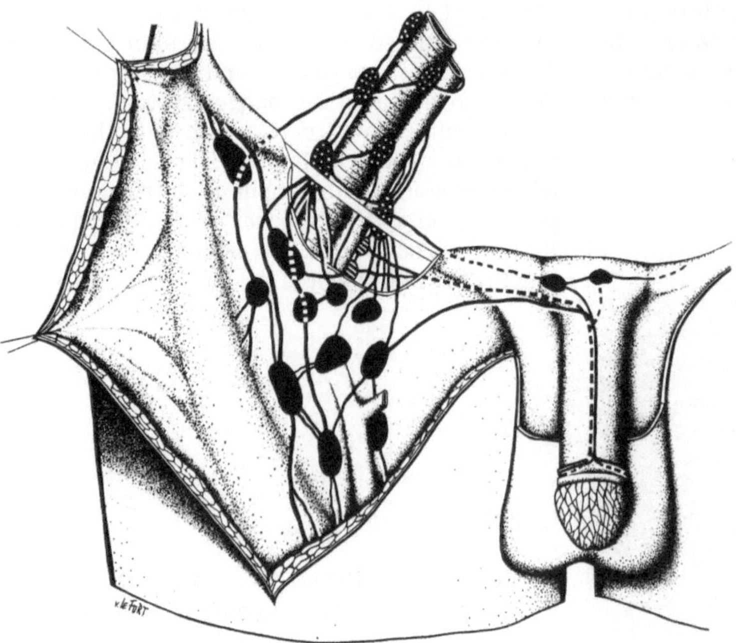

Fig. 5.16. Lymphatics of the penis. (Courtesy of The University of Texas System Cancer Center, M.D. Anderson Hospital and Tumor Institute, Houston, Texas)

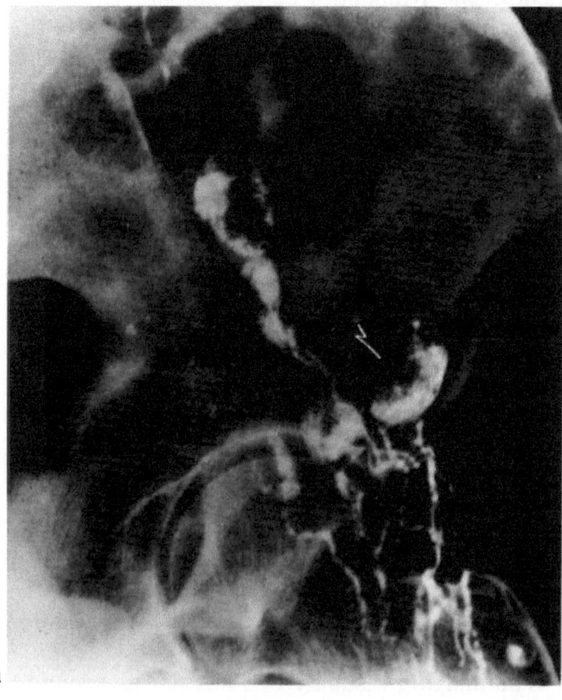

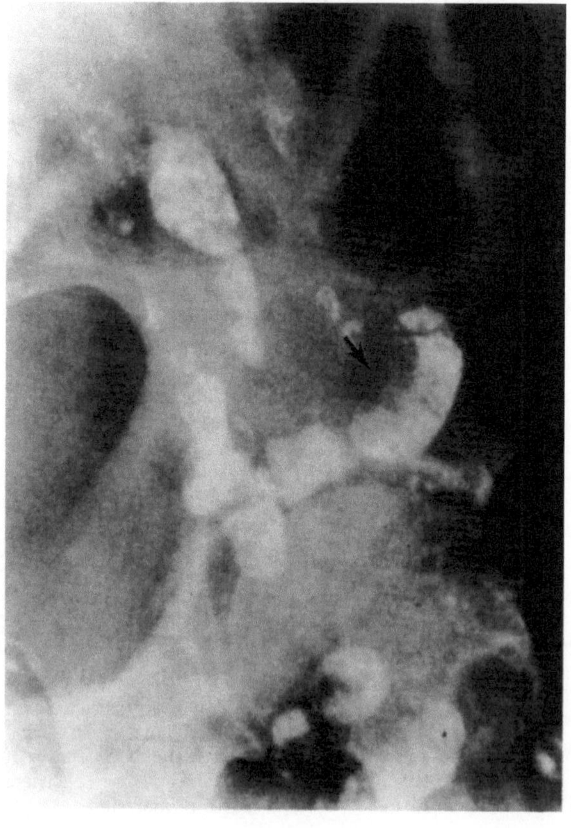

Fig. 5.17a,b. Carcinoma of the penis. Lymphangiogram showing nodal metastasis to left retrocrural node. a Lymphatic phase; b nodal phase.

References

1. Rouviere H (1938) Anatomy of the human lymphatic system. Translated by JM Tobias. Edwards, Ann Arbor
2. Wallace S, Jing BS (1970) Lymphangiographic diagnosis of nodal metastases from testicular malignancies. JAMA 213:94–96
3. Wallace S, Jing BS (1976) Testicular malignancies and the lymphatic system. In: DE Johnson (ed) Testicular tumors, 2nd edn. Medical Examination Publishing Co., Flushing, NY, pp 71–159
4. Chiappa S, Uslenghi C, Bonadonna G (1966) Combined testicular and foot lymphangiography in testicular carcinoma. Surg Gynecol Obstet 123:10–14
5. Husband JE, Peckham MJ, MacDonald JS, Hendry WF (1979) The role of computed tomography in the management of testicular teratoma. Clin Radiol 30:243–252
6. Lee JK, McClennan BL, Stanley RJ, Sagel SS (1976) Computed tomography in the staging of testicular neoplasms. Radiology 130:387–390
7. Dunnick NR, Javadpour N (1981) Value of CT and lymphography: distinguishing retroperitoneal metastases from non-seminomatous testicular tumors. AJR 136:1093–1099
8. Ellis JH, Bies JR, Kopecky KK, Klatte EC, Rowland RG, Donohue JP (1984) Comparison of NMR and CT imaging in the evaluation of metastatic retroperitoneal lymphadenopathy from testicular carcinoma. J Comput Assist Tomogr 8:709–719
9. von Eschenbach AC, Jing BS, Wallace S (1985) Lymphangiography in genitourinary cancer. Urol Clin North Am 12:715–723
10. Arger PH (1985) Computed tomography of the lower urinary tract. Urol Clin North Am 12:677–686
11. Koss JC, Arger PH, Coleman BC, Mulhern CB, Pollack HM, Wein AJ (1981) CT staging of bladder carcinoma. AJR 137:359–362
12. Jing BS, Wallace S, Zornoza J (1982) Metastases to retroperitoneal and pelvic lymph nodes: computed tomography and lymphangiography. Radiol Clin North Am 20:511–530
13. Castellino RA (1975) The role of lymphangiography in "apparently localized" prostatic carcinoma. Lymphology 8:16–20
14. Prando A, Wallace S, von Eschenbach AC, Jing BS, Rosengren JE, Hussey DH (1979) Lymphangiography in staging of carcinoma of the prostate. Radiology 131:641–645
15. Van Engelshoven JMA, Kreel L (1979) Computed tomography of the prostate. J Comput Assist Tomogr 3:45–51
16. LeVine MS, Arger PH, Coleman BG, Mulhern CB, Pollack HM, Wein AJ (1981) Detecting lymphatic metastases from prostatic carcinoma: superiority of CT. AJR 137:207–211
17. Rifkin MD (1985) Ultrasonography of the lower urinary tract. Urol Clin North Am 12:645–656
18. von Eschenbach AC, Zornoza J (1982) Fine-needle percutaneous biopsy. Urology 20:589–590
19. Jing BS, Wallace S (1985) Lymphatic imaging of solid tumors. In: Clouse ME, Wallace S (eds) Lymphatic imaging: lymphography, computed tomography and scintigraphy, 2nd edn. Baltimore, Williams and Wilkins, pp 290–451

Chapter 6

Experience with Modified Chiba Needle in Staging of Urological Cancer

L. Luciani

Introduction

Although sporadic information of fine needle aspiration biopsy (FNAB) on various tissues and organs has been reported since 1847 [1, 2], the technique has been more widely used as a diagnostic procedure only since the 1930s [3]. FNAB was initially employed to obtain samples of tissue from superficial lesions or palpable organs. However, increased expertise in its use and the introduction of new imaging techniques as guidance for the needle, as well as the increased experience of cytopathologists, have allowed the extension of this procedure to deep sited organs and lesions.

In the field of urological neoplasms the transrectal FNAB, as originally described by Franzèn et al. (1960) [4], has been widely applied in diagnosing prostatic carcinoma because of its safety and accuracy. Encouraging results have been obtained by several authors [5–30] in the preoperative "N" categorisation of urological neoplasms by means of aspiration biopsy cytology (ABC) of the regional nodal chains. The same technique has also been performed for the cytological diagnosis of equivocal primary and metastatic renal and adrenal masses, always showing high reliability and effectiveness. In this chapter I report our experience at the S. Chiara Hospital of using a modified Chiba needle[1] in the staging of bladder and prostatic carcinoma and in diagnosing the nature of primary and metastatic renal and adrenal masses.

Material and Methods

Our series includes 31 patients with bladder carcinoma ranging in age from 35 to 72 years (mean 61 years), 37 patients with biopsy-proved prostatic carcinoma aged from 51 to 75 years (mean 65 years) and 15 cases of renal (7) and adrenal masses (8). All patients with prostatic and bladder carcinoma underwent fluoroscopic guided ABC of the pelvic nodes for staging purposes.

The needle we used was derived from the standard Chiba needle. The original 45° bevel was reduced to 15°, and two small lateral faces improved penetration ability. Two or four elliptical fen-

[1] Cyto-Aspir, patent pending. Cook Urological Inc., Spencer, Ind., USA.

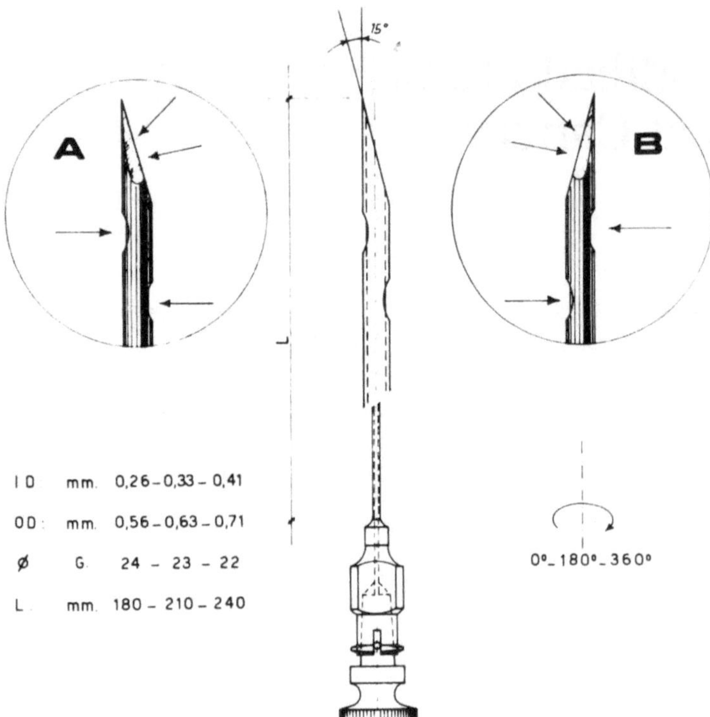

Fig. 6.1. Modified Chiba needle, details of the tip. The bevel angle of 15° and the multiple aspirating sites (*A* and *B*) are illustrated. (Luciani and Piscioli [27]; reproduced by kind permission of the Editor of *Urology* and the publisher, Hospital Publications Inc.)

estrations were made just under the bevel on opposite walls of the needle to permit the aspiration of a large quantity of cytological material. During penetration the holes were closed by a removable obturator which did not affect the flexibility of the instrument and its guidability. The needle was 22, 23 and 24 gauge with an outer diameter of 0.71, 0.63 and 0.56 mm, an inner diameter of 0.41, 0.33 and 0.26 mm, and a length of 240, 210 and 180 mm, respectively (Fig. 6.1).

Both cytological and lymphographic findings were classified only as positive or negative for metastatic disease. No equivocal diagnostic category was used. The nodes were aspirated regardless of their lymphographic appearance.

After bland sedation and local anaesthesia had been administered to the patient the needle was inserted through the anterior abdominal wall, up to the target node. Because of the sharpness of the needle tip, no previous skin incision was needed. When possible, the nodal chains on the same side as the tumour were aspirated first, starting from the medial margin of the nodes. Dislocations of the nodes synchronous with the movements of the needle indicated the proper placement of the tip. Subsequently, the node was penetrated and up and down movements and rotations were made with the needle in order to ensure a more adequate sampling.

The aspiration manoeuvre was accomplished using a plastic syringe handle[2] which has a more forceful vacuum aspiration than a metal-glass syringe. The four or five drops obtained from the aspiration were extruded onto separate slides and thinly spread with another slide. Subsequently, one slide was immediately fixed in 95% ethanol for 2–3 min and stained by the rapid Papanicolaou technique. If the smear was unsatisfactory the aspiration was immediately repeated. Usually, no more than three aspirations were necessary to obtain adequate cytological material. The remaining slides were then processed according to the standard Papanicolaou method.

The same needles were used to aspirate the adrenal and renal masses. In these cases the needle was passed from a posterior or anterior approach and computed tomography (CT) and ultrasonography were the guidance mechanisms. Cytological specimens from such lesions were stained with haematoxylin and eosin and processed according to standard methods. No significant complications were encountered.

[2] Cameco, Sweden.

Table 6.1. Lymphographic, cytological and histological results in the staging of 68 patients with prostatic and bladder carcinoma

Carcinoma	Primary tumour	Patients	Lymphography N+	Lymphography N−	Acc.	Sens.	Spec.	Cytology N+	Cytology N−	Acc.	Sens.	Spec.	Histology N+	Histology N−	Surgical stage
Prostatic	T1	9	6(FP6)	3				9	0				9	0	9 T1 N+M0
	T2	15	4(FP3)	11(FN5)				3	12(FN3)				6	9	6 T2 N+M0 9 T2 N0M0
	T3	13	13	0				12	1(FN1)				13	6	13 T3 N0M0
					62%	74%	50%			89%	86%	100%			
Bladder	T1	3	0	3				0	3				0	3	3 T1 N0M0
	T2	8	2(FP1)	6(FN1)				2	6				1	7	1 T2 N+M0 7 T2 N0M0
	T3	20	9(FP6)	11(FN1)				3	17(FN1)				3	11	3 T3 N+M0 11 T3 N0M0
	T4												2	4	2 T4 N+M0 4 T4 N0M0
					71%	67%	72%			97%	83%	100%			

N+, Nodal involvement; N−, no nodal involvement; FP, false positive finding; FN, false negative finding; Acc., accuracy; Sens., sensitivity; Spec., specificity.

Results

The results of lymphographic, cytological and histological investigations on the 68 patients with prostatic and bladder carcinoma are summarised in Table 6.1. A total of 233 lymph nodes were punctured and histological verification was accomplished in 166. There were no false positive results and the false negative findings numbered 11. The presence of malignant cells both on histological and cytological examination was shown in 27 nodes. The overall accuracy of the method in determining the nodal status was 93%, with a sensitivity of 71% and specificity of 100%.

The cytological findings were consistent with bilateral adrenal metastasis from renal cell carcinoma in one case, with right adrenal metastasis from prostatic adenocarcinoma in one case and from bladder carcinoma in another case, and with adrenal metastasis from lung carcinoma in one case. Diagnosis of adrenal adenoma was made in two cases, and two adrenal cysts were also punctured. As regards the renal masses the diagnosis was renal cyst in four cases, of cystic adenocarcinoma in one case and of renal adenocarcinoma in the remaining two cases.

Discussion

Accurate staging is essential in determining the prognosis and assessing the therapy of patients with urological neoplasms. Pedal lymphography is adversely affected by a high percentage of false negative and false positive results [31, 32], and staging lymphadenectomy cannot be routinely practised because of the related morbidity and even mortality [33–37]. In this connection ABC has proved to be a safe and reliable method which can improve the accuracy of pedal lymphography in determining the presence or absence of nodal metastatic disease with minimal discomfort for the patients.

In our experience and according to reports in the literature no significant complications have even been documented in the use of fine needles, nor can implantation of tumour cells along the needle track be considered of any clinical importance. Despite the fact that such an event has never been reported, consideration should be given to the general adoption of more aggressive therapeutic procedures in the case of positive findings so that the risk of metastatic dissemination by means of the puncture remains negligible.

The presence of malignant cells in the aspirate is always considered diagnostic of nodal metastasis and may spare patients with prostatic and bladder carcinoma a pelvic node dissection or even an unnecessary radical operation. False positive findings have rarely been reported [21]. On the other hand, negative cytological findings are of no value in excluding metastatic disease. False negative results are strictly related to the failure in obtaining representative and adequate cytological material from the nodes, the occurrence of micrometastasis and the possibility of errors in microscopic interpretation.

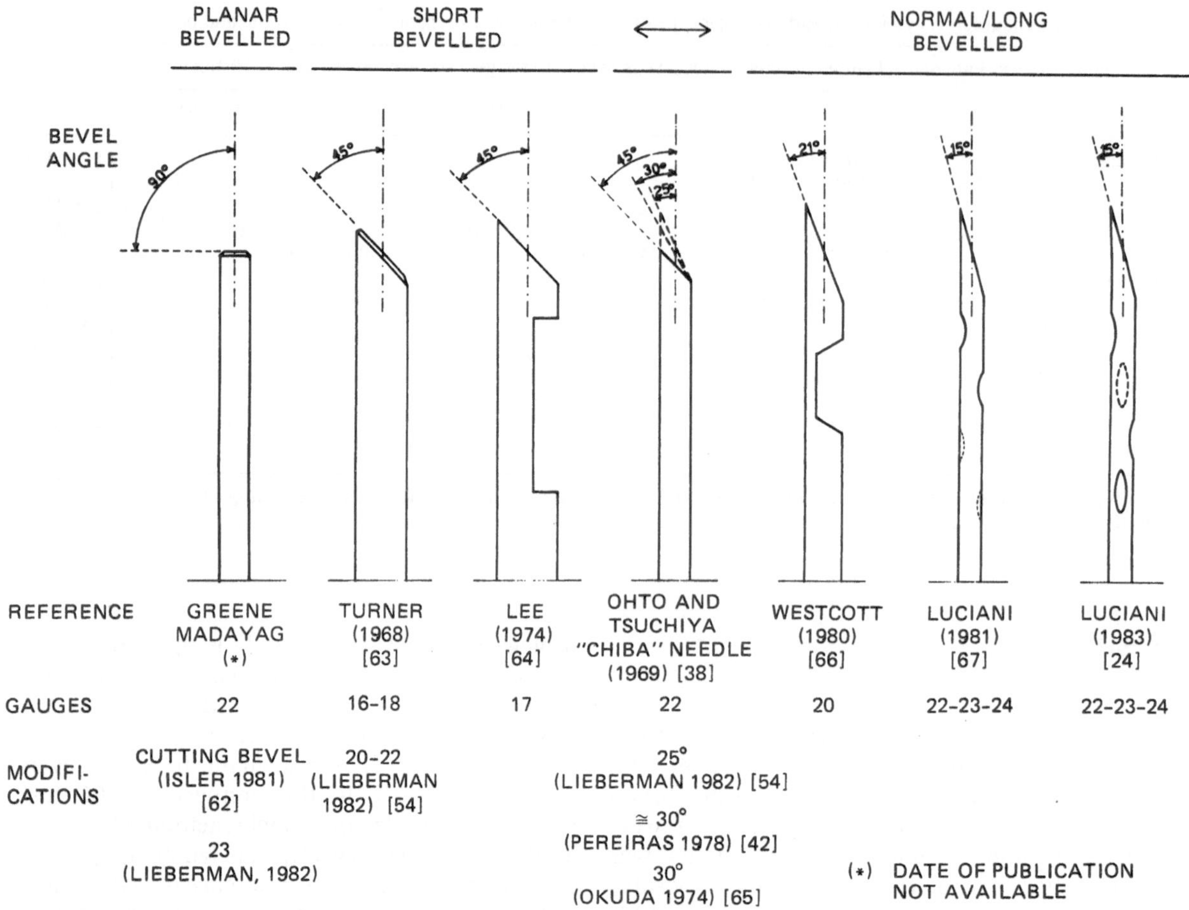

Fig. 6.2. Series of aspiration biopsy needles showing the different devices of bevel angle and aspirating openings. (Luciani and Piscioli [27]; reproduced by kind permission of the Editor of *Urology* and the publisher, Hospital Publications Inc.)

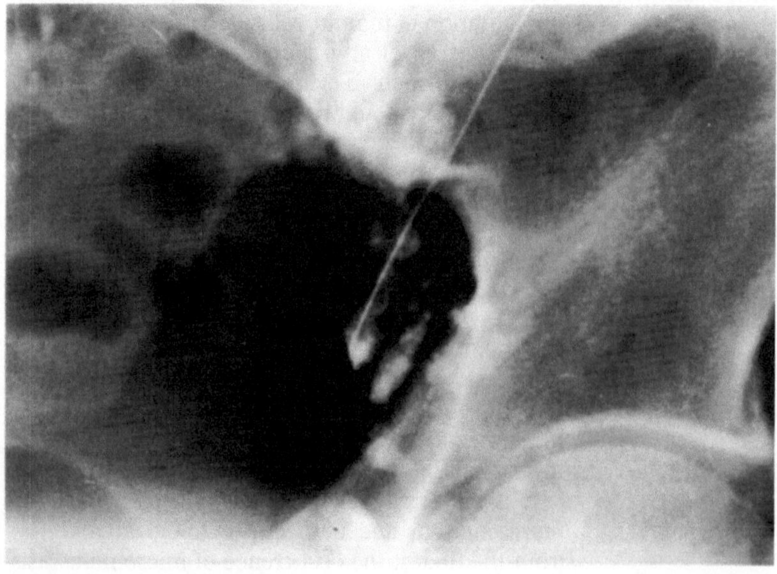

Fig. 6.3. Fine needle aspiration biopsy of lymph node. The needle tip is placed throughout the entire nodal chain. (Luciani and Piscioli [27]; reproduced by kind permission of the Editor of *Urology* and the publisher, Hospital Publications Inc.)

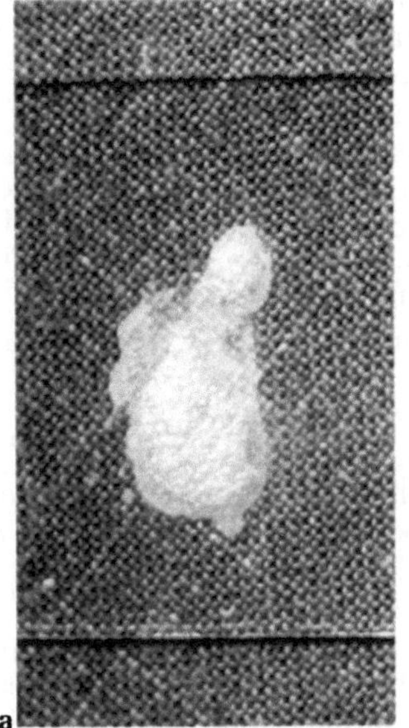

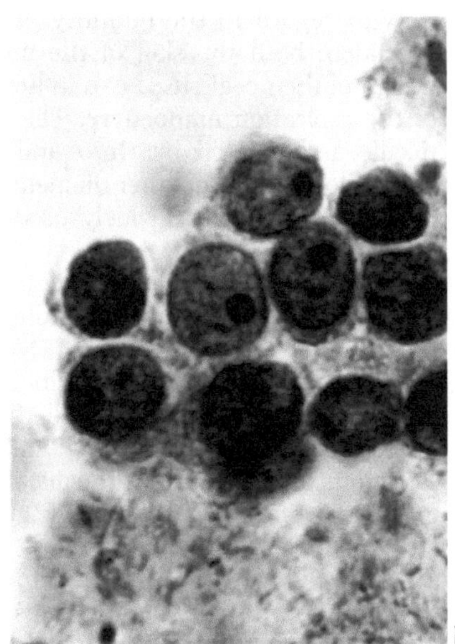

Fig. 6.4. a Drop of cytological material from prostatic biopsy extruded onto slide. **b** Light microscopy view showing malignant cells consistent with prostatic carcinoma.

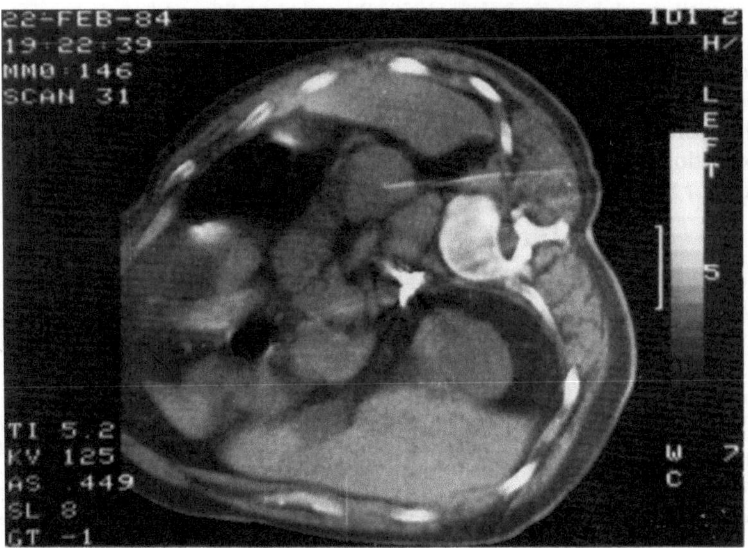

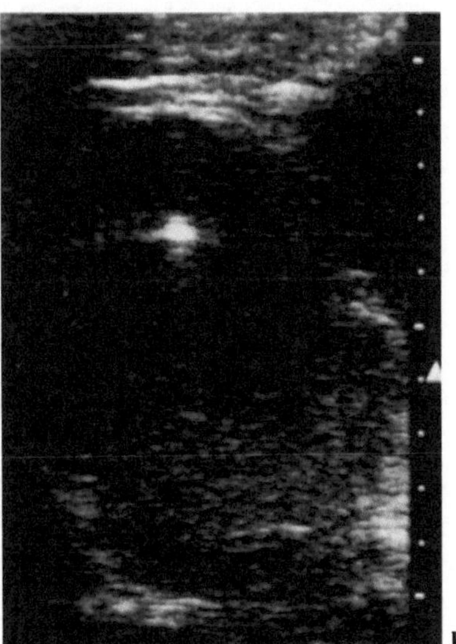

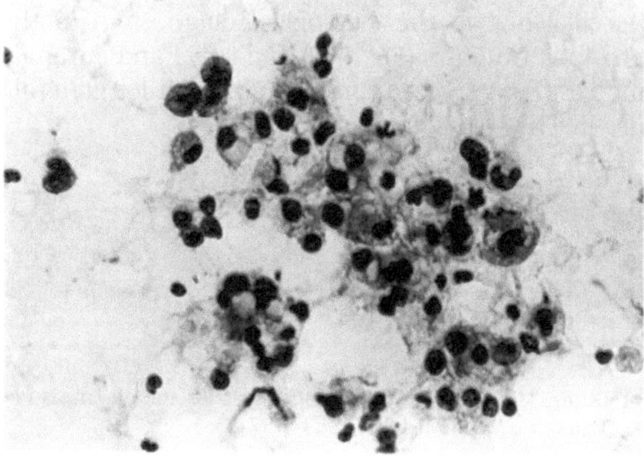

Fig. 6.5. a CT scan showing needle tip properly placed in left adrenal mass. **b** Sonogram of the right kidney showing large suprarenal mass adjacent to the upper pole of the kidney. Needle tip echo seen inside the mass. **c** Aspirate from the mass. Cluster of malignant cells consistent with adenocarcinoma. (**a** and **c** from Luciani et al. [68]; reproduced by kind permission of the Editor of *Journal of Urology* and the publisher, Williams and Wilkins Co., Baltimore, Maryland)

With regard to the quantity of the cytological specimen, both the skill of the operator and the design of the needle used can influence the success of the aspiration manoeuvre. The standard Chiba needle as proposed by Ohto and Tsuchiya [38] (15 cm long, 0.7 mm outer diameter, 0.5 mm inner diameter) is the most widely used needle in ABC (Fig. 6.2) [5–8, 10, 13–15, 16, 18–26, 28–30, 39–61]. Nevertheless, other fine needles have been devised in order to obtain better diagnostic samples (Fig. 6.2) [62–67].

In our series we used a new fine needle which is characterised by a thicker wall than that of the Chiba needle, a longer bevel and multiple fenestrations on either side of the distal portion. These improvements confer a greater guidability, penetration and aspiration ability to the instrument. As a result of lowering the bevel angle to 15° and the provision of four side openings, the aspirating surface of the needle is increased. By turning and withdrawing the needle much more of the nodal tissue around the needle can be aspirated and examined. Furthermore, the faceted lateral margins of such an acute bevel and the thicker wall, which gives greater strength to the needle tip, ensure better penetration ability and provide a quick, reliable access into the entire nodal chain (Fig. 6.3). Such characteristics of the needle reduce the percentage of insufficient material, the time of execution of the procedure and the exposure of the operator to direct radiation.

As mentioned previously, in our series of 166 aspirated lymph nodes with histological control the overall accuracy was 93%, sensitivity was 71% and specificity was 100%. The main source of error with ABC is the presence of micrometastases, which are difficult to detect even at the histological examination since serial section is routinely impractical. ABC of the pelvic nodes is an accurate, minimally invasive procedure which provides sufficient information on nodal status of patients with urological neoplasms without resorting to inopportune surgery. In conclusion, since our needle has proved its efficiency in the cytological diagnosis of both prostatic tumours (Fig. 6.4) and renal and adrenal masses (Fig. 6.5), we can recommend it for general use in TNM categorisation.

References

1. Kunn MA (1847) A new instrument for the diagnosis of tumours. Mth J Med Sci 7:853–857
2. Paget J (1853) Lectures on tumours. Longman, London. Quoted by Webb AJ (1974) Bristol Med Chir J 89:59–64
3. Martin HE, Ellis EB (1930) Biopsy by needle puncture and aspiration. Ann Surg 92:169–172
4. Franzèn S, Giertz G, Zajicek J (1960) Cytological diagnosis of prostatic tumours by transrectal aspiration biopsy: a preliminary report. Br J Urol 32:193–196
5. Göthlin JH (1976) Post-lymphographic percutaneous fine needle biopsy of lymph node guided by fluoroscopy. Radiology 120:205–207
6. Zornoza J, Wallace S, Goldstein H, Lukeman JM, Jing BS (1977) Transperitoneal percutaneous retroperitoneal lymph node aspiration biopsy. Radiology 122:111–115
7. Wallace S, Jing BS, Zornoza J (1977) Lymphangiography in the determination of the extent of metastatic carcinoma. The potential value of percutaneous lymph node biopsy. Cancer 39:706–708
8. Zornoza J, Johnson K, Wallace S, Lukeman JM (1977) Fine needle aspiration biopsy of retroperitoneal lymph nodes and abdominal masses. An updated report. Radiology 125:87–88
9. Göthlin JH (1978) Percutaneous transperitoneal fluoroscopic guided fine needle biopsy of lymph nodes. Acta Radiol [Diagn] (Stockh) 20:660–664
10. Wein AJ, Ring EJ, Freiman DB, Oleaga JA, Carpiniello VL, Banner MP, Pollack HM (1979) Applications of thin needle aspiration biopsy in urology. J Urol 121:626–629
11. Rupp N, Rothenberger KH, Bajer-Pietsch E, Feuerbach ST, Esch U (1979) Die perkutane Feinnadelbiopsie von Lymphknoten. Fortschr Geb Roentgenstr Nuklearmed 130:328–331
12. Rothenberger KH, Hofstetter A, Pfeifer KJ, Rupp N (1979) Transperitoneal Feinnadel Biopsie retroperitonealer. Lymphknoten in der Karzinomadiagnostik. Prog Med Virol 97:2218–2222
13. Göthlin JH, Hoeim L (1980) Percutaneous transperitoneal fine needle biopsy of normal looking lymph nodes and small lesion at lymphography: a preliminary report. Urol Radiol 1:237–239
14. Correa RJ, Kidd RC, Burnett L, Brannen GE, Gibbons RP, Cummings KB (1981) Percutaneous pelvic lymph node aspiration in carcinoma of the prostate. J Urol 126:190–191
15. Efremidis SC, Dan JS, Nieburgs HI, Mitty HA (1981) Carcinoma of the prostate: lymph node aspiration for staging. AJR 136:489–492
16. Göthlin JH, Hoeim L (1981) Percutaneous fine needle biopsy of radiographically normal lymph nodes in the staging of prostate carcinoma. Radiology 141:351–354
17. Göthlin JH, Rupp N, Rothenberger KH, MacIntosh PK (1981) Percutaneous biopsy of retroperitoneal lymph nodes: a multicentric study. Eur J Radiol 1:46–50
18. Rothenberger KH, Hofstetter A, Rupp N, Pfeifer KJ (1981) Transabdominal fine needle biopsy of lymph nodes for urologic cancer staging. In: Schulman CC (ed) Advances in diagnostic urology. Springer, Berlin Heidelberg New York, pp 113–117
19. Dan SJ, Wulsohn MA, Efremidis SC, Mitty HA, Brendler H (1982) Lymphography and percutaneous lymph node biopsy in clinically localized carcinoma of the prostate. J Urol 127:695–698
20. von Eschenbach A, Zornoza J (1982) Fine-needle percutaneous biopsy: a useful evaluation of lymph node metastases from prostate cancer. Urology 6:589–590
21. Wajsman Z, Gamarra M, Park JJ, Beckley SA, Pontes JE, Murphy GP (1982) Fine-needle aspiration of metastatic lesions and regional lymph nodes in genitourinary cancer. Urology 4:356–360
22. Wajsman Z, Gamarra M, Park JJ, Beckley SA, Pontes JE

(1982) Transabdominal fine-needle aspiration of retroperitoneal lymph nodes in staging of genitourinary tract cancer (correlation with lymphography and lymph node dissection findings). J Urol 128:1238–1240

23. Khan O, Pearse E, Bowley N, Williams G, Krausz T (1983) Combined bipedal lymphangiography, CT scanning and transabdominal lymph node aspiration cytology for node staging in carcinoma of the prostate. Br J Urol 55:538–541
24. Luciani L, Piscioli F (1983) Accuracy of transcutaneous aspiration biopsy in definitive assessment of nodal involvement in the prostatic carcinoma. Report of 24 cases and review of the literature. Br J Urol 55:321–325
25. Boccon-Gibod L, Katz M, Cochand B, Le Portz B, Steg A (1984) Lymphography and percutaneous fine needle node aspiration biopsy in the staging of bladder carcinoma. J Urol 132:24–26
26. Luciani L, Scappini P, Pusiol T, Piscioli F (1985) Comparative study of lymphography and aspiration cytology in the staging of prostatic carcinoma. Urol Int 40:181–189
27. Luciani L, Piscioli F (1985) Experience with modified Chiba needle in staging of genitourinary tumors. Urology 25:568–572
28. Piscioli F, Scappini P, Luciani L (1985) Aspiration cytology in the staging of urological cancers. Cancer 56:1173–1180
29. Roger B, Cochand-Priollet B, Katz M, Laval-Jeantet M, Boccon-Gibod L, Leduc A (1985) La cytoponction percutanée à l'aiguille fine des ganglions opacifiés par lymphographic, en pathologic urogenitale. A propos de 174 cas. J Urol (Paris) 91:565–573
30. Luciani L, Piscioli F, Pusiol T, Scappini P (1986) The value of aspiration cytology in the definitive staging of bladder carcinoma. Br J Urol 58:26–30
31. Schubert J, Heidl G (1983) The value of various methods for determining lymphatic metastases in prostatic carcinoma. Eur Urol 9:257–261
32. Taddei L, Fuochi C, Menichelli E, Luciani L (1984) Accuracy of lymphography in staging of prostatic and bladder carcinoma: 88 cases with aspirative cytological and post-lymphadenectomy histological verification. Diagn Imaging 53:91–98
33. Babcock JR, Grayhack JT (1979) Morbidity of pelvic lymphadenectomy. Urology 13:483–486
34. Fowler JE Jr, Barzell W, Hilaris BS, Whitmore WF Jr (1979) Complications of ^{125}Iodine implantation and pelvic lymphadenectomy in the treatment of prostate cancer. J Urol 121:447–451
35. Freiha FS, Pistenma DA, Bagshaw MA (1979) Pelvic lymphadenectomy for staging of prostatic carcinoma: is it always necessary? J Urol 122:176–177
36. Herr HW (1979) Complications of pelvic lymphadenectomy and retropubic prostatic I^{125} implantation. Urology 14:226–229
37. Paul DB, Loening SA, Narayaha AS, Culp DA (1983) Morbidity from pelvic lymphadenectomy in staging carcinoma of the prostate. J Urol 129:1141–1144
38. Ohto M, Tsuchiya Y (1969) Nonsurgical percutaneous transhepatic cholangiography techniques and cases. Medicina 6:735–738
39. Ruttiman A (1968) Iliac lymph node aspiration biopsy through paravascular approach. Radiology 90:150–151
40. Goldstein HM, Zornoza J, Wallace S, Anderson JH, Bree RL, Samuels BI, Lukeman J (1977) Percutaneous fine needle aspiration biopsy of pancreatic and other abdominal masses. Radiology 123:319–322
41. Scheible W, Coel M, Siemers PT, Siegel H (1977) Percutaneous aspiration of adrenal cysts. Am J Roentgenol Radium Ther Nucl Med 128:1013–1016
42. Pereiras RV, Meiers W, Kunhardt B, Troner M, Huston D, Barkin JS, Viamonte M (1978) Fluoroscopically guided thin needle aspiration biopsy of the abdomen and retroperitoneum. Am J Roentgenol Radium Ther Nucl Med 131:197–202
43. Kard W, Ekelund L (1979) Ultrasound angiography and fine needle aspiration biopsy in diagnosis of renal neoplasms. Acta Radiol [Diagn] (Stockh) 20:649–659
44. Bonfiglio TA, MacIntosh PK, Patten SF, Jr, Cafer DJ, Woodworth FE, Kim CW (1979) Fine needle aspiration cytopathology of retroperitoneal lymph nodes in the evaluation of metastatic disease. Acta Cytol 23:126–130
45. Haaga JR (1979) New techniques for CT-guided biopsies. Am J Roentgenol Radium Ther Nucl Med 133:633–641
46. MacIntosh PK, Thomson KR, Barbaric ZL (1979) Percutaneous transperitoneal lymph node biopsy as a means of improving lymphographic diagnosis. Radiology 131:647–649
47. Ferrucci JT, Wittenberg J, Mueller PR, Simeone JF, Harbin WP, Kirkpatrick RH, Taft PD (1980) Diagnosis of abdominal malignancy by radiologic fine-needle aspiration biopsy. Am J Roentgenol Radium Ther Nucl Med 134:323–330
48. Levin NP (1981) Fine needle aspiration and histology of adrenal cortical carcinoma: a case report. Acta Cytol 25:421–424
49. Buonocore E, Skipper GJ (1981) Steerable real-time sonographically guided needle biopsy. AJR 136:387–392
50. Zornoza J, Ordonez N, Bernardino ME, Cohen MA (1981) Percutaneous biopsy of adrenal tumors. Urology 18:412–416
51. Heaston DK, Handel DB, Ashton PR, Keorobkin M (1982) Narrow gauge needle aspiration of solid adrenal masses. AJR 138:1143–1148
52. Nosher JL, Amorosa JK, Leiman S, Plafker J (1982) Fine needle aspiration of the kidney and adrenal gland. J Urol 128:895–899
53. Halvorsen RA Jr, Heaston DK, Johnston WW, Ashton PR, Burton GM (1982) CT guided thin needle aspiration of adrenal blastomycosis. J Comput Assist Tomogr 6:389–391
54. Lieberman RP, Hafez GR, Crummy AB (1982) Histology from aspiration biopsy: Turner needle experience. AJR 138:561–564
55. Lüning M, Neuser D, Kursawe R, Pötschke B (1983) CT guided percutaneous fine needle biopsy in the diagnosis of small adrenal tumours. Eur J Radiol 3:358–364
56. Pagani JJ (1983) Normal adrenal glands in small cell lung carcinoma: CT-guided biopsy. AJR 140:949–951
57. Price RB, Bernardino ME, Berkman WA, Sones PJ Jr, Torres WE (1983) Biopsy of the right adrenal gland by the transhepatic approach. Radiology 148:566
58. Juul N, Torp-Pederson S, Holm HH (1984) Ultrasonically guided fine needle aspiration biopsy of retroperitoneal mass lesions. Br J Radiol 57:43–46
59. Pagani JJ (1984) Non-small all lung carcinoma adrenal metastases. Computed tomography and percutaneous needle biopsy in their diagnosis. Cancer 53:1058–1060
60. Katz RL, Patel S, Mackay B, Zornoza J (1984) Fine needle aspiration cytology of the adrenal gland. Acta Cytol (Baltimore) 28:269–282
61. Flanigan RC, Mohler JL, King CT, Atwell R, Umer MA, Loh FK, McRoberts JW (1985) Preoperative lymph node evaluation in prostatic cancer patients who are surgical candidates: the role of lymphangiography and computerized tomography scanning with directed fine needle aspiration. J Urol 134:84–87
62. Isler RJ, Ferrucci JT Jr, Wittenberg J, Mueller PR, Simeone JF, van Sonnenberg E, Hall DA (1981) Tissue core biopsy of abdominal tumors with a 22 gauge cutting needle. AJR 136:725–728

63. Turner AF, Sargent EN (1968) Percutaneous pulmonary needle biopsy—an improved needle for a simple direct method of diagnosis. Am J Roentgenol Radium Ther Nucl Med 104:846–850
64. Lee LH (1974) A new biopsy needle and its clinical use. Am J Roentgenol Radium Ther Nucl Med 121:854–859
65. Okuda K, Tanikawa K, Emura T, Kuratomi S, Jinnouchi S, Urabe K, Sumikoshi T, Kanda Y, Fukuyama Y, Musha H, Mori H, Shimokawa Y, Yakushiji F, Matsuura Y (1974) Nonsurgical percutaneous transhepatic cholangiography diagnostic significance in medical problems of the liver. Dig Dis Sci 19:21–36
66. Westcott JL (1980) Direct percutaneous needle aspiration of localized pulmonary lesions: results in 422 patients. Radiology 137:31–35
67. Luciani L, Menichelli E, Fuochi C, Taddei L (1981) Lo staging linfonodale del carcinoma della prostata. Minerva Med 72:789–800
68. Luciani L, Scappini P, Pusiol T, Piscioli F (1985) Aspiration cytology of simultaneous bilateral adrenal metastases from renal cell carcinoma. A case report and review of the literature. J Urol 134:315–318

Chapter 7

Aspiration Biopsy of Lymph Nodes in Urological Malignancies: Its Value in Different Diagnostic Procedures in Tumour Staging

K. H. Rothenberger

The planning of treatment in urological cancer requires exact tumour staging. The prognosis of the disease depends on the absence of metastases. It is especially difficult to determine lymph node status. Lymphatic metastases are not as rare as suspected and their growth is independent of haematogenous metastases. The frequency of lymph node metastases is highly dependent on tumour grade and stage.

Lymph Drainage of Urological Tumours

Lymph node metastases proceed from the ilio-inguinal and pelvic area to the para-aortic region and finally the thoracic duct. Primary lymph node sites of *tumours of the bladder and prostate* are regional nodes surrounding these organs: the external, internal and common iliac nodes and the promontory and obturator nodes.

The first lymph node sites of *testicular tumours* are the high lumbar and para-aortic nodes. In some rare cases the lymph first drains into the common iliac nodes [1].

Lymph node drainage of *penile tumours* terminates in the superficial inguinal lymph nodes. There seems to be a "sentinel" lymph node medial to the superficial epigastric vein. There may be anastomosis with contralateral inguinal nodes and, in rare cases, a lymph vessel enters the pelvis and terminates in the internal and external iliac lymph nodes.

The lymph drainage of *kidney tumours* terminates in lumbar lymph nodes. They are of no interest for aspiration biopsy because in every case the kidney will be removed together with the lymph nodes.

We can see from this short look at lymph drainage that some nodes cannot be visualised by pedal lymphography. In the case of penile cancer, penile lymphography can be performed, although we ourselves do not have any experience in this method. There could be further progress in lymph node diagnosis if we had a reproducible method of indirect lymphography. Our trials, like those of other authors, have been unsuccessful.

Staging Methods

Histology

Histology is by far the best method of detecting metastases. It requires operative lymph node dissection, an extended operative procedure sometimes leading to morbidity and mortality. It must be pointed out that this operation is carried out solely for diagnosis, not treatment. Some lymph nodes remain in place even after careful operation (about 25% in the estimate of some authors), so that metastases may be missed.

Lymphography

Lymphography is a slightly invasive method with few complications. It has less diagnostic reliability for metastases because it relies on purely macroscopic findings [2–4]. If strict criteria for diagnosis of metastases are applied [5], such as marginal lymph node defects of more than 1 cm with or without stasis in the corresponding lymph vessels, the diagnostic accuracy in demonstrating metastases can be as high as 85% or more. In negative lymphography, however, up to 30% of metastases are overlooked. Another problem with lymphography is that some of the first nodal sites of bladder, prostatic and penile cancer will not be opacified because these lymph nodes lie beside the pathway of the contrast medium injected during pedal lymphography [6, 7]. High para-aortic nodes (the first site of testicular tumours) are represented constantly only to the level of L-2 on the left side and L-3 on the right side [1, 8, 9]. Upper sites will not be visible constantly because of the inflow from renal lymph nodes. A lymph node metastasis may not take up any contrast medium and thus may not be visualised.

Computed Tomography

Computed tomography (CT) is a non-invasive method and may only show mass growth without differentiation of the interior structure. Internal structure can only be seen in very large lymph node metastases, which could be differentiated by all methods. The sensitivity and specificity of documentation of small lymph node metastases in the pelvis are not very high with this method. On the other hand, CT is preferred for high lumbar areas because of well-defined landmarks, e.g. the aorta and vena cava [10].

Sonography

Sonography is a non-invasive method which may only show large metastases. Intestinal formations and gas diminish the quality of the test [11].

Aspiration Cytology

Aspiration cytology of lymph nodes is a simple technique to complete staging procedures before histological (operative) examination. In general, we combine aspiration cytology with lymphography and, especially in the case of large tumour masses, with sonography or CT [10,12–14].

Technique of Aspiration Cytology

Immediately after the storage phase of lymphography several lymph nodes along the lymphogenous pathway of metastases are punctured under fluoroscopic guidance with the patient lying supine. In general, we do not use general anaesthetics. The patient is given only a sedative combined with an analgesic. The pain of local anaesthesia is similar to that produced by the puncture of a lymph node. To reduce radiation exposure for both patient and examiner we have developed an instrument giving good guidance of the Chiba needle (23 gauge). The instrument consists of a handle and a transparent (visible light and X-ray) cylinder to compress the abdominal wall and to guide the tip of the needle perpendicularly into the lymph node (Fig. 7.1). The correct position of the tip of the needle in the node can be recognised by a to-and-fro movement of the node when the needle is slightly shaken. In thin nodes a slight rim where contrast disappears can be seen as a result of the compression effect (Fig. 7.2). The stylet is then removed from the needle and a pistol handle with syringe (e.g. Franzen or Cameco) is connected to

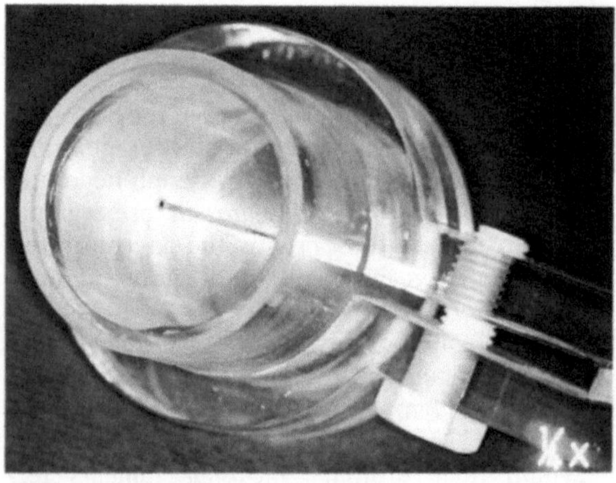

Fig. 7.1. Guiding instrument for aspiration biopsy.

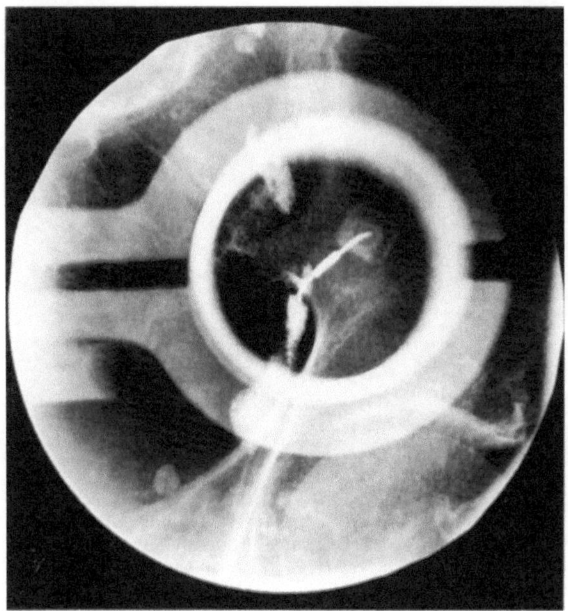

Fig. 7.2. X-ray view of guiding instrument with Chiba needle positioned in a lymph node.

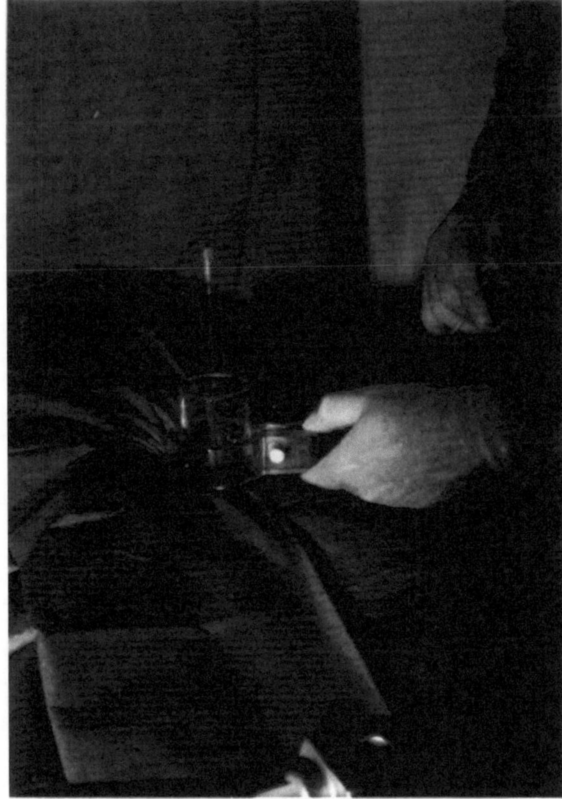

Fig. 7.3. Technique of puncture.

create a negative pressure. After slightly rotating and moving the needle back and forth over a few millimetres the negative pressure is released and the needle is removed (Fig. 7.3). The content of the needle is expressed onto glass slides to prepare smears. The air-dried material is stained by May–Grünwald–Giemsa for cytological examination. We examine about 10 lymph nodes in every patient [14–16].

Complications

In accordance with other authors [12,17] we did not encounter any severe complications, although the needle passes intestine and vessels. We consider severe coagulopathies, ileus and local inflammation as contraindications.

Results

Only the positive finding of tumour cells in one or more lymph nodes proves the presence of metastatic disease. In contrast, the lack of tumour cells cannot prove a lack of nodal metastases as it is possible that the metastatic lymph node was not punctured or that the aspirated material was inadequate. In about 70% of cases we are able to obtain nodal material, although it is no problem to repeat the procedure if necessary. A general problem is the interpretation of the cytological findings. Reactions to the contrast medium are easily recognised. Cells of the mesothelium may cause problems (Figs. 7.4–7.6).

Cancer of the Prostate (Table 7.1)

We examined a series of 25 patients with G2 and G3 carcinomas. We found lymph node metastases in 10 patients by aspiration cytology, while lymphography revealed metastases in only 5 patients. Six of these patients underwent lymph node dis-

Table 7.1. Lymph node metastases in prostatic cancer patients

Lymphography	Cytology	$n=25$	Histology ($n=6$)
+	+	5	1 × positive
+	–	3	0
–	–	12	4 × negative
–	+	5	1 × positive

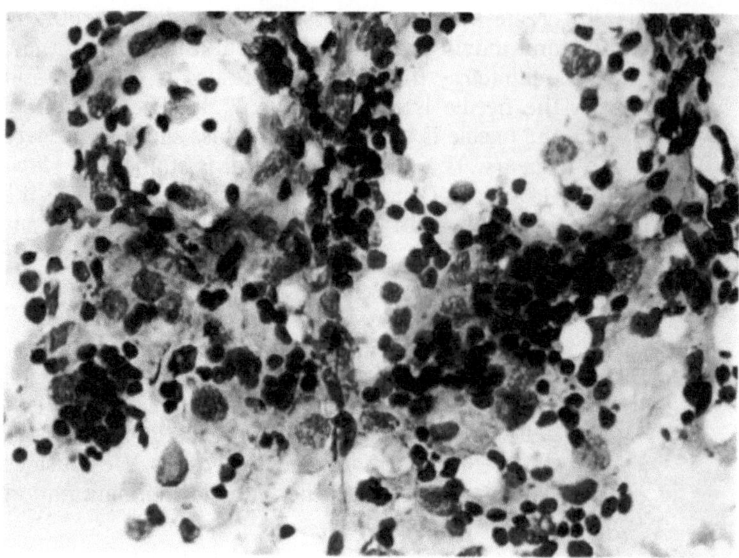

Fig. 7.4. Epithelioid and inflammation cells in a smear from a lymph node.

section and the results were the same as in aspiration cytology.

Bladder Cancer

In a series of 61 patients initially treated by transurethral resection (TUR) we found tumour cells in lymph nodes in 17, so we treated these patients additionally with radiotherapy or cytostatics. In four patients who underwent cystectomy with lymph node dissection, the findings were the same as in aspiration cytology.

Testicular Tumours

After introducing the technique of aspiration cytology in 1977, we examined patients with testicular tumours the day before therapeutic lymph node dissection in order to obtain experience and to see possible damage during operation. We only saw small retroperitoneal effusions of blood in 2 out of 10 patients, and in 9 out of 10 patients the histological results were the same as the cytological findings. One smear showed no material. Owing to a lack of results, today we no longer practise lymphography and aspiration cytology in the case

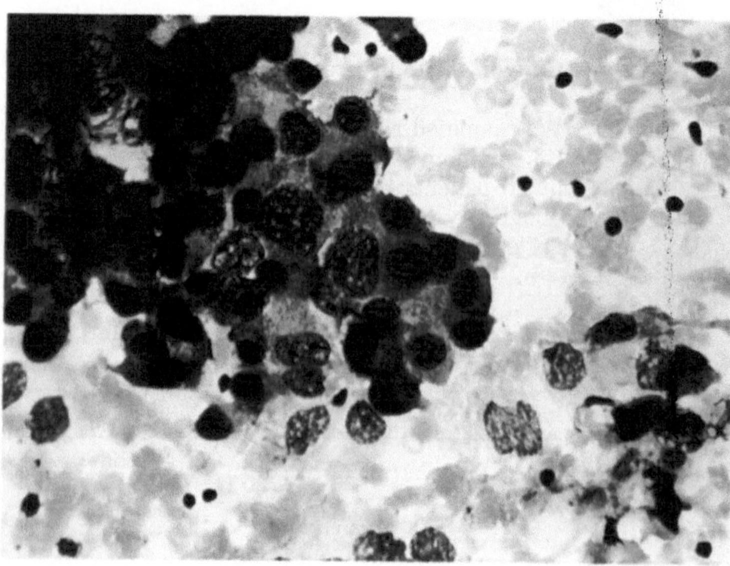

Fig. 7.5. Tumour cells of prostatic cancer in a smear from a lymph node.

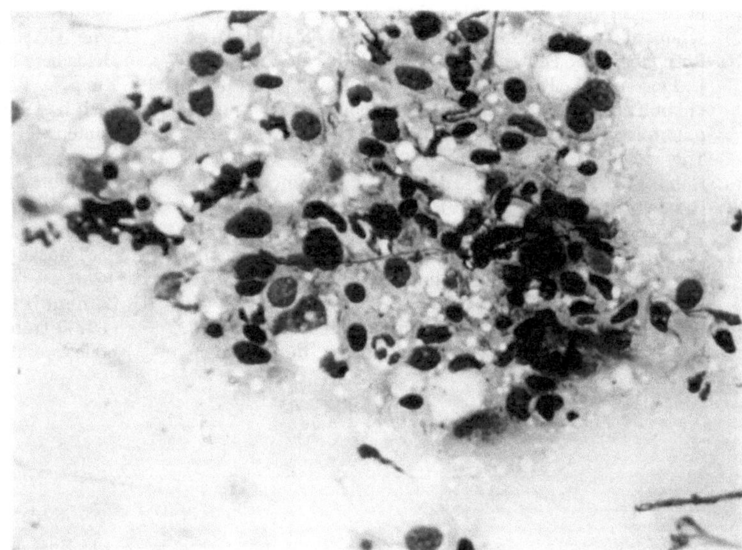

Fig. 7.6. Tumour cells of a squamous cell carcinoma of the penis showing keratinisation.

of testicular tumours, except in some cases of seminomatous tumours undergoing irradiation.

Penile Cancer

In the lymphographic study of penile cancer, besides pedal lymphography, penile lymphography, which visualises the sentinel lymph node, may also be performed. In 14 of our own patients aspiration cytology and histology showed the same results. Nevertheless, because of our experience of the possibility of false negative results on cytological examination, we think that in cases of negative aspiration cytology the lymph staging operation should be performed.

Conclusions

Aspiration biopsy of lymph nodes along the possible pathway of metastases in cancer of the prostate, bladder and penis is a simple method of improvement of tumour staging. Only positive findings are conclusive. If no tumour cells are found, further diagnostic steps (lymph node dissection) are recommended.

In non-seminomatous testicular tumours lymph node dissection is part of therapy, so that only in seminomatous tumours, where most commonly treatment involves abdominal irradiation in addition to orchiectomy, may aspiration biopsy be indicated for better staging. We hope that in future new methods of indirect lymphography will ensure opacification of all initial nodal sites for percutaneous aspiration biopsy.

References

1. Zaunbauer W, Kunz R, Leuppi R (1977) Die diagnostische Zuverlassigkeit der Lymphographie bei Patienten mit malignen Hodentumoren. Fortschr Geb Rontgenstr Nuklearmed Erganzungsband 126 (4):335–338
2. Feuerbach S, Rupp N, Rossmann W, Heller HJ, Rothenberger K, Tauber R, Schmidt G (1979) Lymphknotenmetastasen—Diagnose durch Lymphographie und CT. Fortschr Geb Rontgenstr Nuklearmed 130 (3):323–328
3. Fuchs WA (1965) Lymphographie und Tumordiagnostik. Springer, Berlin Heidelberg New York
4. Zyb AF, Ostapowitsch ON, Luschnikov JF, Zodikova LB (1976) Lymphographische Semiotik benigner Lymphknotenschadigungen im Lichte der morphologischen Angaben. Radiologe 16:200–208
5. Wallace S, Jing BS, Zornoza J (1977) Lymphangiography in the determination of the extent of metastatic carcinoma. Cancer 39:706–718
6. Merrin C, Wajsman Z, Baumgartner G, Jennings E (1977) The clinical value of lymphangiography: are the nodes surrounding the obturator nerve visualized? J Urol 117:762–764
7. Spellmann MC, Castellino RA, Ray GR, Pistenma DA, Bagshaw MA (1977) An evaluation of lymphography in localized carcinoma of the prostate. Radiology 125:637–644
8. Göthlin J, Jonsson K (1976) Lymphangiographic criteria of metastases. An evaluation of patients with malignant testicular teratoma. Acta Radiol [Diagn] (Stockh) 17:321–327
9. Storm PB, Kern A, Loening SA, Brown RC, Culp DA (1977)

Evaluation of pedal lymphangiography in staging non-seminomatous testicular carcinoma. J Urol 118:1000–1002

10. Rothenberger K, Feuerbach S, Friesen A, Hofstetter A, Pensel J, Pfeifer KKJ, Rupp N (1982) Diagnostik von Lymphknotenmetastasen durch Lymphographie, Computertomographie und perkutane Feinnadelbiopsie. Therapiewoche 32:701–703
11. Holm HH, Pedersen JF, Kristensen JK, Rasmussen SN, Hanche S, Jensen F (1975) Ultrasonically guided percutaneous puncture. Radiol Clin North Am 13:493–503
12. Göthlin JH (1976) Post-lymphographic percutaneous fine needle biopsy of lymph nodes guided by fluoroscopy. Radiology 120:205–207
13. Rothenberger K, Rupp N, Feuerbach S, Bayer-Pietsch E (1979) Lymphknoten-Metastasennachweis durch transabdominelle Femnadelbiopsie. Verhandlungbericht. Deutsche Gesellschaft für Urologie 1978. Springer, Berlin Heidelberg New York, p 65
14. Rupp N, Rothenberger K, Bayer-Pietsch E, Feuerbach S, Esch U (1979) Die perkutane Feinnadelbiopsie von Lymphknoten. Fortschr Geg Rontgenstr Nuklearmed Erganzungsband 130 (3):328–331
15. Viamonte M, Ruttimann A (1980) Atlas of lymphography. Thieme, Stuttgart
16. Rothenberger K, Pensel J (1980) Verbesserung der Technik der transkutanen Feinnadelbiopsie. Fortschr Med 98:1283–1286
17. Zornoza J, Wallace S, Goldstein HM, Lukeman M, Jing BS (1977) Transperitoneal percutaneous retroperitoneal lymph node aspiration biopsy. Radiology 122:111–115

Chapter 8

Fine Needle Aspiration Biopsy of Pelvic and Periaortic Lymph Nodes in the Evaluation of Neoplastic Disease

T. A. Bonfiglio and M. A. Fallon

Introduction

Fine needle aspiration biopsy (FNAB) has gained acceptance as a valid method of diagnosing malignant disease. The technique can be easily and accurately used at numerous body sites. Transabdominal needle aspiration biopsy of virtually any intra-abdominal or retroperitoneal tissue can now be readily carried out utilising one of several imaging techniques to localise the lesion and guide placement of the needle. Use of the technique for the evaluation of lymph node pathology has been reported from a number of centres [1–5]. An earlier study from this laboratory attested to the value of FNAB in combination with lymphography in the evaluation of retroperitoneal lymph nodes for possible metastases. Since then our experience has expanded to include a much larger number of cases, in many of which ultrasound or computed tomography (CT) scanning rather than lymphography was used to guide the needle aspiration biopsy. This retrospective evaluation of our experience appraises the accuracy of 109 positive and negative FNABs of pelvic and periaortic lymph nodes in relation to histological and clinical follow-up. This group includes 45 patients reported in the original series from this laboratory who now have a longer period of follow-up.

Materials and Methods

Fine needle transabdominal aspiration biopsies of lymph nodes were performed successfully in 109 cases, a case being defined as the result of one or several aspirations from a site in one patient during one attempt to obtain material. Needle placement was directed by fluoroscopy following lymphography, CT or ultrasonography. In all cases a cytotechnologist in the procedure room prepared slides of the aspirated material. The slides were processed after 95% ethanol fixation using a rapid Papanicolaou stain and after air drying using a Giemsa stain. Special stains were used as indicated. Surgical or autopsy tissue samples were available from 64 of these patients for correlation. These tissues were identified as from the same or different body sites as the aspiration biopsies. If no relevant tissue samples were available, the follow-up clinical history and radiological findings were reviewed. The follow-up period was at least 1 year for all patients.

The cytological interpretation was compared with the histological diagnosis and/or subsequent clinical course in each case. The ability to determine correctly by needle aspiration the presence or absence of malignant disease in the sampled lymph node(s) was evaluated. The sensitivity, specificity, accuracy, positive and negative predictive values and their respective 95% confidence intervals were calculated. The classification of tumour cell types as identified cytologically was compared with that determined histologically in cases where tissue specimens were available.

Results

In 41 cases aspiration biopsy produced a diagnosis of malignancy. In 68 cases the aspiration sample was interpreted as negative for malignancy. Tissue was available for review in 39 of the positive cases, in 13 instances from subsequent removal of the sampled nodal area and in 26 from tissue from another site. In 29 of the negative cases lymph nodal tissue from the sampled area was able to be reviewed. In the remainder of the cases the accuracy of the cytological evaluation was judged on the basis of clinical course and radiological follow-up. No false positive interpretations were made, while four false negative examinations were determined. These figures result in a specificity of 100% with a 100% predictive value for a positive test, an accuracy rate of 96%, a sensitivity of 91% and a 94% predictive value of a negative biopsy. The data are summarised in Tables 8.1 and 8.2.

The majority of the biopsies were performed for the evaluation of possible metastatic disease from a known or suspected lower genitourinary tract malignancy, with the predominant population representing gynaecological patients. This is reflected in the distribution of diagnoses by cell type as noted in Table 8.3, with most of the adenocarcinomas and squamous carcinomas being from these sites. In eight cases, however, suspected or unsuspected lymphoma was diagnosed (Figs. 8.1–8.10).

Table 8.1. Results of correlation with tissue and clinical course

Site	True positive	True negative	False positive	False negative	Total
Periaortic LNs	25	35	0	1	61
Pelvic LNs	16	28	0	3	47
Mesenteric LNs	0	1	0	0	1
All LNs	41	64	0	4	109

LNs, lymph nodes.

Table 8.3. Cytological diagnoses in positive lymph node biopsies

Diagnosis	No. of cases
Adenocarcinoma	15
Squamous cell carcinoma	13
Transitional cell carcinoma	1
Germ cell tumours	1
Sarcoma	1
Lymphoma	8
Undifferentiated malignant tumour	2
Total	41

The four false negative examinations included a periaortic node biopsy in a patient with periaortic lymphadenopathy on lymphangiogram. FNAB was interpreted as negative, but tissue biopsy of the same node was interpreted as nodular, poorly differentiated lymphocytic lymphoma. The other three cases involved pelvic lymph nodes and included two cases in which metastatic squamous cell undetected by FNAB was noted after nodal excision. The final case involved a patient with nodular sclerosing Hodgkin's disease and a positive lym-

Table 8.2. Calculated determinations of accuracy

Site	Sensitivity[a]	Specificity	Accuracy[a]	PV positive result	PV negative result[a]
Periaortic LNs	96% (88–100+%)	100%	98% (94–100+%)	100%	97% (91–100+%)
Pelvic LNs	84% (68–100%)	100%	94% (87–100+%)	100%	90% (81–99%)
All LNs	91% (83–99%)	100%	96% (92–100%)	100%	94% (88–100%)

LNs, lymph nodes; PV, predictive value.
[a] 95% confidence interval given in parentheses.

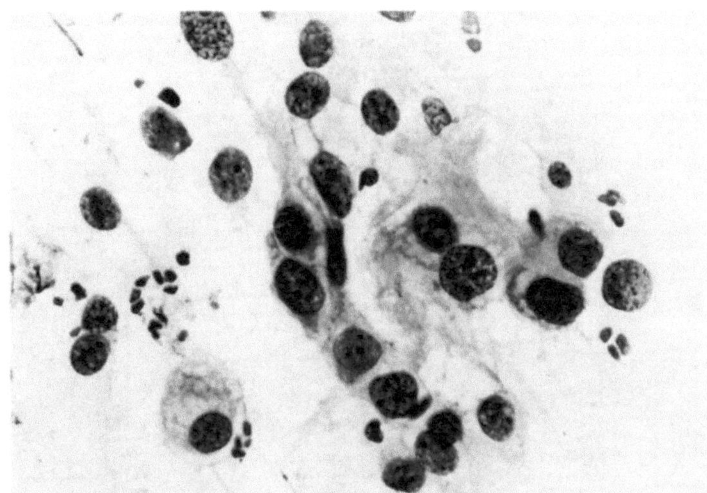

Fig. 8.1. Metastatic squamous cell carcinoma from the uterine cervix in an FNAB of a pelvic lymph node. (Papanicolaou stain, × 120)

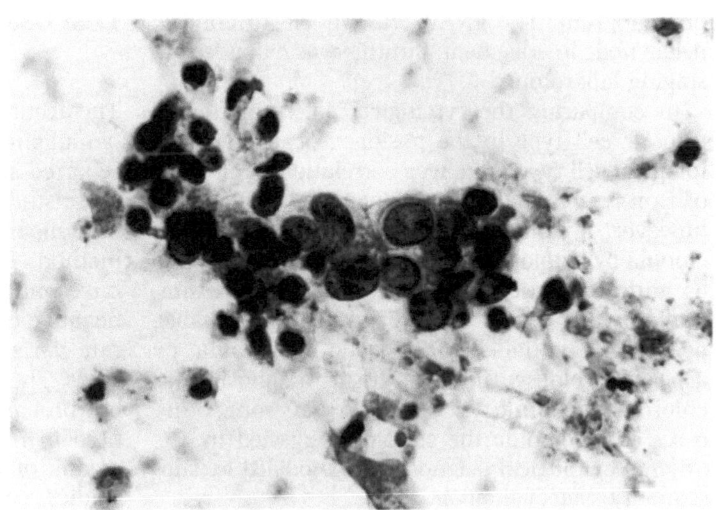

Fig. 8.2. Aspiration of a periaortic lymph node with metastatic, non-keratinising squamous cell carcinoma. (Papanicolaou stain, × 120)

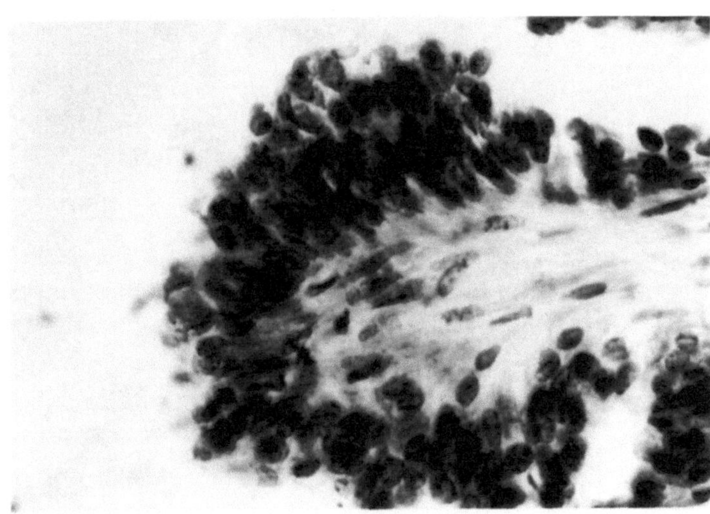

Fig. 8.3. Small tissue fragment in FNAB of a pelvic lymph node in a female with an established diagnosis of papillary adenocarcinoma of the uterine cervix. This biopsy established the diagnosis of lymph node involvement. Note that the papillary nature of the neoplasm is evident. (Papanicolaou stain, × 95)

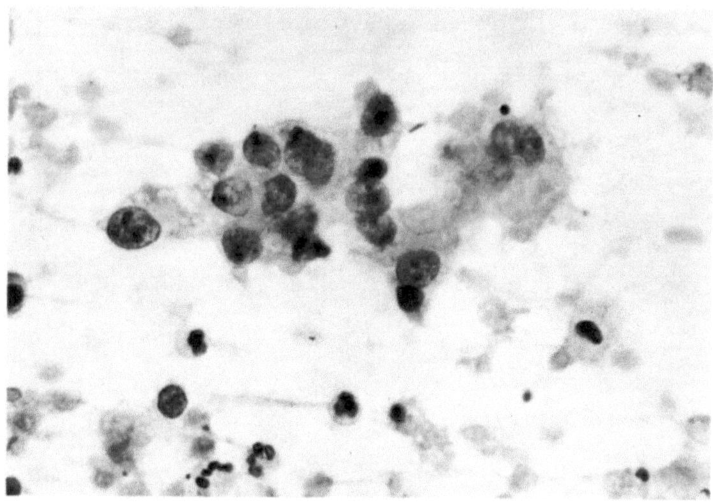

Fig. 8.4. Malignant tumour cells derived from metastatic adenocarcinoma of the endometrium in an FNAB of a pelvic lymph node following lymphography. (Papanicolaou stain, × 120)

phangiogram. The FNAB was unable to confirm pelvic node involvement, but this was diagnosed at staging laparotomy.

In comparing the cytological determination of tumour cell type in the positive FNAB with histological cell type there was correlation in all cases of non-lymphomatous malignancy. In one case, however, a tumour diagnosed as anaplastic carcinoma by cytological examination was interpreted as immunoblastic sarcoma by histological examination. In a second case diagnosed as probable, poorly differentiated lymphocytic lymphoma by aspiration biopsy the final tissue diagnosis was chloroma (granulocytic sarcoma). In four other cases of lymphoma the subtype suggested by the original cytological diagnosis was modified by subsequent tissue evaluation.

Discussion

The accuracy of FNAB in determining the presence of malignant tumour in lymph nodes has varied in reported studies from 65% to 100% [2, 6–14]. In these studies, as in our series, false positive examinations have not been a problem of this diagnostic method. Instances of false positive examinations have only rarely been reported [5, 15, 16]. False negative examinations are well recognised but the rate varies considerably in the reported series of cases. The reasons for this variability are uncertain but probably depend in large part on the methods used in obtaining the biopsies as well as on the means of evaluating the results in the different studies.

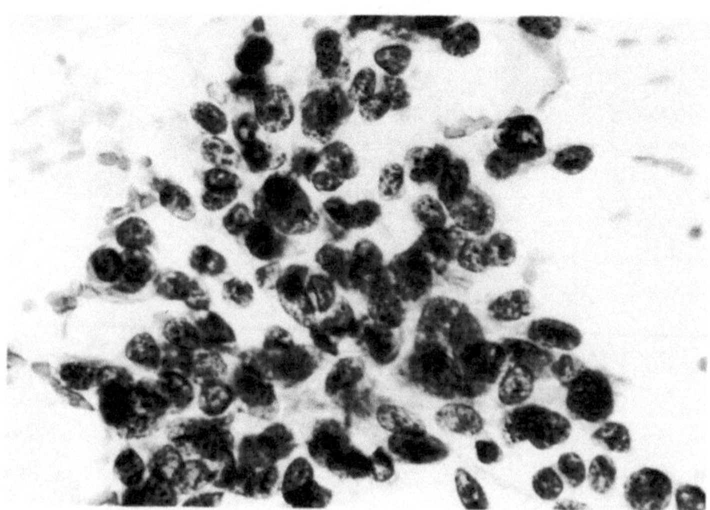

Fig. 8.5. FNAB of a pelvic lymph node with tumour cells derived from metastatic high-grade transitional cell carcinoma of the urinary bladder. (Papanicolaou stain, × 120)

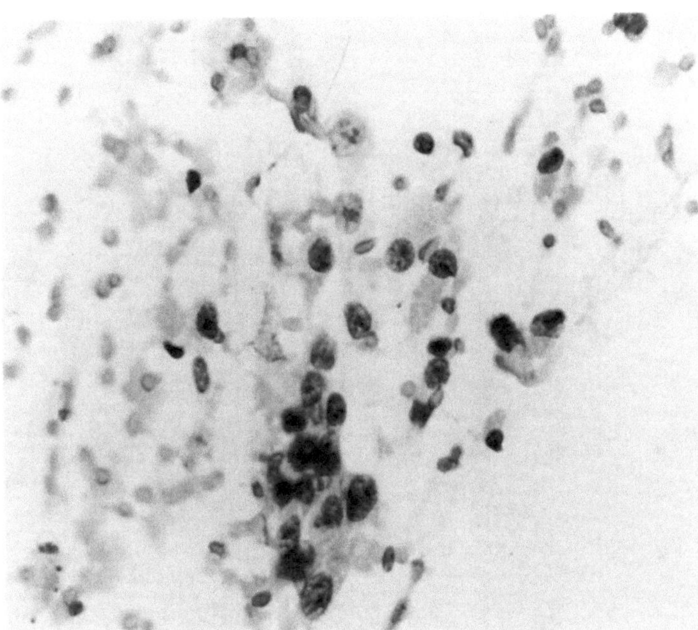

Fig. 8.6. Single cells and small aggregates of cells from prostatic adenocarcinoma aspirated from an enlarged pelvic lymph node. (Papanicolaou stain, × 120)

In this study there was a very low false negative rate and therefore high sensitivity and a high predictive value noted for a negative result. In analysing the results, however, several points must be stressed regarding the negative needle biopsy. In this series, of the 68 negative cases only 29 underwent lymph node dissection. Even though clinical follow-up of the remaining 39 patients revealed no evidence of nodal involvement in the sampled area it is possible that some of these patients had undetected nodal disease. Taking this into consideration and recalculating the figures using only those patients undergoing lymph node dissection in the evaluation, the predictive value of a negative FNAB is still a respectable 86%.

In our procedures multiple samples were often taken, with repeat sampling being performed at the same setting if the initial samples of a suspicious node or nodal group were negative. This, of course, contributed to the low number of false negative examinations.

Although the overall accuracy rates reported are good, it is noted that metastatic carcinoma can be diagnosed with greater accuracy than lymphoma by FNAB [7, 17]. Hodgkin's disease has been correctly identified in 70%–88% of cases in reported

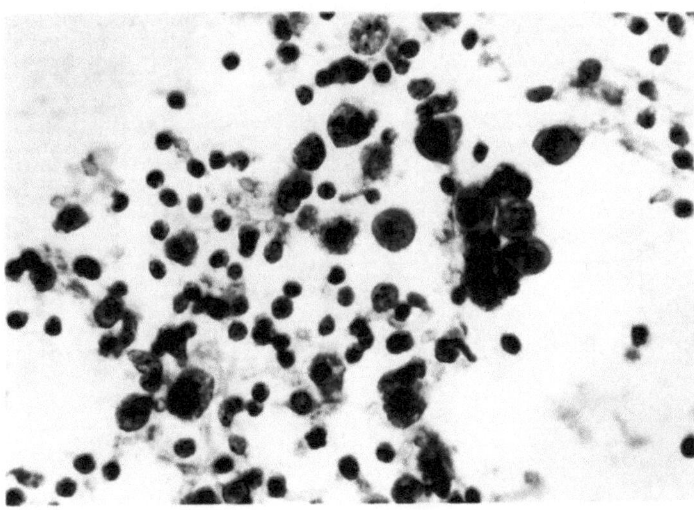

Fig. 8.7. Tumour cells and small lymphocytes in an FNAB of a periaortic lymph node from a young male with testicular seminoma. The cytological findings are diagnostic of lymph node involvement. (Papanicolaou stain, × 120)

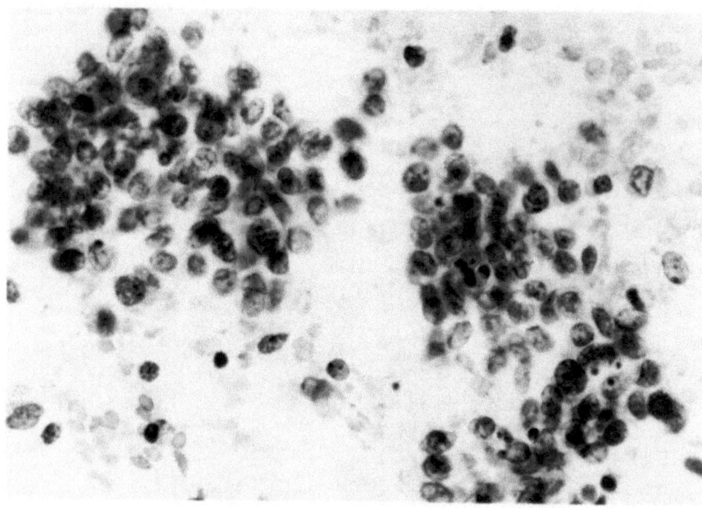

Fig. 8.8. Loose aggregates of tumour cells from serous adenocarcinoma of the ovary metastatic to a periaortic lymph node are present in this example of an FNAB. (Papanicolaou stain, × 95)

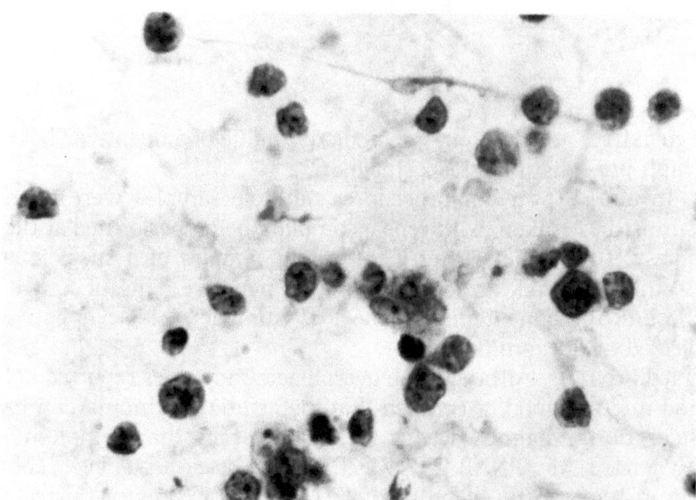

Fig. 8.9. Malignant lymphoma large cell type. These large atypical lymphoid cells were numerous in this periaortic lymph node biopsy. (Papanicolaou stain, × 120)

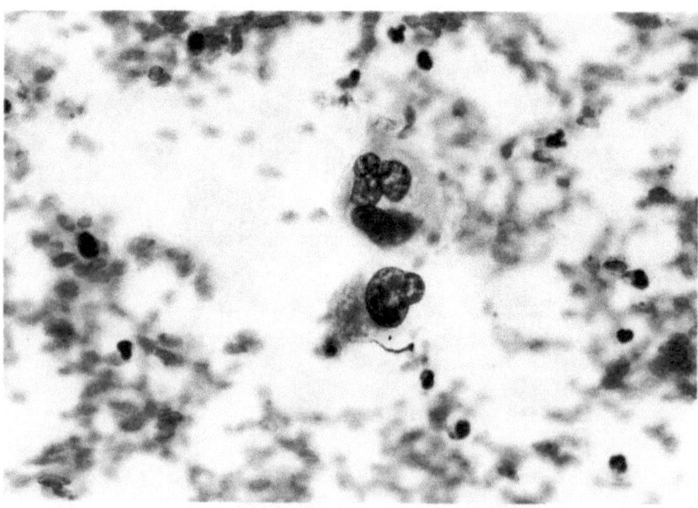

Fig. 8.10. Large atypical lymphoid cells in FNAB of an abdominal lymph node in a patient with Hodgkin's disease, nodular sclerosis type. While not classic Reed–Sternberg cells, these Reed–Sternberg variants are considered diagnostic of nodal involvement in a patient with known disease. (Papanicolaou stain, × 120)

series, but the determination of non-Hodgkin's lymphoma subtypes has been less successful [1, 9]. It is generally appreciated that the precise subclassification of aspirated material of haematological cell types is more difficult than that of epithelial tumours. This is also attested to by our results. However, the identification of the likelihood of a lymphomatous process is not difficult. The additional information now available through the use of immunocytochemical techniques to evaluate cell markers promises to make FNAB even more accurate in this difficult area.

There is considerable debate on the value of FNAB in detecting the presence of metastases in nodes not appearing abnormal by standard imaging techniques. Some authors have advocated its use in this regard, citing a relatively high rate of detection [14–16], while others see little use for the technique in this setting [18, 19]. The methods used in this study do not allow us to speak directly on this point as only nodes that were abnormal on radiological or ultrasonographical examination were biopsied.

In spite of the low false negative rate obtained with careful sampling of apparently abnormal nodes and close coordination between the cytopathology laboratory and the radiologist, we stress, as others have, that a negative biopsy does not exclude the presence of neoplastic nodal disease. While it is possible to reduce the likelihood of false negative examinations by careful evaluation and sampling as demonstrated in this study, the technique cannot definitively insure the absence of microscopic nodal involvement. However, this does not negate the value of the procedure in the evaluation of patients with cancer. Many patients, through the use of this relatively inexpensive technique, are spared unnecessary, complicated surgery. Documentation of nodal involvement allows for better treatment planning, gives valuable prognostic information and provides solutions to problems in patient management.

In this series of cases, FNAB of lymph nodes was used to detect recurrent disease as well as to stage newly diagnosed disease. In both circumstances the method is a useful addition to other diagnostic modalities.

References

1. Bloch M (1967) Comparative study of lymph node cytology by puncture and histopathology. Acta Cytol 11:139–144
2. Bonfiglio TA, Macintosh PK, Patten SF, Cafer DJ, Woodworth FE, Kim CW (1979) Fine needle aspiration cytopathology of retroperitoneal lymph nodes in the evaluation of metastatic disease. Acta Cytol 23:126–130
3. Frable WJ (1976) Thin-needle aspiration biopsy. Am J Clin Pathol 65:168–182
4. Godwin JT (1964) Cytologic diagnosis of aspiration biopsies of solid or cystic tumors. Acta Cytol 8:206–214
5. Lopes Cardozo P (1964) The cytologic diagnosis of lymph node punctures. Acta Cytol 8:194–202
6. Sundaram M, Wolverson MK, Heiberg E, Pilla T, Vas WG, Shields JB (1982) Utility of CT-guided abdominal aspiration procedures. AJR 139:1111–1115
7. Ferrucci JT Jr, Wittenberg J, Mueller PR, Simione JF, Harbin WP, Kirkpatrick RH, Taft PD (1980) Diagnosis of abdominal malignancy by radiologic fine needle biopsy. AJR 134:323–330
8. McDonald TW, Morley GW, Choo YC, Shields JJ, Cordoba RB, Naylor B (1983) Fine needle aspiration of para-aortic and pelvic lymph nodes showing lymphangiographic abnormalities. Obstet Gynecol 61:383–388
9. Zornoza J, Jonsson K, Wallace S, Lukeman J (1977) Fine needle aspiration biopsy of retroperitoneal lymph nodes and abdominal masses: an updated report. Radiology 125:87–88
10. Zornoza J, Cabanillas FF, Altoff TM, Ordonez N, Cohen MA (1981) Percutaneous needle biopsy in abdominal lymphomas. AJR 136:97–103
11. Zornoza J, Wallace S, Goldstein HM, Lukeman JM, Jing B (1977) Transperitoneal percutaneous retroperitoneal lymph node aspiration biopsy. Radiology 122:111–115
12. Piscioli F, Pusiol T, Leonardi E, Luciani L (1985) Role of percutaneous pelvic node aspiration cytology in the management of bladder carcinoma. Acta Cytol 29:37–43
13. Luciani L, Piscioli F (1983) Accuracy of transcutaneous aspiration biopsy in definitive assessment of nodal involvement in the prostatic carcinoma: report of 24 cases and review of the literature. Br J Urol 55:321–325
14. Gothlin JH (1976) Post-lymphographic percutaneous fine needle biopsy of lymph nodes guided by fluorography. Radiology 120:205–207
15. Wajsman Z, Gamarra M, Park JJ, Beckley S, Pontes JE (1982) Transabdominal fine needle aspiration of retroperitoneal lymph nodes in staging of genitourinary tract cancer: correlation with lymphography and lymph node dissection findings. J Urol 128:1238–1240
16. Wajsman Z, Gamarra M, Park JJ, Beckley S, Pontes JE, Murphy GP (1982) Fine needle aspiration of metastatic lesions and regional lymph nodes in genitourinary cancer. Urology 4:356–360
17. Zajdela A, Ennuyer A, Batini P, Poncet P (1976) The diagnostic value of cytologic aspiration biopsy in adenopathies—cytohistological comparison in 1756 cases. Bull Cancer (Paris) 63(3):327–340
18. Kidd R, Crane RD, Dail DH (1984) Lymphangiography and fine needle aspiration biopsy: ineffective for staging early prostate cancer. AJR 142:1007–1012
19. Kidd R, Correa R (1984) Fine needle aspiration biopsy of lymphangiographically normal lymph nodes: a negative view. AJR 142:1005–1006

Chapter 9

Percutaneous Fine Needle Biopsy of Abdominal and Pelvic Lesions: Needle Passes Necessary for Secure Diagnosis with Fluoroscopy and Computed Tomography Guidance

J. H. Göthlin and G. Gadeholt

An extensive number of investigations regarding techniques, results, indications, hazards and complications of fine needle biopsy have been published, but few reports mention the number of passes performed at each biopsy and little is known of the number of passes necessary for securing diagnosis [1–4]. As the use of fine needle biopsy seems to be increasing and as the number of passes necessary to obtain acceptable security in diagnosis is often discussed among colleagues, we present investigation in 296 patients.

Methods and Materials

Over a 9-year period 184 males and 112 females, age range 16–94 years, were subjected to fine needle biopsy of abdominal and pelvic lesions. In 178 patients the biopsy was guided by fluoroscopy (FL) (mainly a retrospective study) and in 118 by computed tomography (CT) (mainly a prospective study). Each aspirate was marked with a number or a letter regardless of how many slides were smeared. The cytology report has been given for every aspirate (pass) and it is thus possible to state the first positive pass and the number of passes performed (which were also recorded by the radiologist for every biopsy performed in each patient). The cytological diagnosis was confirmed by various methods: surgical confirmation, autopsy and clinical course. The main problem has been verification of false negative and inconclusive results.

The biopsy technique with FL has been described elsewhere [3]. With CT, the target was defined on consecutive slices just before the biopsy. The distance from the ventral or dorsal midline of the patient or from the lateral surface vertical plane was measured on the monitor screen. The depth from the skin surface to the target along the intended biopsy route was determined and the insertion point on the skin marked. Techniques using grids on the surface were first used but soon abandoned, as they were considered to be unnecessarily time consuming and not capable of increasing precision. The position of the tip of the needle was verified by single or multiple CT slices.

The needles used varied widely: length was chosen according to the distance from the skin to the target. Stylets were sometimes employed. Only needles with an outer diameter of 1.2 mm or less were used. The Chiba needle has been avoided lately in the CT series for deep targets as its flexibility decreases precision in hitting the intended part of the target organ.

The original cytology reports have been used, as the aim of this investigation was to find out how many passes are required to obtain a diagnosis prior to treatment in the patient.

Results

In Table 9.1 are presented the correct, inconclusive or false negative and non-representative results. The overall results were better for CT than for FL guidance: 91% correct versus 73%. For both modalities adrenal glands, kidneys, ureter and soft tissue masses showed the best results, while correct diagnosis of pancreatic disease with FL was obtained in only 61% and with CT in 81% of patients. Of a total of 120 pancreatic biopsies 17 were false negative or inconclusive and 24 yielded no diagnostic material, mainly in the case of benign disease.

In Table 9.2 the first cytologically positive pass at each biopsy is presented. With FL, 59% of the correct diagnoses were obtained at the first pass, whereas the corresponding figure with CT was 81%. At least three passes were necessary to obtain diagnostic material from the pancreas, while for the other organs and soft tissue masses two passes often were sufficient.

Table 9.3 reflects that in most patients with false negative/inconclusive diagnosis or with non-representative material only two passes had been performed, but similar results were also obtained after four or more passes.

Table 9.1. Comparison of results of diagnosis by aspiration biopsy with FL and CT

	No. of patients		Correct		Inconclusive or false negative		Not representative	
	FL	CT	FL	CT	FL	CT	FL	CT
Liver	22	16	19	14	1	1	2	1
Pancreas	93	27	57	22	14	3	22	2
Bile ducts, gall bladder	11	14	9	13	—	1	2	—
Adrenal glands	—	7	—	7	—	—	—	—
Kidneys (no cysts)	6	11	6	11	—	—	—	—
Ureter, urinary bladder	23	18	23	17	—	1	—	—
Stomach, large bowel	11	5	7	5	—	—	4	—
Soft tissue, abdomen + pelvis	12	20	9	18	1	1	2	1
Total	178	118	130	107	16	7	32	4
% of total	100	100	73	91	8	6	18	3

Table 9.2. First cytologically positive pass at each biopsy

Location	Needle aspiration no.							
	1		2		3		4	
	FL	CT	FL	CT	FL	CT	FL	CT
Liver	10	11	5	3	3	2	1	—
Pancreas	29	16	8	2	11	3	9	1
Bile ducts, gall bladder	5	9	3	3	1	1	—	1
Adrenal glands,	—	6	—	1	—	—	—	—
Kidneys	5	8	—	2	1	—	—	1
Ureter, urinary bladder, gynaecological lesions	19	16	3	2	1	—	—	—
Stomach, large bowel	4	4	—	1	1	—	2	—
Soft tissue in abdomen or pelvis	5	17	3	3	—	—	1	—
Total in each group	77	87	22	17	18	6	13	3
% of all correct diagnoses (Table 9.1)	60	77	17	15	14	5	10	3

Table 9.3. False negative or inconclusive diagnoses and non-representative material

	Number of passes performed											
	False negative, inconclusive						Not representative					
	2		3		4+		2		3		4+	
	FL	CT	FL	CT	FL	CT	FL	CT	FL	CT	FL	CT
Liver	1	1	—	—	—	—	2	1	—	—	—	—
Pancreas	9	1	5	1	—	1	15	1	4	1	3	—
Bile ducts, gall bladder	—	1	—	—	—	—	2	—	—	—	—	—
Ureter, urinary bladder	—	1	—	—	—	—	—	—	—	—	—	—
Stomach, large bowel	—	—	—	—	—	—	1	—	2	—	1	—
Soft tissue	—	—	—	1	1	—	2	1	—	—	—	—
Total	10	4	5	2	1	1	22	3	6	1	4	—

Discussion

In the discussion we will deal with problems which have not gained much attention in the earlier literature. It has to be borne in mind that the results in only about 300 patients will merely indicate the number of passes necessary to secure a cytological diagnosis with reasonable certainty. The differences in type of lesion and organ, guiding modality and skill of the examiner and cytologist mean that the results will differ in varying materials.

The superiority of CT over FL as a guiding modality was not unexpected (Tables 9.1 and 9.2). It gives greater precision in controlling the exact desired position in the lesion and is superior for very small lesions. Especially when using angiography in pancreas and liver lesions it may be difficult to reach the desired region. However, the improved accuracy with CT is gained only with increased time consumption. A high percentage of correct results have been reported earlier [e.g. 5].

Another advantage CT has over FL is that it is easier to avoid traversing organs in the pathway to the target organ. In this connection it should be mentioned that we have not found it worthwhile to use techniques with grids on the skin surface [e.g. 6]. It is easy to measure distances on the CT monitor and to calculate point of entrance, pathway and depth to the lesion. A tape strip mounted on the biopsy needle to show calculated depth to the lesion is of advantage.

We have not been able to demonstrate any superiority of one needle over another [7, 8], with the exception that flexible needles too often deviate from the intended course when deep lesions are to be punctured. A stylet seems to be important only to increase the rigidity of the needle.

In most patients with a correct cytological diagnosis the first needle pass yielded diagnostic material (FL 60% and CT 77% of all correct diagnoses). The organ most likely to yield diagnostic material on the first pass was the pancreas (benign and malignant lesions combined). In a study of 45 malignant and 79 benign pancreatic lesions it was shown that in 72% of the cases carcinoma cells were found in the first aspirate, while in benign lesions the corresponding figure was 63% [4]. In our investigation the pancreatic material is too restricted (79 patients) to break down into malignant and benign conditions (most were malignant lesions), but we feel that we had greater difficulty in obtaining diagnostic material from benign lesions.

If the first pass gave non-representative or inconclusive material it was not possible to predict which of the next passes would be diagnostic (Table 9.2). However, the indication in Table 9.3 is that more than one pass after the initial one is necessary to give a correct diagnosis. Table 9.2 shows that generally there are no large differences between pass 2 and any of the next passes performed in obtaining diagnostic material.

False positive results are unusual. None was demonstrated in our series, which is in accordance with the general experience reported in the literature. False negative results, on the other hand, are not uncommon: in our material 16 out of 178 patients gave false negative results with FL and 7 out of 118 with CT. It is probably possible to decrease the percentage of false negatives by increasing the number of passes, but a limit has to be set for practical reasons. In our study using FL, frequently only two passes (53% of the cases) were employed, which are probably too few. With CT three or more passes were not unusual and may partly have contributed to the overall higher percentage of correct cytological diagnoses. In an earlier study of 20 patients with varying lesions one pass was enough for diagnosis in 15 patients, while for three patients two samples were necessary and in the remaining two patients three and four passes were needed [2]. In material from another 75 biopsies the number of passes required for a correct diagnosis was not recorded, but one single pass was felt to be sufficient if a rapid stain technique was used and cytological interpretation was carried out immediately. Further passes were performed if the diagnoses were not conclusive [6]. Whether there is time enough for such a procedure in most hospitals is doubtful, but the idea of checking every cytology specimen immediately is at least theoretically attractive. In a well-controlled biopsy study of the pancreas it was demonstrated that a minimum of four aspirates per biopsy was required for satisfactory results [4].

In our material there is a considerable percentage of non-representative results (with FL 32/178, with CT 4/118). In most of these cases only two passes were performed (Table 9.3) and our high percentage is probably due to too few passes. It is interesting to note that non-representative material and false negative/inconclusive results are virtually absent in kidney and ureteral lesions (Table 9.3). This may be caused by two factors: the lesions may have been easy to hit and the cells in the lesions were probably easy to dislodge [9]. In other lesions the unrepresentative material may be due to difficulties in dislodging cells rather than problems in hitting the lesions. An abundance of blood in the sample may also make it more difficult to find diagnostic cells. Therefore, at the latest, aspiration should be stopped at the first sign of blood in the syringe, preferably before any material at all can be seen in the syringe.

From our series and the literature some conclusions can be drawn:

1. Most of the correct diagnoses by percutaneous biopsy can be made from the first aspirate.
2. Of the organs most commonly punctured, pancreas required most passes for a correct diagnosis.
3. There is increased certainty of getting a correct diagnosis with an increasing number of aspirations.
4. Urinary pathway lesions frequently give a correct diagnosis on the first pass.
5. CT gives greater precision in guiding the needle than FL.
6. A fast bedside staining technique with immediate cytological interpretation may be of advantage, even if time consuming.
7. The number of passes necessary for secure diagnoses is not established but four passes may represent a suitable standard.

References

1. Kidd R, Freeny PC, Bartha M (1979) Single pass fine needle aspiration biopsy. AJR 133:333–334
2. Ferrucci JT Jr, Wittenberg J, Mueller PR, Simeone JF, Harbin WP, Kirkpatrick RH, Taft PD (1980) Diagnosis of abdominal malignancy by radiologic fine-needle aspiration biopsy. AJR 134:323–330
3. Göthlin JH (1982) Fluoroscopically guided, percutaneous, transperitoneal fine-needle biopsy. Eur J Radiol 2:130–134
4. Lüning M, Kursawe R, Schöpke W, Lorenz D, Menzel A, Hoppe E, Meyer R (1985) CT guided percutaneous fine-needle biopsy of the pancreas. Eur J Radiol 5:104–108
5. Dondelinger R (1982) CT-guided percutaneous biopsy. J Belge Radiol 65:227–243
6. Haaga JR, Alfidi RJ (1976) Precise biopsy localization by computed tomography. Radiology 118:603–607
7. Andriole JG, Haaga JR, Adams RB, Nuncz L (1983) Biopsy needle characteristics assessed in the laboratory. Radiology 148:659–662
8. Torp-Pedersen S, Juul N, Vyberg M (1984) Histological sampling with a 23 gauge modified Menghini needle. Br J Radiol 57:151–154
9. Göthlin JH, Barbaric ZL (1978) Fluoroscopy-guided percutaneous transperitoneal fine needle biopsy of renal masses. Urology 11:300–302

Chapter 10

The Role of Aspiration Cytology in Staging and Monitoring of Malignant Neoplasms

R. Giardini, S. Pilotti, L. Alasio and F. Rilke

Aspiration cytology is a well-established, easy and safe procedure which enables the pathologist to participate actively in the post-diagnostic and follow-up phase of the cancer patient's development [1]. Fine needle aspiration biopsy (FNAB) of a clinically suspicious lymph node or non-vascular mass is one of the first procedures performed in the work-up of a patient with a potentially recurrent disease because it provides accurate assessment with minimal morbidity, mortality and cost [2–5]. To quantify our experience, we reviewed the material pertaining to this area of oncological cytopathology.

In this study we analysed 520 patients followed for known malignant disease. An FNAB was obtained between July 1979 and March 1985 and was used to sample superficial, subcutaneous and deep visceral masses. With regard to the distribution of the anatomical sites of the aspirated masses, superficial lymph nodes and soft tissues were the most frequent sites [6] and were examined in 342 patients (66%). In 178 cases (34%) the diagnostic material was obtained from deep-seated masses and was usually aspirated by radiologists using radiological guidance techniques such as scintiscan, echotomography and computed tomography. The lesion was aspirated using a 22-gauge needle and a hand-activated 20-ml syringe. The material was smeared on non-frosted glass slides and either immediately fixed in 95% ethanol or air dried. The slides were stained by the Papanicolaou technique or the May–Grünwald–Giemsa method. The patients were widely distributed with respect to age (range 1–90 years) and the male-to-female ratio was approximately equal. The FNABs from abdominal and retroperitoneal masses were taken from patients whose mean age was comparatively low. Males predominated in the group with retroperitoneal disease.

Nodal metastatic disease was detected in 193 patients with secondary spread. Most of the nodal aspirates were taken from the supraclavicular location, the next most frequent sites being the cervical, axillary and inguinal areas. There were cases of metastases to predictable regional lymph nodes, such as to the cervical lymph nodes in the course of head and neck malignancies. However, the involvement of supraclavicular lymph nodes by topographically distant tumours was not a rare event [7]: the frequency of gynaecological malignancies metastatic to this area was surpassed only by that of breast and lung cancers and was twice that of gastrointestinal carcinomas. The primary tumours most often encountered in the metastasising phase originated in the breast, head and neck, lung, upper respiratory and digestive tracts and skin, particularly malignant cutaneous melanomas. In contrast, retroperitoneal nodes were most commonly seeded by germinal cell tumours of the testes.

In the analysis of FNABs from soft tissue nodules metastases (60%) should be distinguished from recurrences (40%). It was found that malignant melanoma and breast carcinoma were the malignancies that most frequently metastasised to soft tissues, malignant melanoma showing a clear prevalence of metastases (including those in

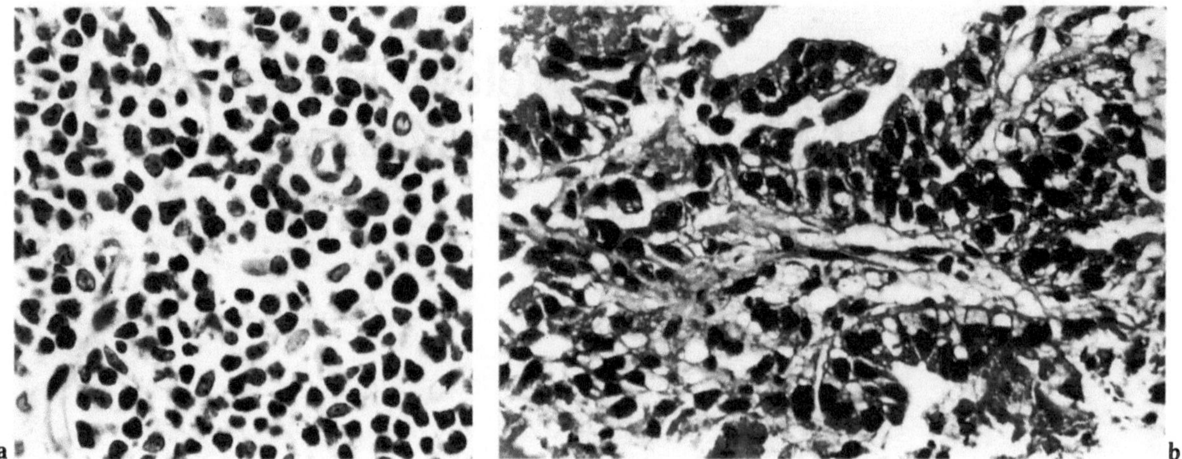

Fig. 10.1a,b. A 50-year-old male with a lymphoplasmacytoid malignant lymphoma. **a** The biopsy of an axillary node shows the complete replacement by an abnormal population made up of lymphocytes, lymphoplasmacytoid cells, immunoblasts, centroblasts and centrocytes (PAS-haematoxylin, × 510). **b** Histological section of an FNAB specimen from a pulmonary nodule showing a second primary adenocarcinoma. (H & E, × 310)

transit) over local recurrences. Of the total number of malignant tumours diagnosed in nodal aspirates, superficial lymph nodes were involved more often by adenocarcinomas from various sites than by squamous cell carcinomas. In general, malignant squamous cells exhibited a remarkable tendency to form sheets and cohesive clusters; adenocarcinoma cells metastatic to lymph nodes showed great variation in size and shape, but also a striking resemblance to the primary neoplasm.

FNAB, which was performed to detect possible recurrences or metastases of known tumours, in some cases led to the diagnosis of a second malignancy [8–10]. In the lungs a second primary tumour was found in 10 cases, with intervals after the first primary ranging from 1 to 168 months. Among lung malignancies almost all histological types were represented; however, adenocarcinoma and bronchiolo-alveolar carcinoma were the most frequent (Fig. 10.1).

In one case a pulmonary squamous-cell carcinoma developed 84 months after the diagnosis of Hodgkin's disease and, in addition, the patient developed a third primary tumour in the urinary bladder 60 months later. Other cases of second malignancies at varying intervals after diagnosis of the first were found in the soft tissues (three cases), kidneys (two cases), liver and thyroid gland.

Of all FNABs made in this study there were 442 cases (85%) of metastases, 58 (11%) recurrences and 20 (4%) second malignancies.

References

1. Schultenover SJ, Ramzy I, Page CP, LeFebre SM, Cruz AB Jr (1984) Needle aspiration biopsy: role and limitations in surgical decision making. Am J Clin Pathol 82:405–410
2. Betsill WL Jr, Hajdu SI (1980) Percutaneous aspiration biopsy of lymph nodes. Am J Clin Pathol 73:471–479
3. Craig ID, Shum DT, Desrosiers P, McLeod C, Lefcoe MS, Paterson NAM, Finley RJ, Woods B, Anderson RJ (1983) Choriocarcinoma metastatic to the lung. A cytologic study with identification of human choriogonadotropin with an immunoperoxidase technique. Acta Cytol (Baltimore) 27:647–650
4. Frable WJ (1976) Thin-needle aspiration biopsy. A personal experience with 469 cases. Am J Clin Pathol 65:168–182
5. Ramzy I, Rone R, Schultenover SJ, Buhaug J (1985) Lymph node aspiration biopsy. Diagnostic reliability and limitations—an analysis of 350 cases. Diagn Cytopathol 1:39–45
6. Kehl A, Nagel GA (1984) Bedeutung der Punktionszytologie für den Nachweis von Metastasen bei bekanntem und unbekanntem Primärtumor aus der Sicht des Onkologen. Pathologe 5:243–246
7. Helmkamp BF, Sevin BU, Greening SE, Nadji M, Ng ABP, Averette HE (1981) Fine needle aspiration cytology of supraclavicular lymph nodes in gynecologic malignancies. Gynecol Oncol 11:89–95
8. Friedman M, Shimaoka K, Fox S, Panahon AM (1983) Second malignant tumors detected by needle aspiration cytology. Cancer 52:699–706
9. Pilotti S, Rilke F, Gribaudi G, Damascelli B, Ravasi G (1984) Transthoracic fine needle aspiration biopsy in pulmonary lesions. Updated results. Acta Cytol 28:225–232
10. Volk SA, Mansour RF, Gandara DR, Redmond J III (1984) Morbidity in long-term survivors of small cell carcinoma of the lung. Cancer 54:25–27

Section III

Prostatic Cancer

Chapter 11

Aspiration Biopsy in the Diagnosis of Prostatic Carcinoma: the Importance of Cytological Malignancy Grading in Prognostic Judgement and Choice of Therapy[1]

P. L. Esposti

Transrectal aspiration biopsy of the prostate, introduced by Franzèn et al. [1] at Radiumhemmet in the late 1950s has been accepted as a fast, safe and accurate procedure. At Karolinska Hospital this procedure has become routine in the last 25 years in the diagnosis of prostatic carcinoma [2].

The instrument consists of a syringe, a needle and a needle guide. The needle is 20 cm long, fine (22 gauge) and flexible. The needle guide is a metal tube having at its distal end a steering ring for the palpating index finger. The syringe has a special handle which permits a one-hand grip during aspiration.

The patient is usually placed in the lithotomy position but the aspiration may also be performed if the patient is standing, slightly bent forwards. Previous preparation of the bowel and anaesthesia are usually not required. Complications are comparatively rare and usually not of a serious nature.

The smear is prepared after aspiration by rapidly spreading the material expressed onto a glass with the help of a coverslip. The smear is then left to dry in the air and successively stained with May–Grünwald–Giemsa stain, or fixed in methanol and stained according to Papanicolaou.

Prostatic Carcinoma: Cytological Diagnosis and Malignancy Grading

In prostatic carcinoma the aspirates are generally rich in cells and only occasionally mixed with fluids and blood. The high cellularity gives a distinctive granulated appearance to the unstained smear, macroscopically appreciable. At microscopy the known characteristics of carcinoma become evident: nuclear atypia with prominent, irregular nucleoli, decreased cytoplasmic/nuclear ratio and reduced mutual adhesiveness to the cells, although microacinar structures typical of adenocarcinoma are as a rule present.

According to the degree of deviation from the normal epithelial structure and the appearance of microglandular structures, prostatic carcinoma has been grouped into three different grades [3]: grade I or highly differentiated, when nuclear polymorphism is of moderate degree and the microadenomatous complexes are frequent; grade II or moderately differentiated, when the nuclear polymorphism is more pronounced and the number of

[1] This contribution is reproduced from *European Urology* 11:361–362, by kind permission of the publisher, S. Karger AG, Basel.

free cells is higher, but the general pattern is similar to that of grade I carcinoma; and grade III or poorly differentiated, when there is predominance of dissociated cells, with strongly polymorphous nuclei and only rare microacinar structures.

The cytological degree of differentiation is usually remarkably constant in the smears of the individual tumours. When several degrees of atypia coexist, the most undifferentiated part determines the grade of malignancy. There is no fundamental difference in criteria of differentiation of prostatic carcinoma in aspirates and in histological sections. The histological diagnosis of carcinoma rests mainly on two criteria: nuclear abnormalities and abnormalities in the size, configuration and arrangement of the acini [4]. When such criteria are adopted, comparisons of malignancy grading of prostatic carcinoma in cytological smears and histological sections become possible.

Prognostic Significance of Cytological Malignancy Grading and Its Therapeutic Consequence

Follow-up studies of 469 patients with hormonally treated prostatic carcinoma cytologically diagnosed and after aspiration biopsy [3] showed that the crude 5-year survival rate was 68% in highly differentiated, 55% in moderately differentiated and 11% in poorly differentiated tumours. The following considerations are possible: (1) the survival rate of patients with grade I tumours is only slightly worse than the expected survival of a population of men around 67 years of age; (2) the rate survival in the group of grade II tumours occupies an intermediate position between highly and poorly differentiated cases; and (3) the poor survival rate of patients with grade III tumours indicates that the given therapy was not effective in these carcinomas. As a consequence grade III tumours are now treated at the Karolinska Hospital with high-voltage irradiation, if proved to be without distant metastases.

Nuclear DNA Levels in Aspirated Cells from Prostatic Carcinoma and Its Prognostic Significance

Increased DNA content in the nuclei of aspirated cells expressing chromosomal abnormality can be assessed by cytophotometry. The cytochemical analysis of individual cells, based on quantitative cytophotometric measurements of Feulgen-stained nuclei, showed that nuclei from benign lesions exhibited the normal diploid amount of DNA. Cells from prostatic carcinomas were characterised by various degrees of heteroploidy. A general correlation seems to exist between the degree of heteroploidy and the degree of cytological and clinical malignancy. In a retrospective study [5] old May–Grünwald–Giemsa-stained smears could be destained and used for quantitative DNA analysis after Feulgen staining. It was then shown that patients with good response to therapy and long survival had carcinoma cells characterised by a diploid or combined diploid-tetraploid DNA distribution pattern, while carcinomas from patients with poor response to therapy, exhibited abnormally increased DNA amounts. At present, research work is in progress using rapid flow cytofluorometric DNA analysis of a large number of cells [6].

It is the opinion of the present author that these two complementary methods, i.e. the quantitative DNA determination of a limited number of Feulgen-stained well-identified cancer cells and the rapid flow analysis of large amounts of cells, will increase knowledge of the biological properties of the prostatic tumours and offer additional information to the cytological malignancy grading.

References

1. Franzèn S, Giertz G, Zajicek J (1960) Cytological diagnosis of prostatic tumours by transrectal aspiration biopsy: a preliminary report. Br J Urol 32:193–196
2. Esposti PL (1966) Cytologic diagnosis of prostatic tumours with the aid of transrectal aspiration biopsy. A critical review of 1,110 cases and a report of morphologic and cytochemical studies. Acta Cytol (Baltimore) 10:182–186
3. Esposti PL (1971) Cytologic malignancy grading of prostatic carcinoma by transrectal aspiration biopsy. Scand J Urol Nephrol 5:199–209
4. Mostofi FK (1969) Carcinoma of the prostate. In: Riches E (ed) Modern trends in urology. Butterworth, London, pp 231–263
5. Zetterberg A, Esposti PL (1980) Prognostic significance of nuclear DNA levels in prostatic carcinoma. Scand J Urol Nephrol [Suppl] 55:53–58
6. Tribukait B, Esposti PL, Rönström L (1980) Tumour ploidy for characterization of prostatic carcinoma: flow cytofluorometric DNA studies using aspiration biopsy material. Scand J Urol Nephrol [Suppl] 55:59–64

Chapter 12

Fine Needle Aspiration and Core Biopsy in Detection of Prostatic Cancer: Comparison with Surgical Biopsy

P. Narayan, R. Stein and P. B. Jajodia

Introduction

Prostatic cancer is the second most common cancer in American men. It is estimated that, in 1987, over 96 000 new cases will be diagnosed [1]. Nearly half of all patients have metastatic disease at the time of diagnosis [2]. The need for early detection of prostatic cancer makes it desirable to establish simple diagnostic procedures that yield accurate information and can be performed on an outpatient basis.

Traditionally, in the USA, digital examination followed by core needle biopsy of the prostate via the transrectal or transperineal route has been the preferred means of establishing the diagnosis. The transperineal core biopsy is performed with the Travenol (Tru-Cut) or Vim Silverman needle following local, spinal or general anaesthesia. Usually, two to four cores are obtained from the suspicious areas of the prostate. If these appear inadequate visually, more cores are obtained. Prophylactic antibiotics are given at the physician's discretion. Each core is fixed immediately in formalin, processed and stained with haematoxylin and eosin. All specimens are reviewed by a pathologist and, when malignant, are either assigned a Gleason grade [3] or graded as well-differentiated, moderately differentiated or poorly differentiated carcinoma. Other grading systems [4–7] are less commonly used.

Transrectal needle biopsy of the prostate employs a more direct approach to the prostatic nodule. Usually, the procedure is performed under intravenous sedation. The patient is given pre-biopsy oral or systemic antibiotics. After a cleansing enema, the site for biopsy is identified with the index finger, which also serves as a guide for the biopsy needle. Penetration is made through the rectal wall and fascia into the prostate. Two to four good cores of tissue are usually adequate. Frequently, biopsy is performed on both lobes of the prostate.

The advantages of core needle biopsy include accuracy of needle placement after digital palpation of a prostatic nodule, preservation of architectural integrity of the specimen obtained and the absence of false positive biopsy results. The accuracy of core biopsies has varied from 50% to 96% [8–10].

The disadvantages of core needle biopsy include a relatively high incidence of septic complications (5%–38%), bleeding from haemorrhoidal vessels and the prostatic urethra, and, rarely, tumour cell implantation along the needle biopsy tract [11, 12]. Intravenous sedation and anaesthesia are required, and, occasionally, hospitalisation. The discomfort associated with the procedure may lead the patient to refuse repeat biopsies when the first is unsatisfactory. The main reason for this discomfort is the 18-gauge size of the Tru-Cut biopsy needle and its rigidity. Finer biopsy needles that are more flexible are currently being tested and their use may circumvent some of the disadvantages of the core biopsy technique as performed at present.

In recent years, fine needle aspiration (FNA) cytology of the prostate is being recognised as a useful, if not superior, alternative. Several studies have noted the simplicity, safety and accuracy of FNA of the prostate and its advantages over core needle biopsy [13–18].

Aspiration biopsy of the prostate was one of the first biopsy techniques described in the USA and was performed by Ferguson as early as 1930 [19]. However, Franzèn and associates are to be credited with popularising it [20]. They described transrectal needle aspiration of the prostate in 1960 as a fully developed, easy-to-perform procedure for cytological diagnosis in cases of clinically suspected carcinoma of the prostate. Since interpretation of prostatic cytology was not well developed in the USA at that time, the technique did not gain wide acceptance. However, the development of radiological and endoscopic fine-needle biopsy techniques and the increasing emphasis on cost-effectiveness in diagnostic techniques have substantially increased the use of FNA cytological diagnosis in all malignancies. FNA of the prostate is therefore becoming more widely used as physicians develop the skills necessary for the performance and interpretation of cytological needle aspiration.

FNA of the prostate is still performed as originally described by Franzèn and associates [20, 21], with a flexible 22- or 23-gauge prostatic aspiration needle, needle guide and aspiration syringe. Aspirates either are air-dried and stained with May–Grünwald–Giemsa [22] stain or are fixed in alcohol and stained by the Papanicolaou method.

The advantages of FNA of the prostate can be summarised as follows:

1. It may be performed without anaesthesia as an outpatient procedure.
2. Multiple aspirations capture cells from a larger area of the prostatic nodule.
3. Four-quadrant biopsies may be performed in all patients since the procedure has minimal discomfort.
4. Patient tolerance allows repeat biopsies at frequent intervals in patients with highly suspicious nodules.
5. The easily controlled needle guide permits specific puncture of minute suspicious nodules.
6. The rate of complications, including haematuria and sepsis, is extremely low.

Because of these advantages, FNA may be more likely to become a screening procedure for prostatic cancer than the core biopsy technique as performed at present.

The major disadvantages of FNA are the need for trained personnel to perform and interpret the results. Even in expert hands, there are occasional false negative and false positive results. In our own early experience, FNA was associated with a 12.9% incidence of false negativity, and we had one instance of false positivity in a patient with atypical hyperplasia (Table 12.1). Other studies have shown similar results, although with increasing experience the false negative rates decrease [13–18, 23–26]. Repeat aspirations in clinically suspicious prostates will reduce the incidence of false negative readings. Currently, we perform both core and FNA biopsies in most patients because we consider that these techniques have complementary roles in increasing the diagnostic accuracy in suspected cases of prostatic cancer (Table 12.1).

Table 12.1. Diagnostic accuracy in prostatic cancer: FNA in comparison with core biopsy

	FNA (%)	Core biopsy (%)
Positive[a]	27 (28.42)	21 (22.10)
Negative	53 (55.78)	62 (65.26)
Inconclusive[b]	9 (9.47)	0
Unsatisfactory[b]	6 (6.31)	12 (12.63)
Total no. of patients	95 (100)	95 (100)
False negative	4	10
False positive	1	0

[a] All positive FNA biopsies were confirmed by repeat core biopsies, radical prostatectomy or evidence of metastatic disease.
[b] Patients with inconclusive or unsatisfactory results underwent at least two attempts at biopsy.
A total of 31 patients were confirmed positive.
Diagnostic accuracy of FNA: 26/31 (83.87%).
Diagnostic accuracy of core biopsy: 21/31 (67.74%).
False negative rate of FNA: 4/31 (12.90%).
False negative rate of core biopsy: 10/31 (32.25%).
False positive rate of FNA: 1/31 (3.22%).

Reasons for Sampling Errors in Prostatic Biopsies

There are several reasons why sampling errors occur in prostatic biopsies. Several benign conditions may mimic prostatic cancer. Adenomatous hyperplasia is often associated with nodularity that is suggestive of malignancy. Other conditions such as fibrosis secondary to prostatic infarction, chronic prostatitis, granulomatous prostatic disease and abnormal seminal vesicles may simulate prostatic cancer.

The differences in training of physicians performing biopsies and those reviewing them influence the ability both to obtain and to interpret representative material from suspicious areas of the prostate. For example, during core biopsy of the prostate, an inadequately trained physician may obtain insufficient tissue or may sample skeletal muscle and rectal mucosa. Accuracy rates for core biopsy range from 50% to 96% [8–10]. However, in a recent survey of 20 000 cases in 420 hospitals, 49% of patients presented with clinically palpable disease (stage B or stage C), but the rate of diagnosis by transrectal and perineal biopsy was only 7% and 13% respectively [27]. Although several explanations are possible for this low sensitivity, it is logical to assume that improving the technique and training of physicians performing biopsy will increase the detection of early-stage prostatic cancer.

Physician training is even more important in the performance of FNA. The aspiration technique, although comfortable for the patient, takes considerable skill to be consistently accurate. False negative and inconclusive results are often obtained during early experience because of scanty material and contamination with blood and urine, both of which occur rarely when aspirations are performed by skilled physicians [21]. Also, clinicians are more experienced in detecting suspicious prostatic nodules, while pathologists are more experienced in preparing smears and interpreting cytological data. The number of false positive and false negative findings will therefore depend upon the skill and training of both the person performing the biopsy and the person preparing and interpreting the smears.

The accuracy of the type of biopsy performed depends on the sample obtained. For example, core biopsy samples a 1- to 2-mm area linearly. Therefore, more than one core is necessary for adequate sampling of lesions not completely in the plane of the biopsy. The number of cores obtained and the length of the core are also important in the number of positive diagnoses made. In a recent report, it was found that a minimum of four core biopsies was necessary to give a 94% likelihood of sampling the cancerous area [8]. This still did not address several problems, such as the number of repeat biopsies required or the fate of patients in whom core biopsies of "inadequate" length were obtained. In patients with false negative core biopsies, careful retrospective reviews have shown that atypical and suspicious areas existed that could not be adequately interpreted as cancer because of lack of sufficient tissue. These patients subsequently developed overt prostatic cancer [9].

Our results suggest that FNA tends to sample a larger amount of prostatic epithelium than does core biopsy, as it relies on aspirating cells that are non-adherent and the needle may be moved back and forth within the suspicious area. We have found larger numbers of cells in FNAs positive for malignancy as compared with core biopsies (Fig. 12.1 a–d). This is especially important when a lesion is confined to small area, and a core biopsy may sample it inadequately.

The degree of discomfort and preparation associated with the procedure definitely influences the number of times a biopsy is repeated, and in turn influences the accuracy of the diagnosis. It is well known that some patients are diagnosed only after the second or third biopsy. In our recent series of 95 patients, when both core biopsies and FNAs were performed, repeat biopsies in 5 patients (5.3%) detected prostatic cancer on second or third biopsy (Table 12.2). Other studies have found similar results [9, 28].

Finally, the degree of stromal hyperplasia around an area of tumour may also cause sampling errors. Specimens obtained at radical prostatectomy clearly

Table 12.2. Biopsy results on patients undergoing more than one core and/or FNA biopsy

Patient	Core biopsy			FNA biopsy		
	1st	2nd	3rd	1st	2nd	3rd
1	Neg	Neg	Pos Grade 2/2	Suspicious	Pos (WD)	—
2	Neg	Pos Grade 3/4	—	Pos (WD)	—	—
3	Neg	Neg	—	Neg	Neg	Pos (MD)
4	Neg	Pos Grade 3/3	—	Neg	—	—
5	Inc	Pos Grade 3/3	—	Uns	Neg	Inc

WD, well-differentiated; MD, moderately differentiated; Inc, inconclusive; Uns, unsatisfactory; Neg, negative; Pos, positive.
In this series we had 14 patients who underwent two to three attempts by core and FNA biopsies, and results of both methods were negative.

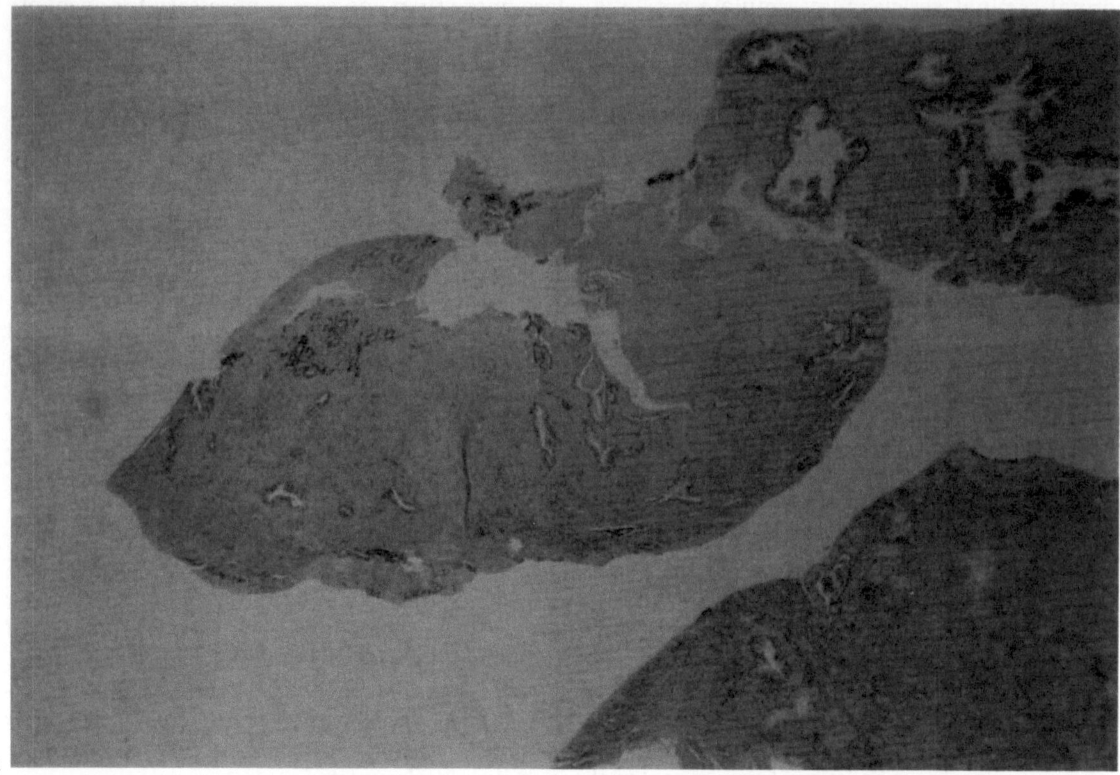

a

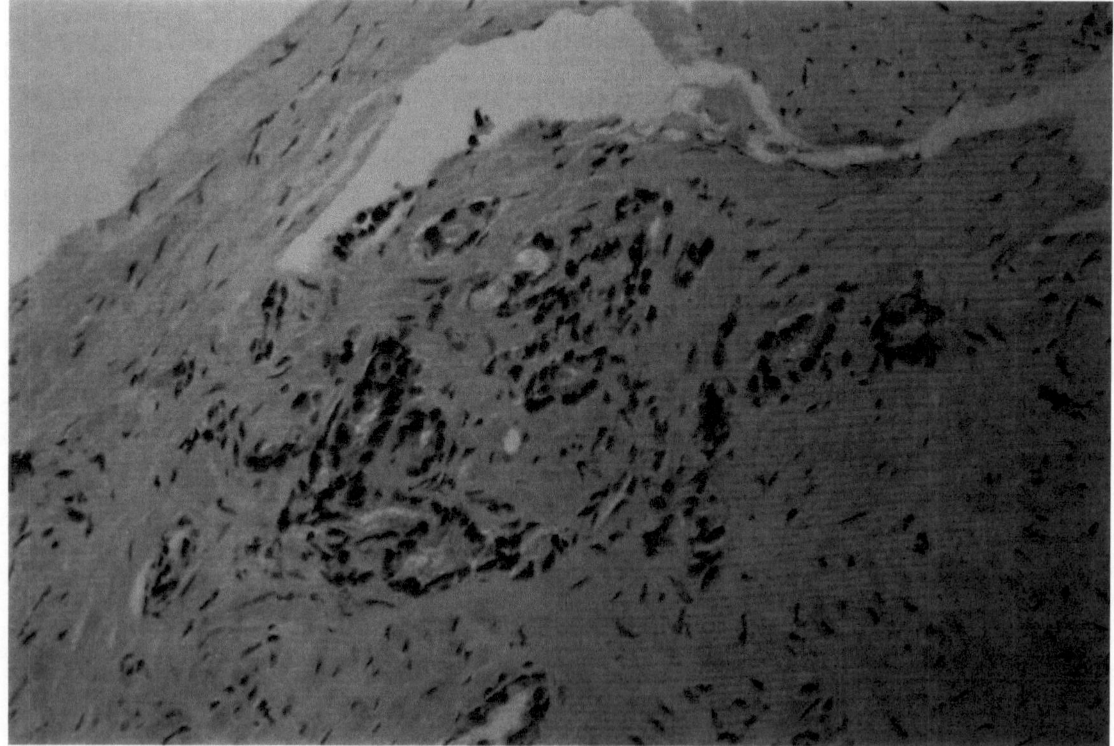

b

Fig. 12.1a–d. Comparison between relatively less cellular core biopsy (**a, b**) and more cellular aspiration material (**c, d**) from the same moderately differentiated adenocarcinoma. (Magnification: **a** × 9; **b** × 65; **c** × 22; **d** × 130)

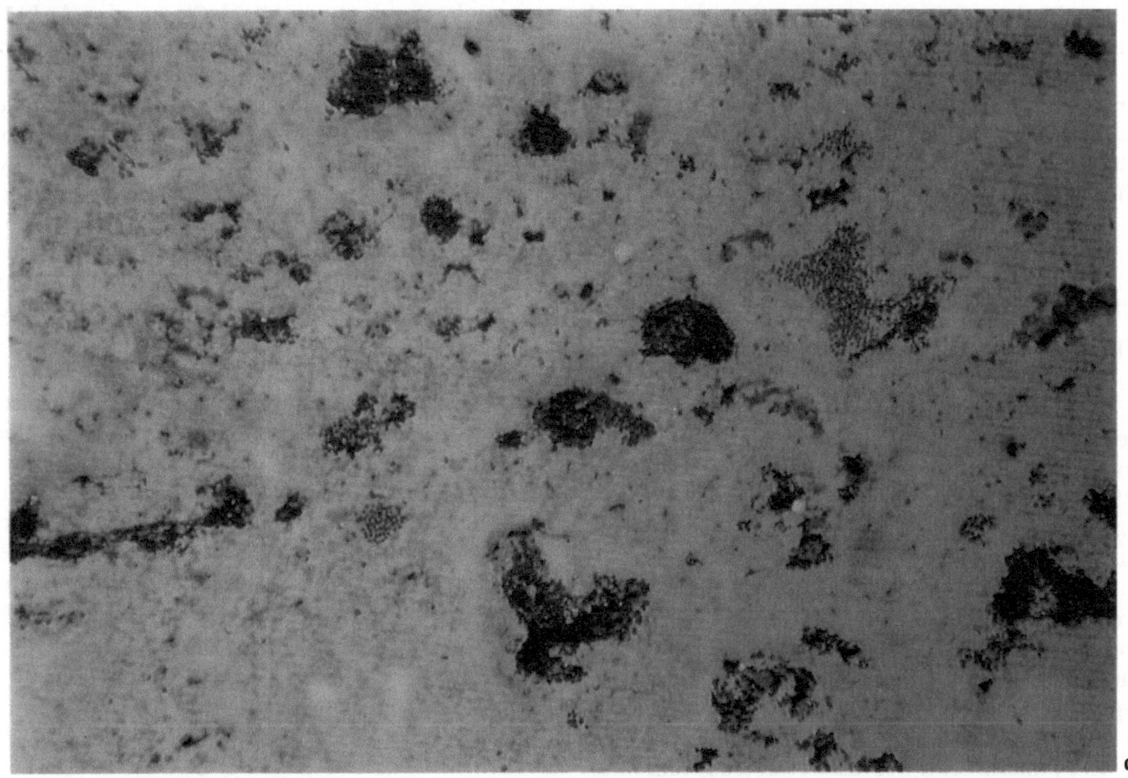
c

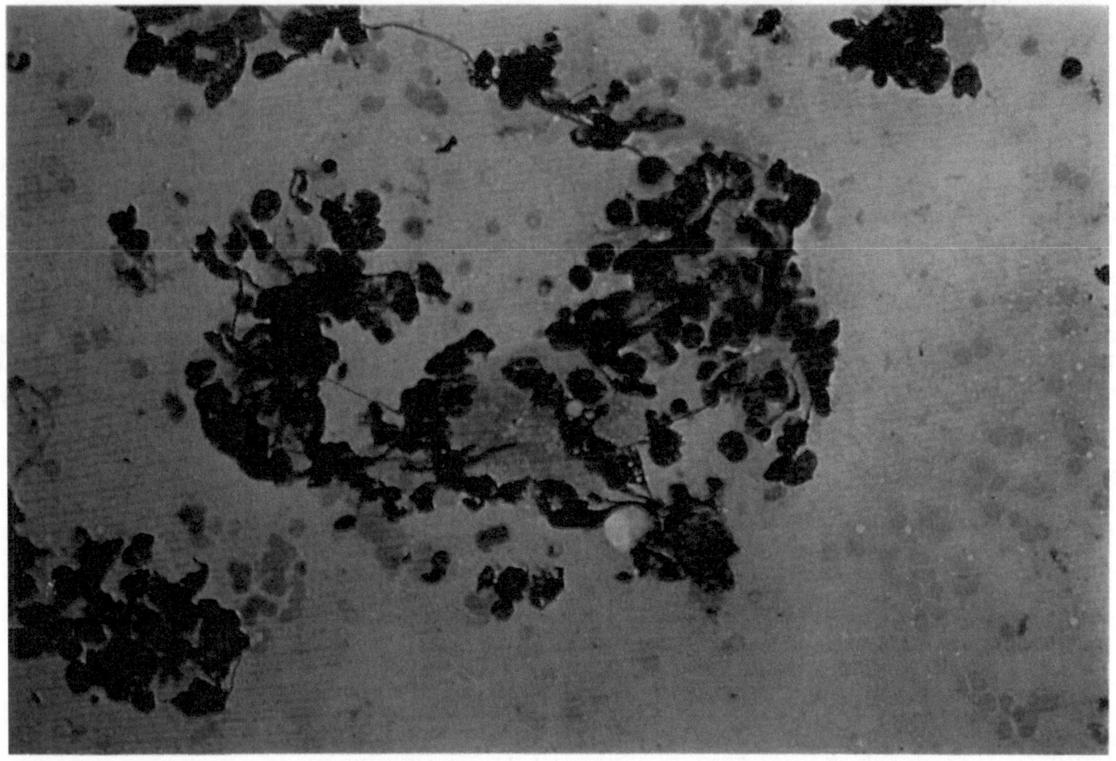
d

▲ **Fig. 12.1** (*continued*)

show that cancerous areas may be interspersed within normal areas and that benign nodules can be present close to areas of tumour cells. Biopsies of suspicious prostatic nodules that yield only fibrous or stromal tissue may actually signify hyperplastic activity around a tumour rather than failure of the physician or technique of biopsy.

Undergrading and Overgrading of Preoperative Biopsies

Grade of prostatic cancer has been used as a prognostic factor that determines biological behaviour [29–32]. A grade of well-differentiated cancer on core biopsy has been used occasionally to direct patients to surveillance-only protocols [33]. Clearly, patients with well-differentiated cancer have a low propensity for metastatic disease (10%–16%) as opposed to patients with poorly differentiated cancer who have a much worse prognosis [33, 34]. Esposti documented the 5-year survival of patients in relation to cytological grade: approximately 70% for 131 patients with well-differentiated carcinoma; only 10% for 73 patients with poorly differentiated carcinoma [35]. It is therefore important to grade the cancer accurately, especially if non-operative treatment is chosen, since results would vary depending on the histology of the tumour. Accurate grading of prostatic carcinoma is also essential to identify patients who have localised, and thus potentially curable, disease since higher-grade lesions tend to be associated with a higher incidence of lymph node metastasis [36–40]. There is a linear association between increasing grade (as determined by biopsy) and stage, and patients with high-grade disease frequently are understaged [29, 37].

In previous studies, we and others have noted that core biopsy of the prostate may under- and overgrade the final combined Gleason grade [41, 42]. In our recent experience with a limited number of patients, we have found that FNA is also associated with significant over- and undergrading (Table 12.3). Preoperative grading, therefore, is inaccurate and may lead to a false sense of security, especially in a patient assigned a grade of well-differentiated carcinoma and placed on a surveillance-only protocol. At the very least, patients on surveillance protocols and those who refuse therapy should undergo frequent repeat biopsies so that they may reconsider their therapeutic options if a higher-grade lesion is detected subsequently.

Inaccurate grading of the biopsy specimen also leads to an incorrect estimate of prognosis. In the Gleason system, patients whose tumours have combined grades of 5 or less appear to have a better prognosis than do those whose tumour grades are over 6 [29]. Our early experience suggests that use of FNA for preoperative grading will be inaccurate for predicting prognosis in about 50%–70% of patients (Table 12.3).

Prostatic Biopsy in Diagnosis of Stage A Prostatic Cancer

Stage A prostatic cancer is an incidental finding in patients undergoing transurethral or suprapubic prostatectomy for benign disease. Stage A_1 is defined as a low-grade, focal lesion present in less than 5% of the total volume of the gland removed, and stage A_2 as a high-grade lesion or any lesion present diffusely or occupying more than 5% of the total volume of the gland removed [43, 44]. The incidence of stage A prostatic carcinoma varies with the patient's age and the pathologist's sectioning technique. Usually, it occurs in 10%–15% of patients undergoing transurethral prostatectomy

Table 12.3. FNA in preoperative grading of prostatic cancer

FNA grade[a]	No. of patients	Combined Gleason[b]	Mean Gleason[b]
1. Well-differentiated	7	5,5,5,7,5,4,4	5
2. Moderately differentiated	14	6,8,4,6,5,7,6,5,8,4,7,8,7,8	6
3. Poorly differentiated	2	10,9	9.5

[a] Well-differentiated = Gleason grades 1–4; moderately differentiated = Gleason grades 5 and 6; poorly differentiated = Gleason grades 7–10.

[b] Grade in radical prostatectomy specimen.

5/7 (71%) of patients in the well-differentiated group were overgraded, although 4/5 (80%) of these patients were within one Gleason grade of the correct diagnosis.

7/14 (50%) of patients in the moderately differentiated group were undergraded, while 2/14 (14.2%) were overgraded.

for suspected benign disease, although rates varying from 4% in the third decade to 80% in the ninth decade have been reported [45]. Also, since incidental carcinomas are often confined to the surgical prostatic capsule, the autopsy incidence is approximately twice the clinical incidence [46, 47].

We attempted to determine if FNA or core biopsy in patients undergoing transurethral prostatectomy for suspected benign hyperplasia would result in a higher detection rate of stage A prostatic cancer. The rationale was that, since stage A disease is more often found in the surgical capsule (which is not always biopsied on transurethral prostatectomy), transrectal biopsies may yield a higher detection rate of these cancers. Also, since FNA biopsies sample a more diffuse area of the prostatic epithelium than do core biopsies, we wanted to determine if FNA is more sensitive. In our limited study, it appears that both FNA and core biopsy are equally unsuccessful in diagnosing stage A cancer of the prostate (Table 12.4). Interestingly, core biopsy of the prostate detected a well-differentiated cancer in one patient in whom a transurethral prostatectomy did not reveal stage A cancer. This patient subsequently underwent radical prostatectomy and was confirmed as having residual cancer in the prostate.

Table 12.4. Diagnostic accuracy of FNA and core biopsy in stage A prostatic cancer

	FNA	Core biopsy	TURP
Negative	24	21	45
Positive	0	1[a]	4[b]
Total	24	21	45

TURP, transurethral prostatectomy.
[a] In this patient, who had cancer detected by core biopsy, pathological findings on the TURP specimen were benign. Radical prostatectomy revealed residual tumour.
[b] None of these stage A carcinomas (3 stage A and 1 stage A_2) were detected by FNA or core biopsy.

References

1. CA–A cancer journal for clinicians (1987) Cancer Stat 37(1):12–13
2. Elder JS, Catalona WJ (1984) Management of newly diagnosed metastatic carcinoma of the prostate. Urol Clin North Am 11(2):283–295
3. Gleason DF (1966) Classification of prostatic carcinomas. Cancer Chemother Rep 50:125–128
4. Broders AC (1925) The grading of carcinoma. Minn Med 8:726
5. Utz DC, Farrow GM (1969) Pathologic differentiation and prognosis of prostatic carcinoma. JAMA 209:1701–1703
6. Gaeta JF (1981) Glandular profiles and cellular patterns in prostatic cancer grading: national prostatic project system. Urology (suppl) 17:33–37
7. Mostofi FK (1976) Problems of grading carcinoma of prostate. Semin Oncol 3:161–169
8. Bonney WW, Robinson RA, Lachenbruch PA et al. (1987) Yield of cancer from prostate needle biopsy. Urology 29:153–156
9. Zinke M, Campbell JT, Utz DC, Farrow GM, Anderson MJ (1973) Confidence in the negative transrectal needle biopsy. Surg Gynecol Obstet 136:78–80
10. Jewett HJ (1956) Significance of the palpable prostatic nodule. JAMA 160:838–839
11. Burkholder GV, Kaufmann JJ (1966) Local implantation of carcinoma of the prostate with percutaneous needle biopsy. J Urol 95:801–804
12. Blackard CE, Soucheray JA, Gleason DF (1971) Prostate needle biopsy with perineal extension of adenocarcinoma. J Urol 106:401–403
13. Hosking DH, Paraskevas M, Hellstein OR et al. (1983) The cytological diagnosis of prostatic carcinoma by transrectal fine needle aspiration. J Urol 129:998–1000
14. Zattoni F, Pagano F, Rebuffi A et al. (1983) Transrectal thin needle aspiration biopsy of the prostate: four years experience. Urology 22:69–72
15. Ljung BM, Cherrie R, Kaufman JJ (1986) Fine needle aspiration biopsy of the prostate gland: a study of 103 cases with histological follow-up. J Urol 135:955–958
16. Esposti PL, Franzèn S (1980) Transrectal aspiration biopsy of the prostate: a re-evaluation of the method in the diagnosis of prostatic carcinoma. Scand J Urol Nephrol [Suppl] 55:49–52
17. Chodak GW, Steinberg GD, Bibbo M, Wied G, Strauss FS, Vogelzang NJ, Schoenberg HW (1986) The role of transrectal aspiration biopsy in the diagnosis of prostate cancer. J Urol 135:299–302
18. Whelan JP, Chin JL, Shapre JR, Davis IR (1986) Transrectal needle aspiration versus transperineal needle biopsy in diagnosis of carcinoma of prostate. Urology 27:410–414
19. Ferguson RS (1930) Prostatic neoplasms—their diagnosis by needle puncture and aspiration. Am J Surg 9:507–511
20. Franzèn S, Giertz G, Zajicek J (1960) Cytological diagnosis of prostate tumors by transrectal aspiration biopsy: a preliminary report. Br J Urol 32:193–196
21. Ljung BM (1985) Fine needle aspiration of the prostate gland: technique and review of the literature. Semin Urol 3:18–24
22. Zajicek J (1979) Aspiration biopsy cytology. I. Cytology of supradiaphragmatic organs. II. Cytology of infradiaphragmatic organs. Basel, Karger (Monographs in clinical cytology, vol 7)
23. Esposti PL (1974) Aspiration biopsy cytology in the diagnosis and management of prostatic carcinoma. Thesis, Stockholm: Stahl & Accidens Tryk, p 46
24. Carter HB, Riehle RA, Koizumi JH, Amberson J, Vaughan ED (1986) Fine needle aspiration of the abnormal prostate: a cytohistologic correlation. J Urol 135:294–298
25. Melograna F, Oertel YC, Kwart AM (1982) Prospective controlled assessment of fine needle prostatic aspiration. Urology 19:47–51
26. Linsk JA, Axilrod HD, Solyn R, Delaverdac C (1972) Transrectal cytologic aspiration in the diagnosis of prostatic carcinoma. J Urol 108:455–459
27. Murphy GP, Natarajan N, Pontes JE et al (1982) The national survey of prostate cancer in the United States by the American College of Surgeons. J Urol 127:928–934
28. Hoskins JH, Millingen GT (1966) Needle biopsy of the pros-

tate. Am Fam Physician-GP 34(2):88–92
29. Gleason DF, Mellinger GT, and VA Cooperative Urological Research Group (1974) Prediction of prostatic adenocarcinoma by combined histological grading and clinical staging. J Urol 111:58–64
30. Pontes JE, Wajsman Z, Huben RP, Wolf RM, Englander LS (1985) Prognostic factors in localized prostatic carcinoma. J Urol 134:1137–1139
31. Frankfurt OS, Chin JL, Englander LS, Greco WR, Pontes JE, Rustum YM (1985) Relationship between DNA ploidy, glandular differentiation and tumor spread in human prostate cancer. Cancer Res 45:1418–1423
32. Evans N, Barnes RW, Brown AF (1942) Carcinoma of the prostate: correlation between the histologic observations and the clinical course. Arch Pathol 34:473–483
33. Epstein JI, Paull G, Eggleston JC, Walsh PC (1986) Prognosis of untreated stage A1 prostatic carcinoma: a study of 94 cases with extended followup. J Urol 136:837–839
34. Bauer WC, McGavran MH, Carlin MR (1960) Unsuspected carcinoma of the prostate in suprapubic prostatectomy specimens. Cancer 13:370–378
35. Esposti PL (1971) Cytologic malignancy grading of prostatic carcinoma by transrectal aspiration biopsy. Scand J Urol Nephrol 5:199–209
36. Paulson DF (1980) Assessment of anatomic extent and biologic hazard of prostatic adenocarcinoma. Urology 15:537–541
37. Kramer SA, Spahr J, Brendler CB, Glenn CB, Paulson D (1980) Experience with Gleason's histopathologic grading in prostatic cancer. J Urol 124:223–225
38. Paulson DF and Uro-oncology Research Group (1979) The impact of current staging procedures in assessing disease extent of prostatic adenocarcinoma. J Urol 121:300–302
39. Golimbu M, Schinella R, Morales P, Kurusu S (1978) Differences in pathological characteristics and prognosis of clinical A2 prostatic cancer from A1 and B disease. J Urol 119:618–622
40. Jewett HJ (1975) The present status of radical prostatectomy for stages A and B prostatic cancer. Urol Clin North Am 2:105–124
41. Lange P, Narayan P (1983) Understaging and undergrading of prostate cancer. Urology 21(2):113–118
42. Tannenbaum M, Tannenbaum S, DeSanctis PN, Olsson CA (1982) Prognostic significance of alveolar surface area in prostate cancer. Urology 19:546–551
43. Carroll PR, Leitner TC, Yen TSB, Watson RA, Williams RD (1985) Incidental carcinoma of the prostate: significance of staging transurethral resection. J Urol 133:811–814
44. Sheldon CA, Williams RD, Fraley EE (1980) Incidental carcinoma of the prostate: a review of the literature and critical reappraisal of classification. J Urol 124:626–631
45. Whitmore WF Jr (1956) Symposium on hormones and cancer therapy. Hormone therapy in prostate cancer. Am J Med 21:697–713
46. Breslow N, Chan CW, Dhom G, Drury RAB, Franks LM, Gellei B, Lee YS, Lundberg S, Sparke B, Sternby NH, Tulinius H (1977) Latent carcinoma of prostate at autopsy in seven areas. Int J Cancer 20:680–688
47. Owen W (1976) Cancer of the prostate: a literature review. J Chronic Dis 29:89–114

Chapter 13

Predictive Value of Histological Grading Systems

D. F. Gleason

Introduction—The Problem

It has proved very difficult to determine the best treatment for prostatic cancer because of its unpredictable, variable and frequently prolonged course. Careful autopsy studies reveal many prostatic cancers, in steadily increasing incidence with age. Most of these were unsuspected and obviously did not need treatment. The increasing incidence with age tells us that many of them were present for many years. Thus, if the incidence of these unsuspected tumours is 10% at age 55, 20% at age 65 and 30% at age 75, then two-thirds of the tumours in the 75-year-old men must have been present for 10 years and one-third must have been present for 20 years! Many would have remained dormant but, unfortunately, they are also found in prostate tissue resected during life, causing a dilemma.

A partial resolution of the uncertainties—at least an estimate of the probability that a certain tumour will progress—is provided by the microscopic appearance of the tumour in a histological grading system.

Histological Grading

Prostatic adenocarcinoma has a unique range of biological and histological malignancy. It varies from rapidly growing, fatal cancers to medical curiosities—"cancers" which never progress—with a complete range of intermediate behaviour patterns. Similarly, its histological structure may vary from anaplastic malignant epithelium to uniform, well-differentiated glands with a complete spectrum of intermediate structures.

There are many well-documented reports going back more than 40 years which prove beyond any doubt that there is strong correlation between the histological structure and the biological malignancy of these tumours. Well-differentiated tumours progress slowly, poorly differentiated tumours progress rapidly, suggesting that both structure and function are dependent upon common, shared genetic and physiological control mechanisms in the chromosomes of the tumour cells.

Most of these reports classified the histological patterns into histological "grades". The US Veterans Administration Cooperative Urological Research Group (VACURG) studies of prostate carcinoma developed a useful histological grading system based upon these correlations.

The VACURG Studies

The VACURG studies were prospective, controlled, randomised comparisons of the common treatments available at that time (1960s). Treatments included

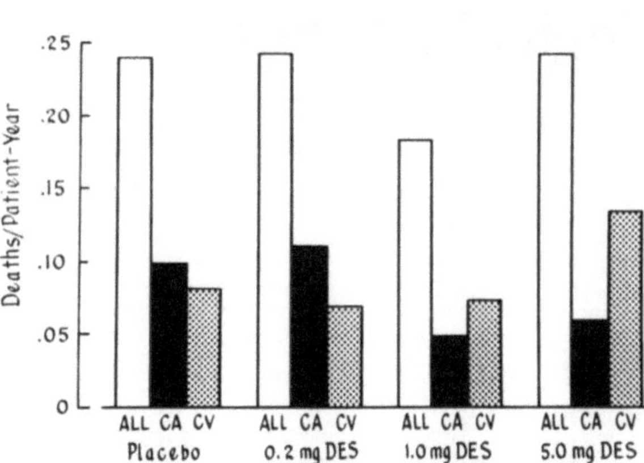

Fig. 13.1. Effect of three dose levels of DES on death rates. *CA*, cancer deaths; *CV*, cardiovascular (thromboembolic) deaths; *figures in parentheses*, number of patients in the dosage group.

diethylstilboestrol (DES), orchiectomy, both together and a placebo treatment.

By 1967 it was clear that there were no significant survival differences between the four types of treatment. However, more detailed analysis revealed that 5 mg DES per day had *reduced cancer deaths* substantially in stages III and IV, but this favourable effect had been almost exactly offset by *increased thromboembolic deaths* (coronary heart disease, strokes and pulmonary embolism) [1, 2, 3].

A second study [4], which randomised DES dose levels in stage III and IV patients, showed that 1 mg DES per day *reduced cancer death rates* as effectively as 5 mg per day but *did not increase thromboembolic deaths* (Fig. 13.1).

The VA Histological Grading System

The large number of carefully studied and carefully followed patients in the VACURG studies yielded much new information and provided the data base upon which the VA histological grading system was developed.

The VA histological grading system [5,6,7] identified five grades of malignancy. Histological patterns were categorised at relatively low magnifications by the degree of glandular differentiation and the pattern of growth in the prostatic stroma (Fig. 13.2). The system was standardised by means of a simplified drawing of the histological patterns (Fig. 13.3). Several different histological patterns were grouped into some of the five grades because they were associated with very similar cancer death rates and because they were often found together in the same tumour. The simplified drawing of the histological patterns has enabled other pathologists to adopt the VA grading system independently and successfully (see below).

Even after grouping the patterns into five grades, it was found that about half of the VACURG patients had two different histological grades in their tumours. We were surprised to find that patients with two different grades of tumour had cancer death rates which were intermediate between the cancer death rates of those patients with the single, pure forms of those two grades. Cancer of the prostate was *not* "as bad as its worst part". Its biological behaviour was more closely related to the *average* histological grade.

Therefore, the VA grading system added the two different grades together, obtaining the scaling effect of averaging, but division by two was omitted, creating a new scale, the "histological score", which could range from 2 to 10. (If only one histological grade was present, that grade was multiplied by 2). The histological score proved to be the strongest single factor correlated with other measures of biological malignancy in the VA studies. It correlated very strongly with death rates (Fig. 13.4).

This histological score on the initial diagnostic biopsy also correlated, to some degree, with almost every other variable which one might expect to be related to the degree of biological malignancy of the tumours, such as the number of organs containing metastases at autopsy many years later, or with the initial, pretreatment findings, such as the presence of hydronephrosis, elevation of serum acid phosphatase, presence of metastases and amount of pain [7]. Byar and Corle [8] documented strong correlation between the histological score and the rate of progression from stages I and II to stage IV in the VACURG patients.

Others have adopted the VA grading system successfully and independently. Corriere et al. confirmed the correlations with survival rates [9]. Paulson et al. [10], Kramer et al. [11] and Thomas et al. [12] found strong correlations between VA histological scores and the incidence of lymph node metastases in staging laparotomies. Luciani and Piscioli [13] and Piscioli et al. [14] confirmed the strong correlation between the histological scores and regional lymph node metastases detected by percutaneous fine needle aspiration cytology. Some of these authors even suggested that the predictions might be strong enough to consider foregoing the risks of the laparotomy and node dissection in

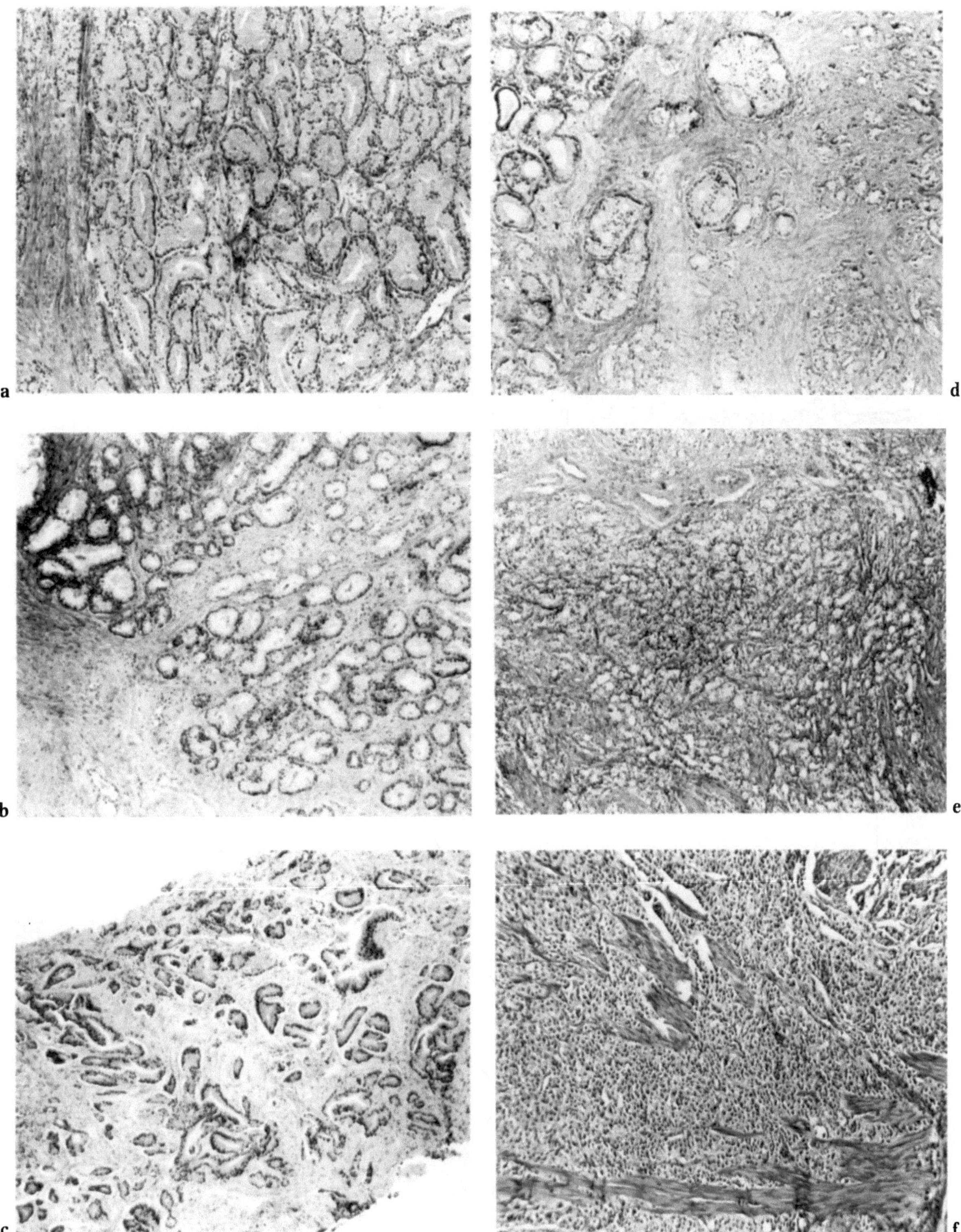

Fig. 13.2a–f. Histological grading system. **a** Grade 1. **b** Grade 2. **c** Best differentiated pattern of grade 3. **d** Grade 3—all three patterns: medium glands, small glands, smooth rounded cribriform masses. **e** Grade 4—ragged, fused glands. **f** Grade 5, anaplastic type.

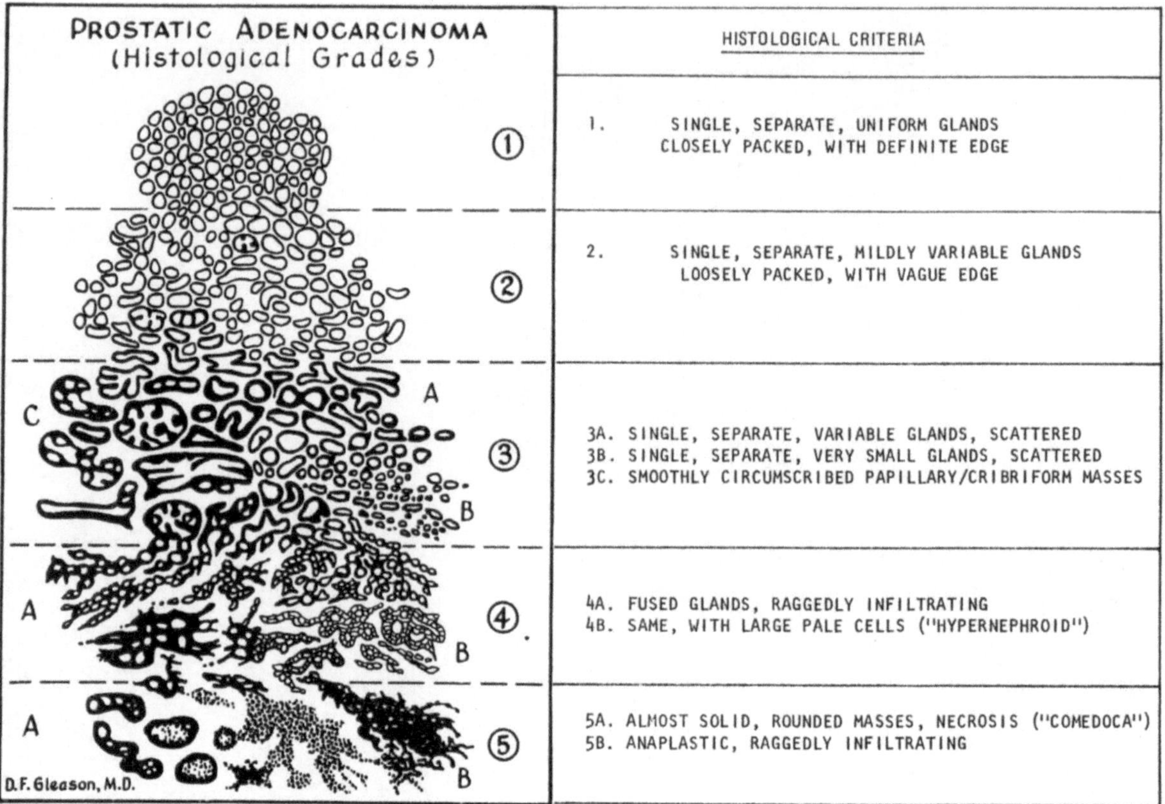

Fig. 13.3. Standardising drawing and description of patterns.

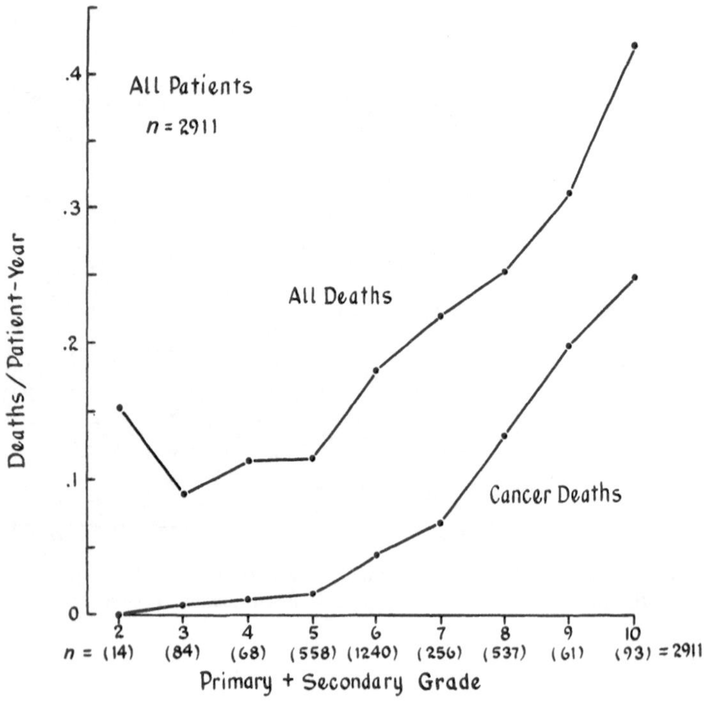

Fig. 13.4. Cancer and total death rates by histological score.

patients with the highest and the lowest histological scores.

Oota showed that minced prostate cancer cells were uniformly permeable to testosterone, but the ability of the cells (rate) to convert testosterone to dihydrotestosterone decreased as the VA histological score increased [15].

Recently, a group at the Memorial Sloan-Kettering Cancer Center confirmed the strength of the VA histological scores as predictors of survival [16]. They also confirmed the superior strength of the combined VA grading-staging "category score" [6, 7], which never attracted any proponents, being passed over in the enthusiasm for subdividing the clinical stages according to the extent of the tumour—A_1, A_2, B_1, B_2, etc.

Some Mathematical Considerations for Predictions of Survival

The survival data for at least the first 10 years after diagnosis in the VACURG studies of prostatic cancer showed striking linearity in logarithmic plots [7], which suggested strongly, but did not prove, that the deaths were occurring in exponential distribution (Fig. 13.5). That is, it appeared that the number of patients dying during each time interval was about the same *fraction* of the patients remaining from the previous interval, and that the data fitted the general (negative) exponential equation:

$$F_t = e^{-kt}$$

in which F_t = the fraction (of patients) remaining at time t, e = the base of the natural logarithms, k = a constant for the group under consideration, and t = elapsed time (years) since the diagnosis of cancer.

The statistic "deaths per year of follow-up" used in the VA studies as one measure of the degree of malignancy in a group of patients is also the actuary's "death rate"—the fraction of individuals dying during each year. It is also the most likely estimate of the exponential slope constant k if the deaths are actually occurring in exponential distribution. Thus, the VA data provide values for k for patients with various grades of tumour [6, 7] for both total death rates and cancer-specific death rates.

However, it must be pointed out that most cancer death rates *decrease* gradually with time after diagnosis, partly because of the favourable effects of treatment and partly because of competing risks (the longer a patient survives his cancer, the more likely he is to die of some other cause). Prostatic cancer is no exception but the gradually decreasing cancer death rate is approximately offset by the slowly increasing "normal" causes of death in these older men (see below). This yields a very close approximation to a constant death rate—an exponential death rate.

Effect of the Age of the Patient at the Time of Diagnosis

It is easy to forget that, in older men, the age of the patient may be the most powerful predictor of the

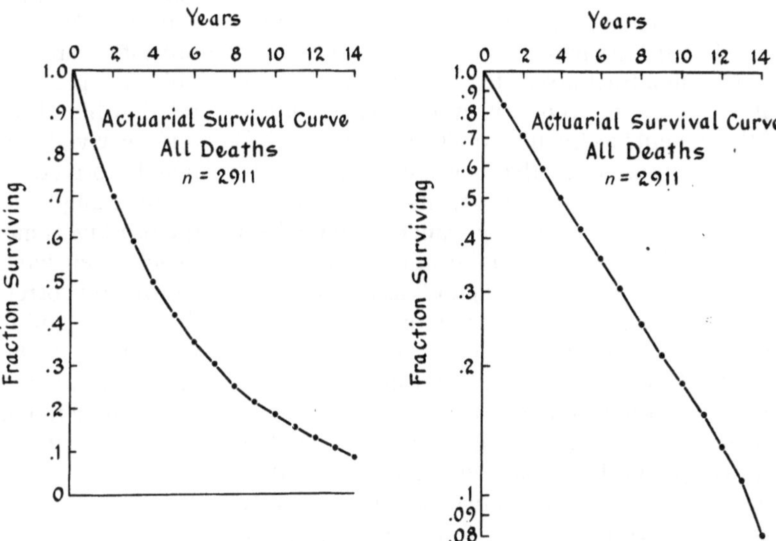

Fig. 13.5. Survival curves on linear and logarithmic scales.

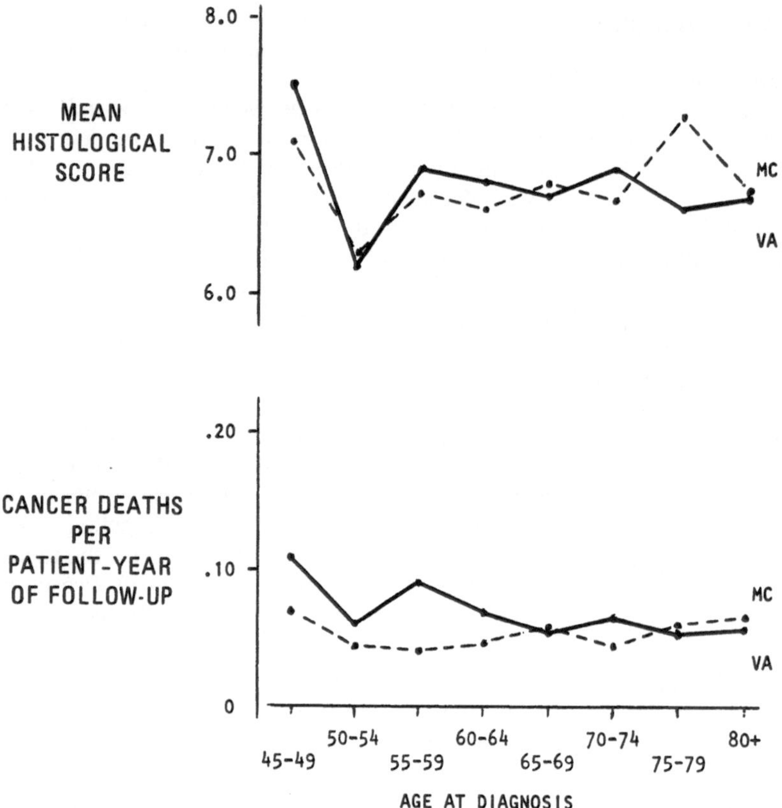

Fig. 13.6. Mean histological scores and cancer death rates by age of patients at time of diagnosis. No trends. *MC*, Mayo Clinic patients.

subsequent death rates. Other factors being equal, a group of older men will die at a faster rate than a group of younger men. The "normal" death rate is not an exponential or linear rate, but is a steadily *increasing* rate which doubles about every 8 years after about age 35 [17].

A very interesting finding in the VACURG studies was that the proportions of the different tumour grades—and their associated, cancer-specific death rates—were quite constant over the range of ages of the patients at the time of diagnosis (Fig. 13.6). The cancer-specific death rate was about 0.06–0.07 (6%–7% of the remaining patients dying of cancer each year) over the entire age range, with no definite trend. Of course, the total death rates increased with advancing age after diagnosis as the "normal" causes of death increased, but the cancer death rates were virtually independent of the age of the patients. Thus, the cancer death rates can be considered separately from all other causes of death.

One can, then, with some confidence combine the age-specific death rates from standard mortality tables with the observed, cancer-specific death rates associated with different histological grades to predict total death rates for groups of patients of specified age and tumour grade. A few examples are presented here (Fig. 13.7), using the 1969–71 US mortality tables [17] and reducing those annual figures sequentially with different added risk factors equal approximately to the cancer-specific death rates associated with several histological grades of prostatic carcinoma. These calculations presume only to provide some insight into the interrelationships between tumour grade, patient age and survivial rates, but they should be approximately correct.

For example, it is widely believed that cancer of the prostate is more malignant in young men than in older men, whereas the VACURG data cited above showed no correlation between the cancer-specific death rates and the ages of the patients. However, when the calculations described above are carried out, a logical explanation for the perceived severer malignancy in younger men is revealed (Fig. 13.7). That is, even the low cancer-specific death rate associated with well-differentiated prostatic carcinoma causes a very significant relative increase over the low normal death rate of a young man.

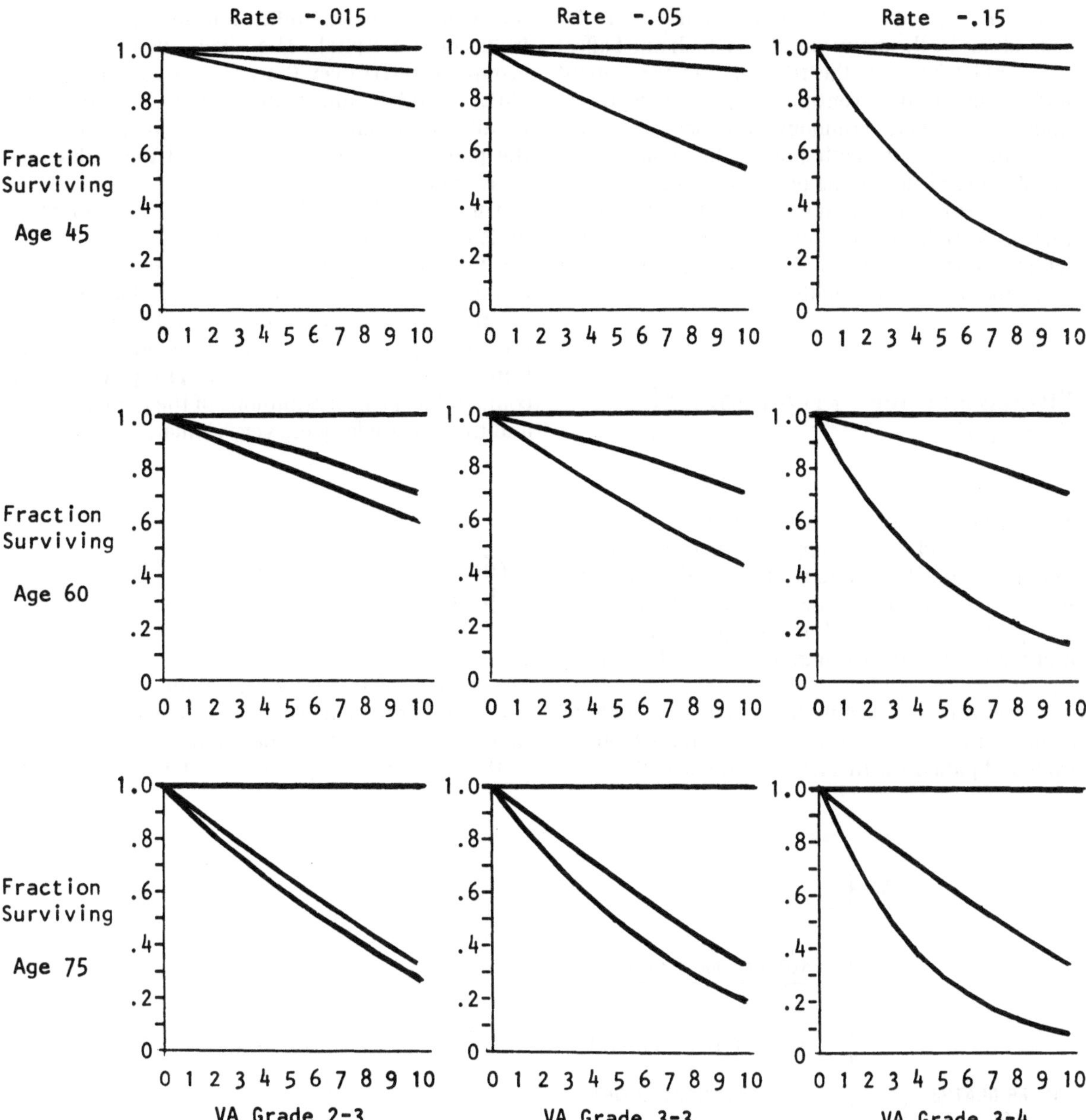

Fig. 13.7. Theoretically calculated survival curves for three different ages. Each death rate increased by three different risk factors, each approximately equal to cancer-specific death rates for histological scores 5, 6 and 7. *Upper curve* "normal"; *lower curve* includes "cancer" death rate.

The same increase over the higher normal death rates in older men is relatively smaller. This age differential is less striking for high-grade tumours because they carry high death rates for all ages.

These curves provide mathematical explanations for the intuitive conclusion that radical treatment is not justified for a 90-year-old patient with prostatic cancer but may be "the only hope" for a 50-year-old man with the same tumour.

The Fallibility of Predictions for Individuals

Although these histological correlations are powerful and the differences between the groups are "statistically strong", they yield only probabilities for the individual patients, whom we must treat one at a time. The probabilities become "randomised" for individuals by the many variables which must affect the conflict between the tumour and the patient and so modify the eventually expressed degree of malignancy. These random variables guarantee that some patients with low-grade tumours will die of cancer before some patients with high-grade tumours. However, this does not confound the conceptual significance of the observed correlations, which are the very basis of the decisions we must make to manage these patients.

"Retrospective" Predictions of Survival

To test the predictive strength of these correlations formally would require applying them to patients in a new clinical trial and waiting for the outcome. Predictions can also be tested, less rigorously, on groups of patients who have already been studied and followed, if their original biopsy slides can be examined.

The author was fortunate enough to be able to examine the microscopic slides of the initial biopsies from 324 patients with adenocarcinoma of the prostate who had been studied and followed carefully in another hospital. The slides were graded in groups of 8–10 over the course of several years, with no other information about the patients. Detailed information on the subsequent course of the patients was received after all the slides had been graded.

The survival data for these patients and comparison with the experience ("predictions") from the VACURG patients is shown in Fig. 13.8. The results are quite similar and provide another example of the strength of the correlation between the histological and biological malignancy of the tumours, suggesting that the two groups were very comparable. (The distributions of the patients' ages and tumour grades were very similar.)

Role of Histological Grading in Selection of Treatment

Selection of treatment for carcinoma of the prostate is a difficult problem. There is no single "best treatment". The choices depend upon the clinical stage of the tumour, the age of the patient, the general

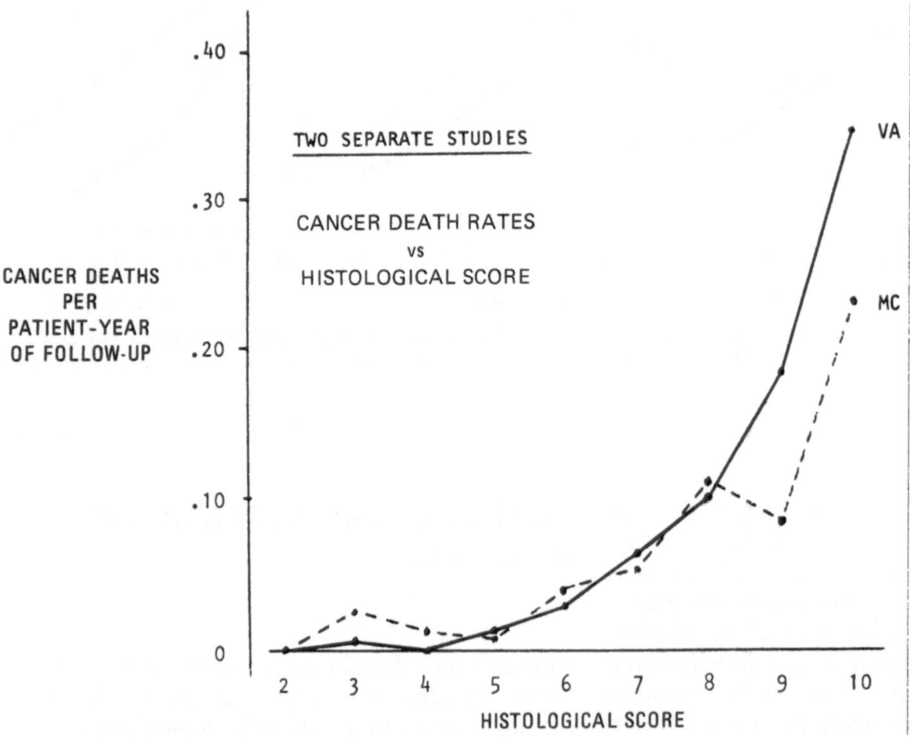

Fig. 13.8. Cancer-specific death rates by histological scores in two different studies. VA, 1200 cases; MC, 324 cases.

health of the patient and the histological grade of the tumour.

These four factors are useful mainly in the negative sense. Most US urologists now recommend close follow-up but no treatment for small, well-differentiated, accidentally discovered tumours (stage A_1). One does not perform radical prostatectomy if the patient already has distant metastases, is 90 years old or is seriously ill with some other disease. Oestrogen treatment must be used very cautiously in patients with coronary heart disease or other thromboembolic disease or tendency. Chemotherapy is probably appropriate, at this stage, only for those patients whose tumours have already extended outside the prostate, are causing serious symptoms or complications and do not respond to hormone therapy.

Summary data such as percentages, death rates and averages may give a false sense of precision. Physicians would like to have their patients sorted into discrete sets: "needs treatment A", "needs treatment B", "needs no treatment". This is, unfortunately, unattainable because biological behaviour is a continuous spectrum, not made up of discrete boxes or steps.

With these reservations firmly in mind, the author presumes to offer his own simplified, incomplete, "cookbook" recommendations for individual patients, mainly in regard to suitability for radical prostatectomy, to be modified according to the clinical stage of the tumour and the patient's general health and age:

Histological score	Recommendation	Comment
2–3–4	Follow only—until evidence of progression	Very few progress
5–6	Best candidates for surgery after signs of progression?	Low, but definite risk of progression
7–8	Radical surgery may be attempted but often stage C or D	High risk of progression
9–10	Usually stage C or D, will eventually need systemic therapy	Very high risk of progression, metastases

Discussion

The observation that prostatic cancers with two different histological grades behave in proportion to the average of the two grades was an important factor in the development of the VA grading system. It also contradicted the old aphorism "a cancer is as bad as its worst part". That aphorism was intuitively satisying because it seemed reasonable that any highly malignant tumour cells should express their degree of malignancy independently of any less malignant cells. However, the observed results indicate that the additional presence of lower grade tumour is usually associated with less malignant overall behaviour. This is probably true for most cancers. It was recently described in cancer of the breast, based on objective morphometric data [18].

The apparent contradiction is partly explained by some relatively recent observations, but it is more completely and simply explained by a very old concept. Sophisticated biological markers have shown that there is substantial heterogeneity in many tumours, involving not only different clones of cells but also varying somatic differentiation, including "maturation" to less malignant cells. Broders [19] had recognised, in 1926, that his various histological grades and the varying degrees of malignancy associated with them must be due to maturation of the tumour cells: "... some carcinomas, by a process of differentiation, put the brakes on themselves, so to speak ... by differentiating ... beyond the point of reproduction ..."—that is, matured beyond the ability to divide again.

Maturation of cancer cells has been demonstrated quite precisely by tracing the "migration" (with time) of newly formed, labelled DNA from poorly differentiated cells into the mature squamous cells of a mouse carcinoma [20]. Induced maturation of leukaemic cells may explain the favourable effects of some non-specific agents [21]. Thus, the mixture of histological grades in some tumours may be considered to be a "snapshot" image of mixtures of actively proliferating cells, areas which have differentiated to slowly proliferating cells and areas of non-proliferating, "post-mitotic" cells. Once that concept is considered, it seems quite easy to find supporting evidence with the microscope (Fig. 13.9).

The combination of the cancer-specific death rates with actuarial death rates yields an individualised prediction for a specific patient (of known age) with a specific histological grade tumour. It could be used to discuss the risk factors with the patient—to be balanced against the probabilities and risks of various treatments. Other risk factors could be added to the calculations and evaluated by computer to yield detailed probability considerations—if the data needed to quantify those risk factors can be accumulated.

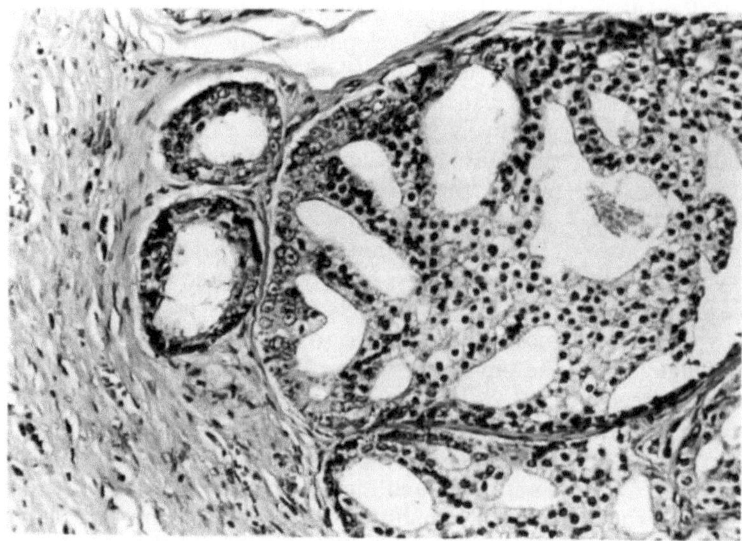

Fig. 13.9. Cribriform carcinoma of the prostate. Note cells with dark cytoplasm, large vesicular nuclei, and large nucleoli in basal layers of tumour and transition to cells with clear cytoplasm, small, dark nuclei and tiny nucleoli, towards the centre of the tumour. Differentiation and maturation?

For example, the simple calculations presented here suggest that a 45-year-old man with a low-grade tumour (histological grade 2–3, with a cancer-specific death rate of about 0.015 per year) is a much more logical candidate for radical prostatectomy than a 75-year-old man with the same tumour. The low-grade tumour more than doubles the death rate for the 45-year-old man and shortens his 50% life expectancy from about 28 years to about 17 years. His long life expectancy gives even the low-grade tumour time to progress and cause trouble. The same tumour in the 75-year-old man increases his death rate by about 20% and shortens his life expectancy from about 7 to about 6 years. It may become difficult to justify the risks and sacrifices of radical prostatectomy for the 75-year-old. The exact point at which those judgements might reverse is a very subjective matter.

Obviously, all such probability predictions can only be proved for *groups* of patients. Nevertheless, all of our patient management decisions are based on such predictive probabilities. We can never hope to predict the exact result for any one patient, but if our clinical policies are firmly based on the most careful observations we can believe that we are doing our best for the patient, and doing the least harm.

References

1. The Veterans Administration Cooperative Urological Research Group (1967) Carcinoma of the prostate: treatment comparisons. J Urol 98:516–522
2. Blackard C, Doe R, Mellinger G, Byar D (1970) Incidence of cardiovascular disease in patients receiving diethylstilbestrol for carcinoma of the prostate. Cancer 26:249–256
3. Byar D (1972) Treatment of prostatic cancer: studies by the VACURG. Bull NY Acad Med 48:751–766
4. Byar D (1973) The Veterans Administration Cooperative Urological Group's studies of cancer of the prostate. Cancer 32:1126–1130
5. Bailar JC, Mellinger GT, Gleason DF (1966) Survival rates of patients with prostatic cancer, tumor stage, and differentiation—preliminary report. Cancer Chemother Rep 50:129–136
6. Gleason DF, Mellinger GT, VACURG (1974) Prediction of prognosis for prostatic carcinoma by combined histological grading and clinical staging. J Urol 111:58–64
7. Gleason DF (1977) Histologic grading and clinical staging of carcinoma of the prostate. In: Tannenbaum M (ed) Urologic pathology: the prostate. Lea and Febiger, Philadelphia, pp 171–197
8. Byar D, Corle D, VACURG (1981) VACURG randomized trial of radical prostatectomy for Stages I and II prostatic cancer. Urology 17 (Suppl): 7–11
9. Corriere J, Cornog J, Murphy J (1970) Prognosis in patients with carcinoma of the prostate. Cancer 25:911–918
10. Paulson D, Piserchia P, Gardner W (1980) Predictors of lymphatic spread in prostatic adenocarcinoma. Uro-oncology Research Group study. J Urol 123:697–699
11. Kramer S, Spahr J, Brendler C, Glenn J, Paulson D (1980) Experience with Gleason's histopathologic grading in prostatic cancer. J Urol 124:223–225
12. Thomas R, Lewis RW, Sarma DP, Coker GB, Rao MK, Robert JA (1982) Aid to accurate staging—histopathologic grading in prostatic cancer. J Urol 128:726–728
13. Luciani L, Piscioli F (1983) Accuracy of transcutaneous aspiration biopsy in the definitive assessment of nodal involvement in prostatic carcinoma. Br J Urol 55:321–325
14. Piscioli F, Leonardi E, Reich A, Luciani L (1984) Percutaneous lymph node aspiration biopsy and tumor grade in staging of prostatic carcinoma. Prostate 5:459–468
15. Oota K (1981) Pathomorphological studies on carcinoma of the prostate in Japan. Prostate 1:125–134

16. Sogani PC, Israel A, Lieberman PH, Lesser M, Whitmore W (1985) Gleason grading of prostate cancer: a predictor of survival. Urology 25:223–227
17. Department of Health, Education and Welfare Publication # (HRA) 75–110. US life tables 1969–1971
18. Sharkey FE (1982) Morphometric analysis of differentiation in human breast carcinoma. Tumor heterogeneity. Arch Pathol Lab Med 307:343–347
19. Broders AC (1926) Carcinoma. Grading and practical application. AMA Arch Pathol 2:376–381
20. Pierce GB, Wallace C (1971) Differentiation of malignant into benign cells. Cancer Res 31:127–134
21. Ross DW (1985) Leukemic cell maturation. Arch Pathol Lab Med 109:309–313

Chapter 14

The Role of Staging in the Management of Prostatic Cancer

L. Giuliani

The knowledge of the true stage of prostatic carcinoma is essential for the appropriate management of the disease [1, 2]. Intracapsular, extracapsular or widespread prostatic carcinoma is to be clearly defined before treatment, as an underestimated evaluation of the extension of the disease may lead to inadequate therapy along with a poor cost/benefit ratio.

Discrepancy between the clinical and the surgical stage of the disease has been widely reported in the literature. Among clinically intracapsular prostatic carcinomas, 23% of stage A_2, 18% of stage B_1 and 37% of stage B_2 showed nodal metastases, and 54% of locally infiltrating tumours had nodal metastases with systemic metastatic disease already present in 30% of the cases. Therefore, the main problem in the choice of an appropriate therapy of prostatic carcinoma is the selection of patients with a true intracapsular disease and no metastatic spread. For this reason, the determination of the local extension of the neoplasm is to be performed according to well-defined criteria. Digital rectal examination, transrectal ultrasonography, computed tomography (CT) and magnetic resonance imaging (MRI) may be used. Skeletal radiographs and bone scans are also necessary in order to find distant metastases, and lymphangiography, MRI, CT, percutaneous fine needle biopsy and staging lymphadenectomy are to be performed to disclose nodal involvement.

Local Staging

Digital rectal examination allows easy recognition of intracapsular and extracapsular neoplasms, but errors are frequent. In fact, by using transrectal ultrasonography it has been shown that capsular discontinuity was present in 40% of clinical stage B prostatic carcinomas [2]. Echography, CT and MRI may be performed in the evaluation of extracapsular extension of the tumour.

Nodal Evaluation

After correct evaluation of the extension of local disease, the determination of the nodal status is mandatory. In our opinion, the diagnostic algorithm shown in Fig. 14.1 is useful in the investigation of the lymph nodes

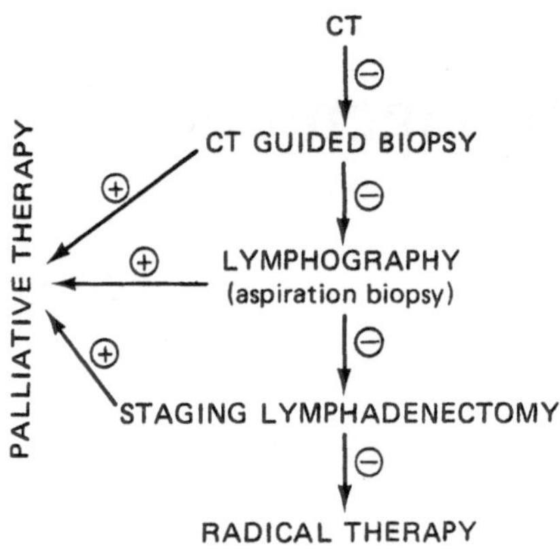

Fig. 14.1. Diagnostic algorithm for nodal evaluation.

Computed tomography has a resolution power of 1.5–2 cm, a specificity of 90%, a sensitivity of 25%–50% and an overall accuracy of 60%–77%. The main limitation of this method is that it only evaluates the volume of the lymph nodes, not their structure [3, 4]. Bipedal lymphangiography has a specificity of 80%, a sensitivity of 35%–75% and an accuracy of 48%–80%, but it does not clearly visualise the obturatory lymph nodes [1, 5, 6, 7]. Ultrasonography has a poor diagnostic accuracy because of its low resolution. The diagnostic accuracy of lymphoscintigraphy has not yet been assessed [1, 3, 4, 8]. Another method to evaluate the nodal status is post-lymphographic fine needle biopsy under CT or fluoroscopy guidance [9, 10, 11, 12].

However, the diagnostic trial must be continued, even if all these procedures have given negative results. Only a staging lymphadenectomy can ensure definitive information on the stage of the disease, which is indispensable to the provision of an effective therapy [1, 6, 13]. In fact, in our series of 74 staging lymphadenectomies performed during the period 1979 to 1986, nodal metastases were found in 2 out of 8 clinical stage B_1 tumours (25%), in 21 out of 53 clinical stage B_2 tumours (39%) and 12 out of 13 clinical stage C tumours (92%). Lymphadenectomy must be bilateral and include all the regional nodes.

In our opinion, the diagnostic trial is essential because it allows us to provide the correct therapy at each stage, avoiding in several cases a useless surgical "radical" therapy.

References

1. Paulson DF and Uro-Oncology Research Group (1979) The impact of current staging procedures in assessing disease extent of prostatic adenocarcinoma. J Urol 121:300–302
2. Pontes JE, Eisenkraft S, Watanabe H, Ohe H, Saitoh M, Murphy GP (1985) Preoperative evaluation of localized prostatic carcinoma by transrectal ultrasonography. J Urol 134: 289–291
3. Mamtora H, Gowland MR, Isherwood I (1981) Diagnostic imaging. In: Duncan W (ed) Prostate cancer. Springer, Berlin Heidelberg New York, pp 76–96 (Recent results in cancer research, vol 78)
4. Martorana G, Damonte P, Oneto F, Giberti C, Biggi E, Rollandi GA, Scopinaro G (1984) Valutazione per immagini dell'interessamento linfonodale nei tumori della prostata. Riv Radiol 24: 56–66
5. Macdonald JS (1981) Lymphography. In: Dundan W (ed) Prostate cancer. Springer, Berlin Heidelberg New York, pp 97–107 (Recent results in cancer research, vol 78)
6. Middleton AW (1981) Pelvic lymphadenectomy with modified radical retropubic prostatectomy as single operation: technique used and results in 50 consecutive cases. J Urol 125:353–356
7. O'Donoghue EPN, Shridhar P, Sherwood T, Williams JP, Chisolm JD (1976) Lymphography and pelvic lymphadenectomy in carcinoma of the prostate. Br J Urol 48: 689–696
8. Schubert J, Heidl G (1983) The value of various methods for determining lymphatic metastases in prostatic carcinoma. Eur Urol 9:257–261
9. Correa RJ, Kidd R, Burnett L, Brannen GE, Gibbons RP, Cummings KH (1981) Percutaneous pelvic lymph node aspiration in carcinoma of the prostate J Urol 126:190–191
10. Flanigan RC, Mohler JL, King CT, Atwell JR, Umer MA, Loh F, McRoberts JW (1985) Preoperative lymph node evaluation in prostatic cancer patients who are surgical candidates: the role of lymphangiography and computerized tomography scanning with directed fine needle aspiration. J Urol 134:84–85
11. Luciani L, Piscioli F (1983) Accuracy of transcutaneous aspiration biopsy in the definitive assessment of nodal involvement in prostatic carcinoma. Br J Urol 55:321–325
12. Sadlowski RW (1978) Early stage prostatic cancer investigated by pelvic lymph node biopsy and bone marrow acid phosphatase. J Urol 119:89–93
13. Golimbu M, Morales P, Al-Askari S, Brown J (1979) Extended pelvic lymphadenectomy for prostatic cancer. J Urol 121:617–620

Chapter 15

Lymphangiography and Fine Needle Aspiration Biopsy of Opacified Lymph Nodes in the Staging of Patients with Carcinoma of the Prostate

R. Kidd and R. Correa Jr.

Introduction

Adenocarcinoma of the prostate frequently metastasises to regional lymph nodes. Because of the prevalence of lymph node metastases and their long-term prognostic significance, in the absence of other evidence of metastatic disease, most staging protocols include some evaluation of pelvic lymph nodes. Many physicians currently favour pelvic lymphadenectomy for this purpose. However, the morbidity (as high as 20%–30% in some series) [1–4] and expense of this procedure have limited its acceptance. Other institutions have advocated lymphangiography (LAG) as an alternative to lymph node dissection (LND) for the staging of carcinoma of the prostate [5–7]. Several factors militate against the use of LAG: (1) the LAG result is frequently equivocal; (2) false positive LAG diagnoses, while much less common than false negative results, do occur; (3) the false negative rate of LAG is disappointingly high, ranging from 33% to 66% in most series [8]; and (4) many radiologists find LAG a tedious and frustrating procedure. Its use in assessing nodal status in patients with carcinoma of the prostate has continued only because LND, the only other widely accepted method for evaluating the pelvic nodes, has substantial morbidity.

Staging Procedures

Some authors [4] believe that a combination of clinical stage, histological grade and serum acid phosphatase level is sufficient for pretreatment staging in most patients with carcinoma of the prostate. The difficulty with this approach is that few patients and physicians can accept the uncertainty of assignment to a treatment group on a statistical basis alone, without histological or cytological knowledge of lymph node status. Such diverse and divergent views indicate that each approach has drawbacks, and none is clearly superior to the others.

In an attempt to define better the status of lymph nodes by a relatively non-invasive method we adapted the use of fine needle aspiration biopsy (FNAB). Our initial experience had suggested that this is a reasonable alternative to the more aggressive and morbid methods of LND [7]. Subsequent experience has tempered our enthusiasm. Our current approach is to use the technique selectively as described in this chapter. The technique of aspiration biopsy we use has been well described before [9, 10] and will not be discussed here.

The most economical approach, in terms of morbidity as well as financial costs, must consider (1) the likelihood of the presence of metastases in lymph nodes, (2) the specificity and sensitivity of the staging method and (3) the morbidity and cost of the staging method. Using these criteria, LAG and FNAB have a limited but definite place in the evaluation of patients with carcinoma of the prostate: namely in patients with a relatively high likelihood of nodal metastases. LAG, even when combined with FNAB of opacified nodes (either normal or abnormal), is ineffective in staging in patients with low-grade, low-stage cancer of the prostate in our experience [11, 12]. We use the clinical stage of disease (as described in Table 15.1) to decide which patients should undergo LAG.

Table 15.1. Clinical staging of carcinoma of the prostate

Stage A	No tumour palpable. Cancer identified only during histological examination of prostatic tissue removed at surgery
A_1 (focal)	Three or fewer microscopic foci of well-differentiated carcinoma
A_2 (diffuse)	More extensive than stage A_1: more than three microscopic foci and/or less than well-differentiated. If diagnosed from an open prostatectomy, more than 5% of the specimen is involved
Stage B	Carcinoma confined within the prostatic capsule. No elevation of serum acid phosphatase level
B_1	Solitary nodule, 1.5 cm or less
B_2	More than one nodule, nodule greater than 1.5 cm or nodule involving more than one lobe
Stage $C_{chemical}$	Carcinoma confined within the prostatic capsule, but with elevated serum acid phosphatase level
Stage C	Carcinoma extending beyond prostatic capsule with or without invasion of adjacent pelvic structures
Stage D	Known distant metastases to bone or lymph nodes

There is no doubt that the incidence of lymph node metastases is related to the clinical stage of disease. Urologists performing pelvic lymph node dissection in patients with newly diagnosed carcinoma of the prostate have reported histologically documented lymph node metastases in 8%–21% of patients with stage B_1 disease, in 23%–38% with A_2 disease, in 20%–40% with B_2 disease and 39%–63% of patients with stage C disease [13–16].

In a previous paper [11], we demonstrated that the sensitivity of LAG is similarly related to clinical stage. We studied 436 patients with carcinoma of the prostate who had LAG as part of their initial evaluation, prior to treatment. FNAB of abnormal opacified nodes was performed routinely. We compared the positivity rate of LAG and FNAB *in each clinical stage* with the positivity rate predicted for that stage, based on published series of patients with carcinoma of the prostate who underwent pelvic LND (Tables 15.2, 15.3). Within each clinical stage, we evaluated the relationship of the outcome of LAG/FNAB to histological tumour grade (Gleason score) and serum acid phosphatase levels (Tables 15.3, 15.4).

Table 15.2. Carcinoma of the prostate: case distribution by clinical stage

A_1	A_2	B_1	B_2	$C_{clinical}$	$C_{chemical}$
4	40	116	73	170	33

Table 15.3. Results of LAG and FNAB by clinical stage

Clinical stage	LAG (−)	LAG (+) or suspicious			FNAB (+)/ total LAG
		FNAB (−)	FNAB (+)	FNAB Not done	
A_1	4	—	—	—	0/4
A_2	31	8	1	—	1/40 (3%)
B_1	95	19	—	2	0/116
B_2	59	12	1	1	1/73 (1%)
$C_{clinical}$	104	30	29	7	29/170 (17%)
$C_{chemical}$	24	5	4	—	4/33 (12%)
Total	317	74	35	10	35/436 (8%)

Table 15.4. Distribution of study cases and positive FNABs by Gleason score

Clinical stage	Gleason score						Not available	Total
	≤5	6	7	8	9	10		
A_1	4	—	—	—	—	—	—	4
A_2	21	7	1	6	—	$\frac{1}{3}$	2	40
B_1	60	41	5	3	2	2	3	116
B_2	34	19	6	$\frac{1}{9}$	2	3	—	73
$C_{chemical}$	12	5	3	$\frac{4}{11}$	2	—	—	33
$C_{clinical}$	35	$\frac{9}{60}$	$\frac{6}{20}$	$\frac{5}{26}$	$\frac{4}{10}$	$\frac{5}{13}$	6	170

Numerators represent those cases with positive FNAB. Categories without numerators contain no positive FNABs.

Relationship of Clinical Stage to Results of LAG/FNAB

As Table 15.3 indicates, only 1 of 40 (3%) LAGs performed for stage A_2 disease, 0 of 116 for stage B_1 disease and 1 of 73 (1%) for stage B_2 disease had nodal metastasis confirmed by FNAB. Stage C disease had a significantly greater percentage of positive FNABs: 29 of 170 (17%) of those with clinical stage C and 4 of 33 (12%) of those patients with Stage $C_{chemical}$ had positive FNAB.

These figures are substantially lower than the 39%–63% positive rate seen at LND, and indicate that LAG, even when coupled with liberal use of FNAB, is hyposensitive even in clinical stage C disease. However, a positive result will avoid the morbidity and expense of a staging LND and allow confident selection of appropriate treatment. A negative LAG/FNAB by itself, on the other hand, is meaningless because of the high false negative rate of LAG.

Relationship of Gleason Score to Results of LAG/FNAB

All patients who had nodal metastases confirmed by a positive FNAB had a Gleason score of 6 or higher (Table 15.4). Said another way, LAG combined with liberal use of FNAB failed to reveal a metastasis in any case with a Gleason score of 5 or less. On the other hand, of the seven patients who had a positive LND, four had a Gleason score of 5 or less.

Influence of Serum Acid Phosphatase

In clinical stage C, 28 patients with *normal* serum acid phosphatase levels had FNAB of abnormal nodes, and FNAB was positive in 10 (36%) of these.

Thirty patients with *elevated* serum acid phosphatase levels had FNAB of abnormal nodes; FNAB was positive in 19 (63%). In 8 of these, the elevation was modest; in the other 11, it was more than twice the highest normal laboratory value.

Other Studies to Evaluate the Sensitivity of LAG

Interinstitutional comparison of LAG in the staging of patients with carcinoma of the prostate has been hampered by (1) lack of lymphangiographical-histological correlation in most studies and (2) a dearth of information regarding case mix. The positivity rate of any nodal staging technique *should be expected* to vary substantially, depending upon the relative proportion constituted by each clinical stage within the study. At least two other investigators have evaluated LAG by clinical stage. Prando and Wallace [17] studied 208 patients with cancer of the prostate by LAG. Of 24 patients with stage B disease, none had a positive LAG. Of 175 patients with stage C disease, 40 (23%) had a positive LAG. Spellman and Castellino [6] compared LAG and LND in 69 patients with carcinoma of the prostate. Of 35 patients with stage B disease, LAG correctly identified 2 patients with histologically proved lymph node metastases and was falsely negative in 5 others. An elevated serum acid phosphatase level did not affect the clinical stage in their study, so it is uncertain whether those patients were stage B by our classification. However, they were quite successful in clinical stage C disease, correctly identifying 14 of 21 patients (64%) with proved lymph node metastases and 11 of 12 patients whose LNDs were negative. We are not sure why our sensitivity (*true positive rate*) is lower than that of Castellino, but we discount the possibility that it is due to inordinately strict criteria for abnormality on LAG since we used FNAB very liberally.

Even in the best of hands LAG is an insensitive test, with a false negative rate of 33%–66% [8]. This insensitivity is *not* due to failure of the LAG to opacify the surgical obturator node, a common site of metastasis for carcinoma of the prostate. Two recent studies [18, 19] conclusively contradict this false but persistent criticism of LAG. The insensitivity of LAG is due instead to its inability to detect small (less than 5–8 mm) nodal metastases and, less commonly, metastases to unopacified or poorly opacified nodes.

Several authors [20–21] have advocated FNAB of lymphangiographically *normal* lymph nodes as a means of increasing the sensitivity of LAG in patients with carcinoma of the prostate. However, we were unable to reproduce their results in a series of 49 patients [12], failing to detect a single metastasis by this technique. We therefore no longer biopsy nodes of normal appearance. However, we continue to biopsy *any* equivocal or suspicious node opacified by LAG.

Discussion

Most authors agree that the likelihood of nodal metastases correlates closely with the tumour grade [4, 13, 22], although others [15, 23] report a disturbingly high incidence of nodal metastases in patients with lower Gleason scores. Our data support the majority view, at least regarding selection of patients for LAG (Table 15.3). Had we restricted the use of LAG in stages A_2 and B_2 to those cases with a Gleason score of 8 or more, and in stage C (including $C_{chemical}$) to those with a Gleason score of 6 or more, we would have performed 255 fewer LAGs (a reduction of 58%) and 45 fewer FNABs without missing a single positive LAG/FNAB. Using these criteria would have improved our positivity rate in patients with stage C disease from 16% to 22%.

In conclusion, lymphangiography is an insensitive means of detecting nodal spread from carcinoma of the prostate in the earlier clinical stages and lower histological grades of this neoplasm. This lack of sensitivity appears to be inherent; LAG cannot reveal small nodal metastases or metastases to small nodes. We had hoped that the liberal use of FNAB, in conjunction with LAG, would improve its sensitivity, but the improvement was minimal. On the other hand, FNAB *did* improve the *specificity* of LAG, by revealing the false positive LAG diagnosis and by resolving the dilemma of an equivocal LAG result.

Based on the data presented here, we currently use LAG and FNAB only in patients with clinical stage C disease and a Gleason score of 6 or more, and in patients with very high-grade (Gleason score of 8 or more) stage A_2 or B_2 tumours, when a positive FNAB will obviate LND or alter the treatment plan. While less sensitive than LND even in this subgroup of patients with carcinoma of the prostate, LAG/FNAB can spare those with a positive result (22% in our series) the morbidity and expense of LND. And if LAG/FNAB is negative, the ability to perform LND is not impaired or significantly delayed.

Finally, the use of FNAB as an adjunct to LAG has no effect whatever on the high false negative rate of LAG, which in most series is 40%–60%. Therefore, a negative LAG or a negative LAG/FNAB result should *not* be taken as evidence that lymphatic metastasis has not occurred.

Acknowledgements. We thank Cindy Soder for her secretarial assistance.

References

1. Wajsman Z (1981) Lymph node evaluation in prostate cancer: is pelvic lymph node dissection necessary? Urology 17(Suppl):80–82
2. Babcock JR, Grayhack JT (1979) Morbidity of pelvic lymphadenectomy. Urology 13:483–486
3. Fowler JE Jr, Barzell W, Hilaris BS, Whitmore WF Jr (1979) Complications of 125 iodine implantation and pelvic lymphadenectomy in the treatment of prostatic cancer. J Urol 121:447–451
4. Frieha FS, Pistenma DA, Bagshaw MA (1979) Pelvic lymphadenectomy for staging of prostatic carcinoma: is it always necessary? J Urol 122:176–177
5. Johnson DE, von Eschenbach AC (1981) Roles of lymphangiography and pelvic lymphadenectomy in staging prostate cancer. Urology 17(Suppl):66–81
6. Spellman MC, Castellino RA, Ray GR, Pistenma DA, Bagshaw MA (1977) An evaluation of lymphography in localized carcinoma of the prostate. Radiology 125:637–644
7. Correa RJ Jr, Kidd R, Burnett L, Brannen GE, Gibbons RP, Cummings KB (1981) Percutaneous pelvic lymph node aspiration in carcinoma of the prostate. J Urol 126:190–191
8. Hoekstra WJ, Schroeder FH (1981) The role of lymphangiography in the staging of prostatic cancer. Prostate 2:433–440
9. Göthlin JH (1976) Post-lymphographic percutaneous fine needle biopsy of lymph nodes guided by fluoroscopy. Radiology 120:205–207
10. Kidd R, Freeny PC, Bartha M (1979) *Single pass* fine-needle aspiration biopsy. AJR 113:333–334
11. Kidd R, Crane RD, Dail DH (1984) Lymphangiography and fine needle aspiration biopsy: ineffective for staging early prostate cancer. AJR 141:1007–1012
12. Kidd R, Correa R Jr (1984) Fine-needle aspiration biopsy of lymphangiographically normal lymph nodes: a negative view. AJR 141:1005–1006
13. Donohue RE, Fauver HE, Whitesel JA, Augspurger RR, Pfister RR (1981) Prostatic carcinoma: influence of tumor grade on results of pelvic lymphadenectomy. Urology 17:435–440
14. Wilson CS, Dahl DS, Middleton RG (1977) Pelvic lymphadenectomy for the staging of apparently localized prostatic cancer. J Urol 117:197–198
15. McLaughlin AP, Saltzstein SL, McCullough DL, Gittes RF (1976) Prostatic carcinoma: incidence and location of unsuspected lymphatic metastases. J Urol 115:89–99
16. Paulson DF, Uro-Oncology Research Group (1979) The impact of current staging procedures in assessing disease extent of prostatic adenocarcinoma. J Urol 121:300–302
17. Prando A, Wallace S, von Eschenbach AC, Jing BS, Rosengren JE, Hussey DH (1979) Lymphangiography in staging of carcinoma of the prostate. The potential value of percutaneous lymph node biopsy. Radiology 131:641–645
18. Merrin C, Wajsman ZM, Baumgartner G, Jennings E (1977) The clinical value of lymphangiography: are the nodes surrounding the obturator nerve visualized? J Urol 117:762–764
19. Zoretic SN, Wajsman Z, Beckley SA, Pontes JE (1983) Filling of the obturator nodes in pedal lymphangiography: fact or fiction. J Urol 129:533–535
20. Göthlin JH, Höiem L (1981) Percutaneous fine-needle biopsy of radiographically normal lymph nodes in the staging of prostatic carcinoma. Radiology 141:351–354
21. Wajsman Z, Gamarra M, Park JJ, Beckley S, Pontes JE (1982) Transabdominal fine needle aspiration of retroperitoneal lymph nodes in staging of genitourinary tract cancer (correlation with lymphography and lymph node dissection findings). J Urol 128:1238–1240
22. Gleason DF, Mellinger GT, VA Cooperative Urological Research Group (1974) Prediction of prognosis for prostatic adenocarcinoma by combined histologic grading and clinical staging. J Urol 111:58–64
23. Zencke H, Farrow GM, Myers RP, Benson RC Jr, Furlow WL, Utz DC (1982) Relationship between grade and stage of adenocarcinoma of the prostate and regional pelvic lymph node metastases. J Urol 128:498–501

Chapter 16

Lymph Node Aspiration in Staging Prostatic Carcinoma: Importance of the Equivocal Lymphogram

S. J. Dan, S. C. Efremidis, J. S. Train and H. A. Mitty

There are many controversies and treatment regimens in managing patients with prostatic carcinoma. The available treatment modalities have differing rates of success and complications. Extent and aggressiveness of the various treatments differ according to the treatment philosophy of the physician. Some are aggressive and aim for cure even in advanced disease, while others opt for symptomatic therapy, leaving curative therapy for those patients proved to have localised disease. In any case, accurate staging of extent of disease allows the most careful matching of treatment to patient. Overstaging localised disease as regionally metastatic may deny potentially curative therapy to those with localised disease, or may greatly increase complications by improperly extending aggressive therapy. Understaging may erroneously restrict aggressive therapy to the prostate or deny more systemic therapy with a better chance at cure or palliation.

In selected patients with apparently localised prostatic carcinoma, we are asked to evaluate the lymphatic system by pedal lymphography and percutaneous needle aspiration of lymph nodes when indicated [1]. The results of these procedures are combined with the patient's total clinical status to plan optimal therapy.

Lymphography and Interpretation

We perform bipedal lymphography in the standard manner using methylene blue to opacify the lymphatics and Ethiodol[1] as the contrast material. X-ray films obtained shortly after introduction of Ethiodol allow evaluation of the lymphatic vascular channels—the angial phase. Additional X-ray films 24 h later are used to evaluate the lymph nodes. The lymphograms are classified as abnormal, equivocal or normal. Abnormal lymphograms are those with nodes containing one or more well-circumscribed filling defects 1 cm or more in diameter (Figs. 16.1, 16.2). They are usually eccentrically placed within the node, and often obliterate the marginal sinus of the node. Angial phase X-ray films are helpful in differentiating large metastatic deposits from prominent lymph node hilar defects or fibrofatty replacement. Hilar defects are most prominent in inguinal and lateral lacunar nodes, are usually placed cephalad or sometimes medial in the node and have lymphatic vessels exiting uniformly from them on angial phase films. Oblique films may be necessary

[1] Savage Laboratories, Melville, NY, USA.

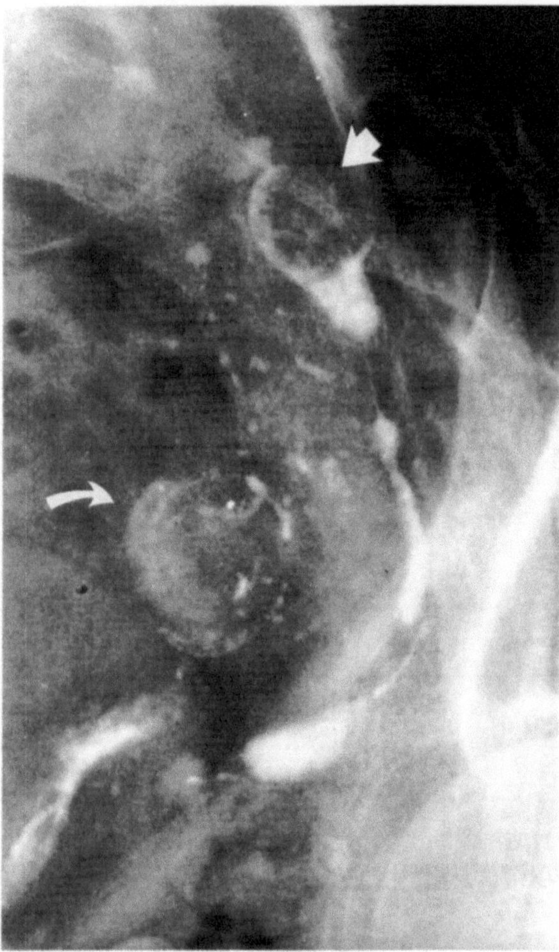

Fig. 16.1. Abnormal defect. Large, well-circumscribed, eccentric filling defect (*arrow*) obliterates marginal sinus of node. Percutaneous needle aspiration confirmed lymph node metastasis. Node which is totally replaced by tumour (*curved arrow*). (Dan et al. [1]; reproduced by kind permission of the Editor of *Urologic Radiology*)

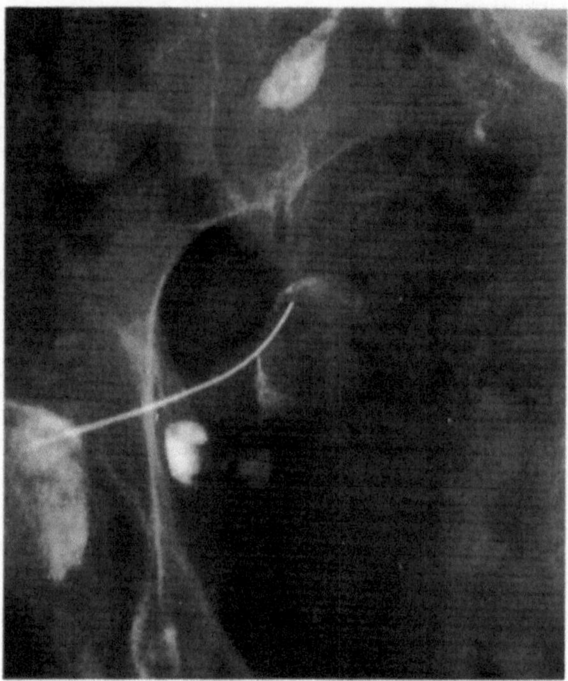

Fig. 16.2. Abnormal defect. Needle tip in periphery of defect. Positive aspiration.

to evaluate the size and borders of the defects and to help visualise the relationship of the defect to the exiting lymphatics.

Equivocal lymphograms contain nodes with suspicious filling defects that do not meet the criteria for abnormality, have nodal irregularities that cannot be deemed normal, or contain areas of absent opacification that cannot be explained on an anatomical basis. They are divided into three groups:

1. Void or gap (Figs 16.3–16.5)—an area of absent opacification of nodes where this is not expected. This may be due to total replacement of nodal tissue by metastasis. It is easier to recognise when it is unilateral and when it involves a group of nodes. Bilateral voids may be taken to be anatomical variations in nodal anatomy or to be more likely due to inflammatory or degenerative changes. When a single node among a group is replaced, this will probably not be noticed. Again the angial phase can be helpful, demonstrating whether the lymphatic vessels cross the void undisturbed, are gently displaced around it, or are obstructed or abruptly deviated at the area of the void. Oblique nodal and angial films can help suggest that an iliac void is due to displacement by tortuous iliac arteries. The remaining nodes are often displaced leftwards, especially in the left common iliac chain, and the lymphatic vessels usually follow the same displaced course.

2. Small defects or fragmentation (Figs. 16.6–16.9). Filling defects similar to definite abnormal defects but less than one centimeter in diameter fall into this category. Small filling defects are less likely to be due to metastasis than those 1 cm or larger and may be the result of inflammatory changes. Oblique films may help resolve apparent small filling defects into spaces between small overlying nodes. Small clusters of opacified nodal tissue, apparently fragmented nodes and nodes that are faintly opacified and have a "ghostly" appearance are also considered equivocal. The abnormalities in this group, especially common in the external iliac chain, may

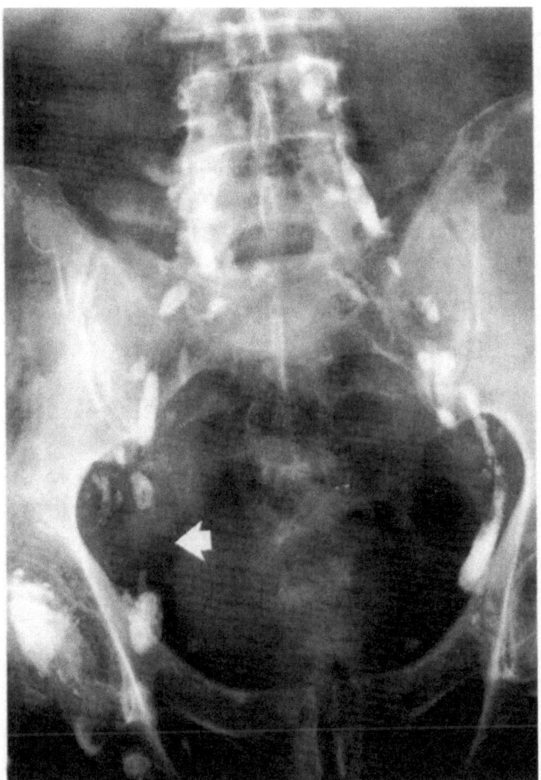

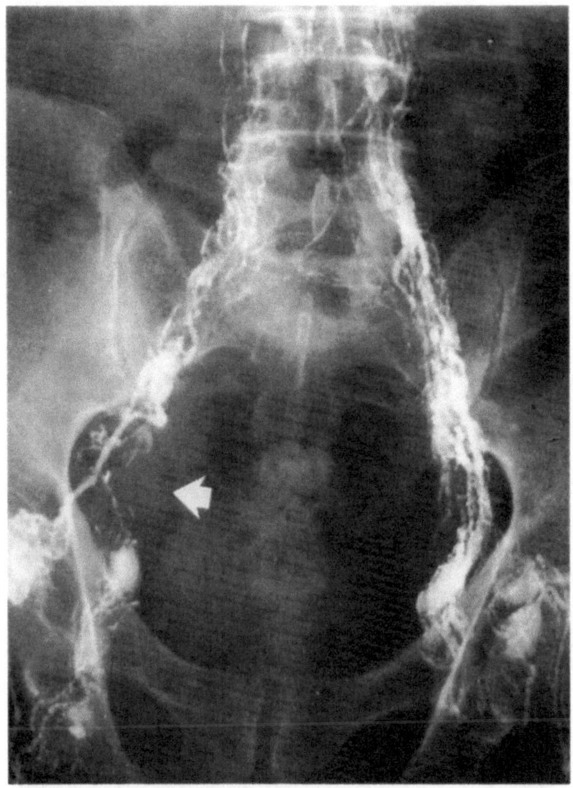

Fig. 16.3. a Equivocal defect—void. Unilateral area of absent nodal opacification in external iliac chain (*arrow*). Positive aspiration. **b** Deviation and local obstruction of lymphatic channels in same area seen during angial phase (*arrow*). (Dan et al. [1]; reproduced by kind permission of the Editor of *Urologic Radiology*)

all be seen as a result of chronic or repeated bouts of inflammation, a common condition in the mostly middle-aged or elderly patients studied for prostatic carcinoma.

3. Irregular nodes or defects (Figs. 16.10, 16.11)—nodes with marginal irregularities and irregular, poorly circumscribed filling defects. This category is less likely to be due to metastasis and may in most cases be inflammatory. An atypical lymph node hilus may sometimes create marginal irregularities that come to aspiration for the exclusion of metastasis.

Materials and Methods

We have analysed a series of 86 cases of prostatic carcinoma with disease localised to the prostate on physical examination (clinical stages A, B and C). Normal bone scan and normal serum acid phosphatase also suggested absence of metastasis. Bipedal lymphograms were performed and interpreted according to the above criteria. Thirty-six of the 86 were considered normal and were not aspirated. In the remaining 50 cases, 89 nodal defects were aspirated.

Following local anaesthesia of the anterior abdominal wall with 1% lidocaine, a 15- or 20-cm, 23-gauge flexible (Chiba) needle was advanced with fluoroscopic guidance into the defects. Multiple needle passes were often required to enter the defects, particularly in external iliac nodes where the majority of aspirations were performed. These nodes surround the iliac vessels which often deflected the tip of the needle from the target. Position of the tip of the needle in the defect was confirmed either by turning the patient into an oblique position and noting that the needle tip remained in the defect or by moving the needle and noting the appropriate motion of the node. When a void was to be aspirated, the needle was advanced into the expected position of the node. Correct depth of the tip of the needle was confirmed by a jiggling motion of the needle. Corresponding motion of opacified

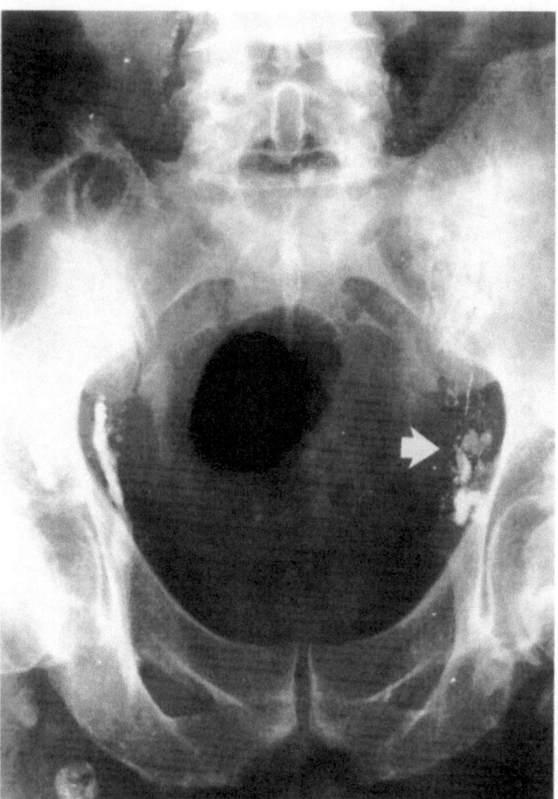

Fig. 16.4. Equivocal defects. Absent nodal opacification above external iliac areas bilaterally without definite positive defects. Also, several poorly opacified external iliac nodes (*arrow*). Positive aspirations in areas of absent opacification bilaterally. (Dan et al. [1]; reproduced by kind permission of the Editor of *Urologic Radiology*)

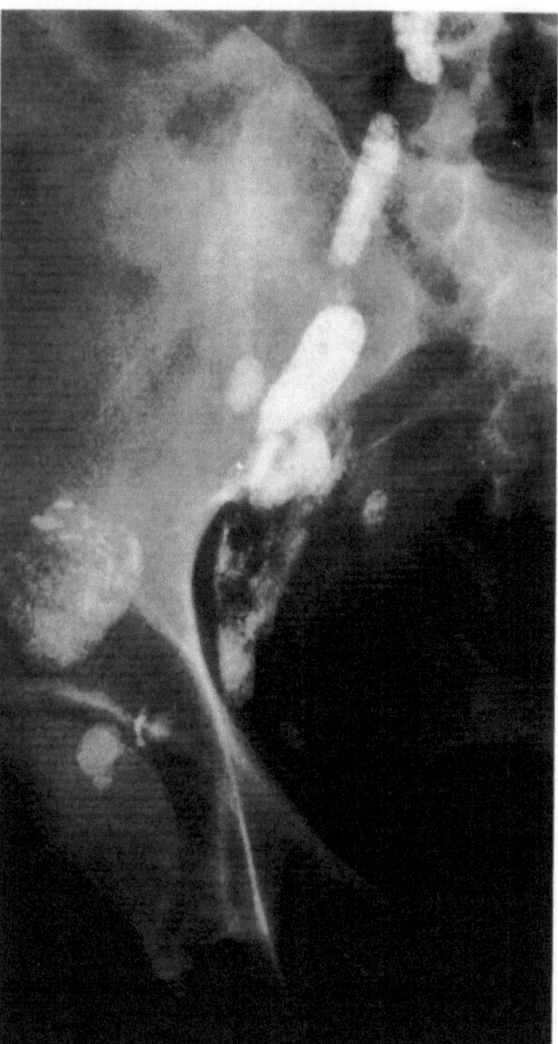

Fig. 16.6. Equivocal defect. Multiple small defects in external iliac node. Positive aspiration. (Dan et al. [9]; reproduced by kind permission of the Editor of *Journal of Urology* and the publisher, Williams and Wilkins Co., Baltimore)

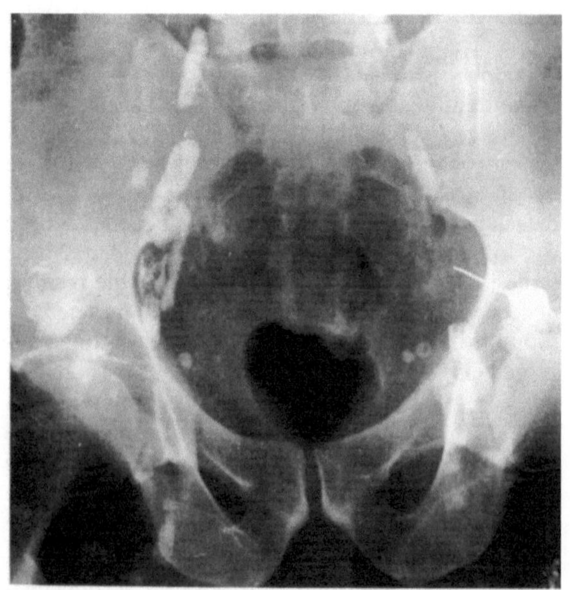

nodes cephalad and caudad to the void indicated correct needle placement. Specimens were obtained by manual suction of a 30 ml syringe while the needle tip was moved to and fro in the defect. Suction is released when the needle tip is no longer in the defect, in order to decrease the aspiration of extraneous tissue and blood. The specimens are

◄

Fig. 16.5. Equivocal defect. Needle tip in external iliac void. Positive aspiration. (Dan et al. [9]; reproduced by kind permission of the Editor of *Journal of Urology* and the publisher, Williams and Wilkins Co., Baltimore)

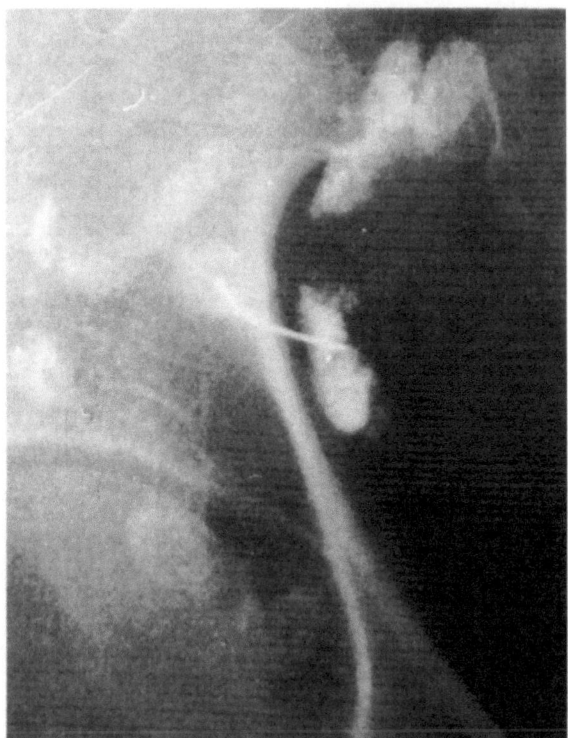

Fig. 16.7. Equivocal defect. Needle tip in small defect in external iliac node. Positive aspiration.

spread on microscope slides immediately after aspiration and fixed with an aerosol alcohol fixative[2]. The slides are examined using cytological criteria following staining with a modified Papanicolaou stain. The cytologist is an essential member of the aspiration team and must be willing to give helpful advice and especially to give as definite an opinion as is consistent with good cytological practice.

Results

Of our 50 patients, 13 had abnormal lymphograms while 37 had equivocal studies (Table 16.1). Of the 13 abnormal lymphograms, 12 (92%) had positive aspiration cytology, and of the 37 equivocal lymphograms, 18 (49%) had a positive aspiration.

Table 16.1. Aspiration results in 50 patients

	No. of patients	Positive biopsies
Abnormal lymphogram	13	12 (92%)
Equivocal lymphogram	37	18 (49%)
Total	50	30 (60%)

The breakdown into separate nodal defects is shown in Table 16.2: 90% of the 20 abnormal defects and one-third of the 69 equivocal defects were positive. Breaking down the latter group, 11 of 13 voids (group 1) were positive (85%). Of 38 small or fragmented defects (group 2), 10 were positive (26%), while only 2 of 18 irregular defects (group 3) were positive (11%).

No complications were encountered, and 13 aspirations were performed on an outpatient basis. This is now done routinely.

2 Cyto-Prep, Fischer Scientific Co., Pittsburg, Pa, USA.

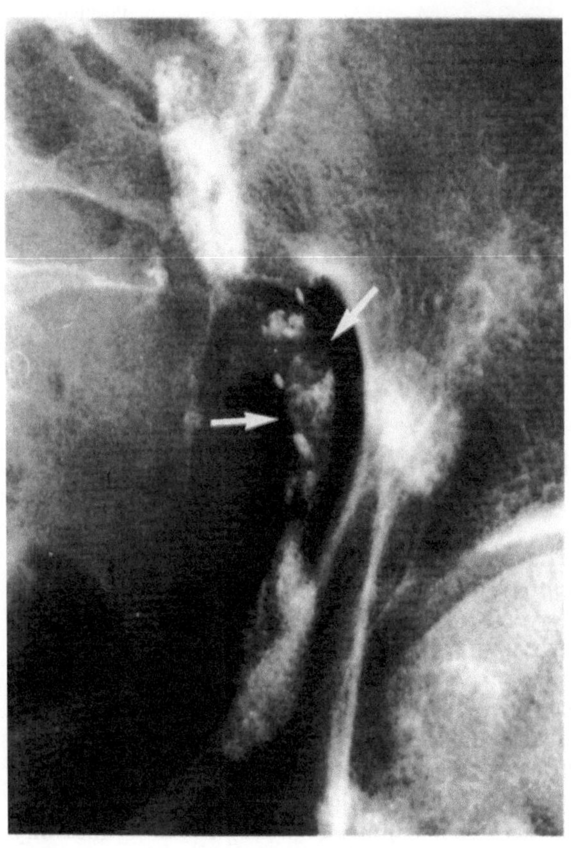

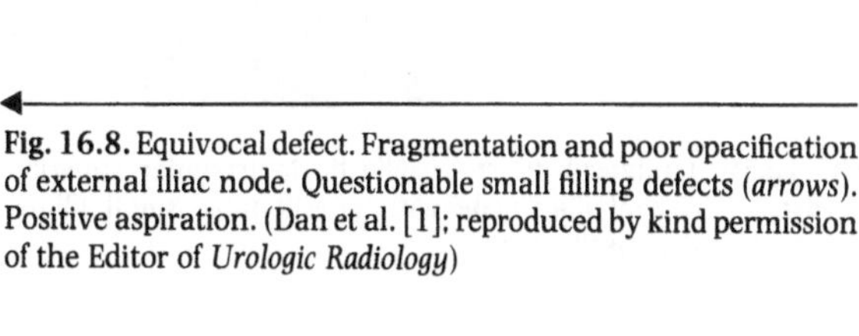

Fig. 16.8. Equivocal defect. Fragmentation and poor opacification of external iliac node. Questionable small filling defects (*arrows*). Positive aspiration. (Dan et al. [1]; reproduced by kind permission of the Editor of *Urologic Radiology*)

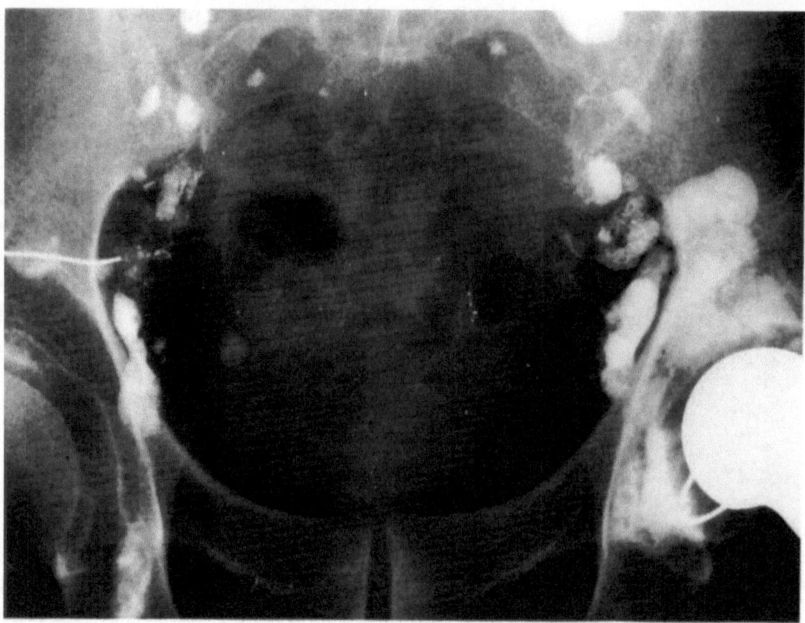

Fig. 16.9. Equivocal defects. Poorly opacified external iliac nodes bilaterally. Needle tip in right node. (Dan et al. [9]; reproduced by kind permission of the Editor of *Journal of Urology* and the publisher, Williams and Wilkins Co., Baltimore)

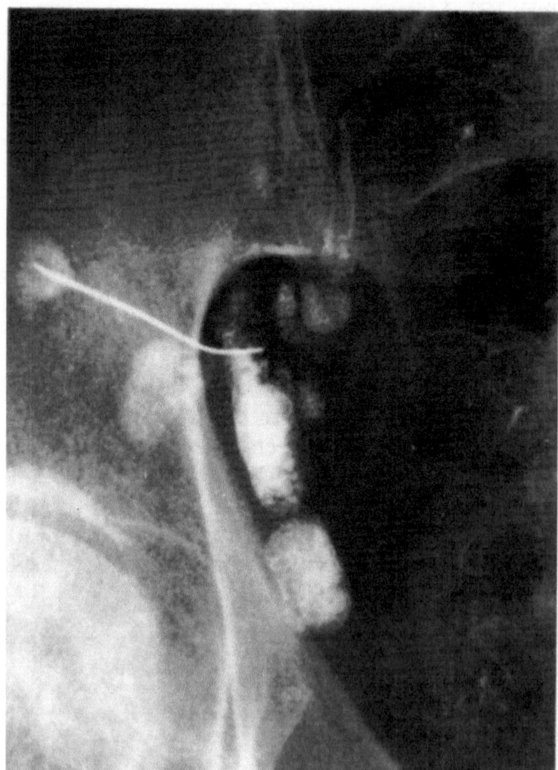

Fig. 16.10. Equivocal defect. Needle tip in irregular superior border of external iliac node with questionable small filling defects. Negative aspiration.

Table 16.2. Aspiration results in 89 nodal defects

	No. of defects	Positive biopsies
Equivocal	69	23 (33%)
1. Voids	13	11 (85%)
2. Small defects, fragmentation or poor opacification	38	10 (26%)
3. Nodal irregularity or questionable defects	18	2 (11%)
Abnormal	20	18 (90%)
Total	89	41 (46%)

Discussion

The goal in lymph node staging of prostatic carcinoma is to define the status of the lymph nodes correctly with minimal morbidity. Pathological examination of the nodes after pelvic lymphadenectomy gives the greatest accuracy, but with considerable morbidity and even mortality [2, 3]. Ultrasound has not been found accurate in lymph node staging. Computed tomography (CT) scanning uses size criteria to detect adenopathy and in good hands can be quite accurate, but will miss any

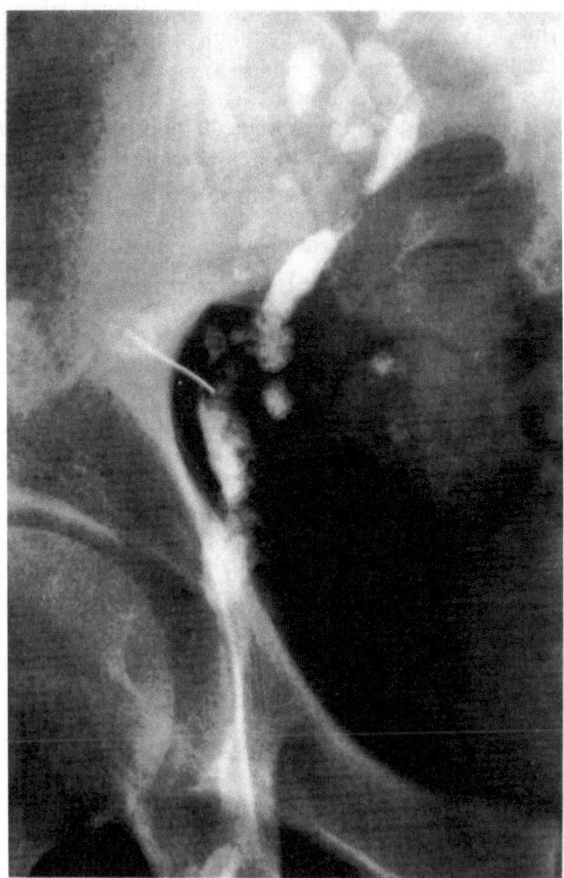

Fig. 16.11. Equivocal defect. Needle tip in irregular medial border of external iliac node. Negative aspiration. (Dan et al. [9]; reproduced by kind permission of the Editor of *Journal of Urology* and the publisher, Williams and Wilkins Co., Baltimore)

metastases that do not result in enlargement of the nodes [4, 5]. Pedal lymphography is semi-invasive, and though it does not evaluate all pelvic nodes (e.g. internal iliac nodes are often not visualised), it has been shown to opacify the major metastatic node groups and can also evaluate the internal nodal architecture, e.g., for small filling defects or faint opacification.

The reported accuracy rate of lymphography in prostatic carcinoma varies widely [3, 6]. This is most likely due to differences in diagnostic criteria for a positive lymphographic finding. Strict criteria for a metastatic deposit are necessary to eliminate a large number of false positive lymphographic findings. These would result if lesser defects which are often due to inflammatory and other changes were called positive. False positives may relegate potentially curable localised disease to systemic chemotherapy or solely symptomatic treatment. On the other hand, not all metastases behave ideally, and all begin as microscopic metastases. Some may present as small filling defects and others may cause fragmentation, etc. Totally replaced nodes may not be recognised as an abnormality on the lymphogram. When these equivocal changes are due to metastasis, strict criteria for a positive lymphographic finding will result in false negatives, subjecting patients with more advanced disease to aggressive local therapy with unnecessary morbidity and mortality.

In an attempt to control false negative and false positive rates, we make liberal use of the middle ground, the equivocal or suspicious group of defects. Of course, our system has an element of subjectivity in separating the equivocal defects from the normals, just as there would be in strictly separating normals from abnormals (e.g. When is a space between nodes called a void? When do degenerative changes become suspicious?). Since "equivocal" is of little or no use for clinical staging and selecting therapy, percutaneous needle aspiration of all equivocal defects is used to give a cytological diagnosis.

Definite abnormal defects are usually aspirated to obtain cytological confirmation of the extent of disease before selecting therapy. The rare case of a benign process (e.g. granuloma) causing a large filling defect will also be suggested by a negative aspiration. Because a negative aspiration is not conclusive, repeat aspiration or surgical confirmation of lack of metastasis may be indicated before aggressive local therapy is undertaken.

False negatives in optimally interpreted lymphograms may be due to metastasis in nodes not opacified or to microscopic metastatic deposits in opacified nodes. Equivocal lymphogram defects can be elucidated by needle aspiration. We do not aspirate normal lymphograms even though microscopic metastasis cannot be excluded. There have been reports of needle aspiration of lymphographically normal nodes yielding positive cytological findings [7, 8]. This has potential implications for all non-lymphadenectomy staging of prostatic carcinoma and points out the limits of lymphographic staging and the inconclusiveness of a negative aspiration. Increasing the number of needle passes to sample all nodes is not practical. Also, no number of aspirations would approach the accuracy of pelvic lymphadenectomy and make a negative aspiration truly diagnostic.

Some clinicians feel that the presence of regional lymph node metastasis in prostatic carcinoma is not necessarily an "all or none" phenomenon and does not automatically change the course of the disease or the prognosis [3]. They would consider the presence of microscopic metastases not detected on a normal lymphogram to be evidence of minimal nodal disease. Perhaps metastasis detected by percutaneous aspiration of some of the equivocal defects would also be considered in this group. In some studies, minimal nodal disease has a prognosis similar to patients with localised disease when definitive therapy is extended to the regional lymph nodes. Clinicians who do not share this view would require pelvic lymphadenectomy to document the absence of lymph node metastasis before undertaking aggressive local therapy.

In summary, our results confirm that lymphographic interpretation cannot be confined by strict criteria for metastasis without significant numbers of false negatives. Of the 37 patients with equivocal lymphographic findings that would presumably be called negative by some, almost half had positive aspiration cytological findings confirming regional lymph node metastasis. These patients had their disease accurately staged without pelvic lymphadenectomy with its morbidity and occasional mortality. Of our three groups of equivocal defects, the voids had positive cytological findings in the greatest number of defects (85%). The group containing small defects and fragmentation was positive in about a quarter of defects and the irregular or questionable group in a smaller number. Even with correct placement of the needle tip in the lymph node defect, needle aspiration is subject to sampling error. Aspiration from a necrotic area in a metastatic deposit may give a false negative. In a void, needle placement is not in an actual defect, but in an area where nodes are expected to have been. Even with these uncertainties, positive results were very common in voids and abnormal defects. When these two groups yield negative cytological findings, we suggest repeating the aspiration. Patients with negative aspiration can be dealt with according to the clinician's treatment philosophy. One schema for the detection of lymph node metastases is indicated in Fig. 16.12. Following lymphography and percutaneous aspiration, the

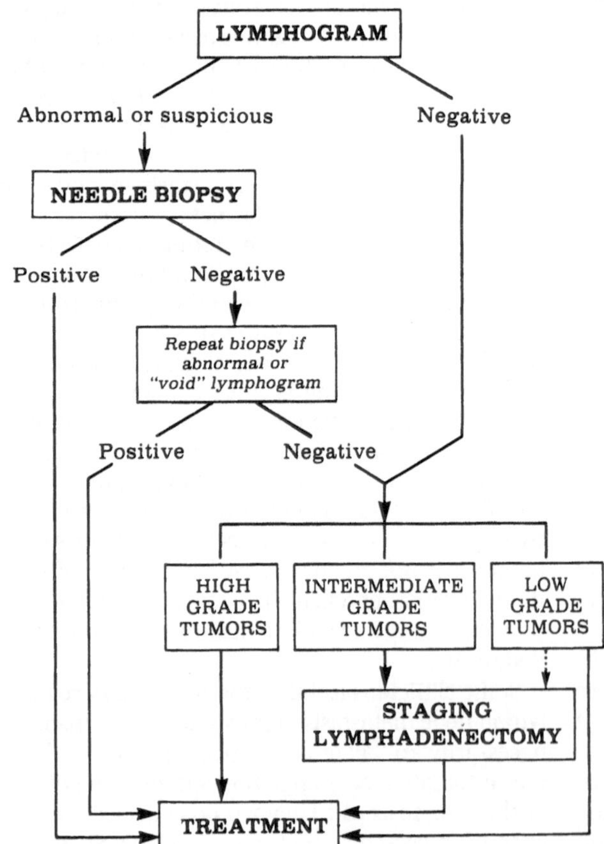

Fig. 16.12. One schema suggested for detection of regional nodal metastasis. (Dan et al. [9]; reproduced by kind permission of the Editor of *Journal of Urology* and the publisher, Williams and Wilkins Co., Baltimore)

pathological grade of the primary tumour is used to determine the aggressiveness of the search for lymph node metastases [9].

Acknowledgement. We thank Ms Louisa Haigler for assistance in preparing this manuscript.

References

1. Dan SJ, Efremidis SC, Train JS, Cohen BA, Mitty HA (1982) Equivocal lymphogram and lymph node aspiration: importance in staging carcinoma of the prostate. Urol Radiol 4:215–219
2. Babcock JR, Grayhack JT (1979) Morbidity of pelvic lymphadenectomy. Urology 13:483–486
3. Johnson DE, von Eschenbach AC (1981) Roles of lymphangiography and pelvic lymphadenectomy in staging prostate cancer. Urology 17(Suppl):66–71
4. Walsh JW, Amendola MA, Konerding KF, Tisnado J, Hazra TA (1980) Computed tomographic detection of pelvic and inguinal lymph node metastases from primary and recurrent malignant disease. Radiology 137:157–166
5. Weinerman PM, Arger PH, Pollack HM (1982) CT evaluation of bladder and prostate neoplasms. Urol Radiol 4:105–114
6. MacDonald JS (1981) Lymphography. In: Duncan W (ed) Prostate cancer. Springer, Berlin Heidelberg New York, pp 97–107 (Recent results in cancer research, vol 78)
7. Gothlin JH, Hoiem L (1981) Percutaneous fine-needle biopsy of radiographically normal lymph nodes in the staging of prostatic carcinoma. Radiology 141:351–354
8. Wajsman Z, Gamarra M, Park JJ, Beckley S, Pontes JE (1982) Transabdominal fine needle aspiration of retroperitoneal lymph nodes in staging of genitourinary tract cancer (correlation with lymphography and lymph node dissection findings). J Urol 128:1238–1240
9. Dan SJ, Wulfsohn MA, Efremidis SC, Mitty HA, Brendler H (1982) Lymphography and percutaneous lymph node biopsy in clinically localized carcinoma of the prostate. J Urol 127:695–698

Chapter 17

Staging Prostatic Carcinomas with Percutaneous Lymph Node Biopsy

J. H. Göthlin

Documentation of regional lymph node involvement is crucial in the prognosis and treatment of genitourinary neoplasm. Traditionally this has required regional lymph node dissection [e.g. 1–10], a procedure with a morbidity as high as 20%–30% [11, 12] and with high costs. A combination of clinical stage, histological grade and serum acid phosphatase level has been reported as sufficient for staging most patients with carcinoma of the prostate [12]. Lymphography without [13–15] or with fine needle aspiration [16–22] has been advocated as an alternative to lymph node dissection. Fine needle biopsy of apparently normal lymph nodes at lymphography in order to verify or exclude micrometastases in early stages of prostatic carcinoma has been reported [17, 20–22].

The probability of lymph node metastasis is related to the clinical stage of the disease. Histologically documented lymph node metastases have been reported (according to the Gleason [23] classification) to be 14%–21% in stage B_1, 23%–28% in stage A_2, 30%–50% in stage B_2 and 39%–63% in stage C disease [5, 7, 24–26].

My own series comprises 54 patients selected from a series of patients with carcinoma of the prostate verified by biopsy. All patients had pedal lymphogram interpreted as normal. Clinically the tumours belonged to group I or II (occasionally III) according to Flocks' [27] classification. Thus none of the patients showed evidence of distant metastases during rectal examination, laboratory tests or radiological work-up.

To minimise the effect of macrophage reaction on the cytological interpretation, percutaneous fine needle biopsy was performed during fluoroscopy the day after lymphography. The technique has been described in detail [17] and will not be dealt with here. It should only be mentioned that in the last 49 patients at least 10 pelvic lymph nodes were biopsied. No clinically evident complications were encountered.

In 11 of the 54 patients one or more positive nodes were found. Positive findings were regarded as true positive, while negative findings were assumed to indicate only absence of disease but not to prove it. Representative material was obtained in only slightly over 80% of the biopsied nodes. Staging surgery was not performed.

Positive cytological findings for metastases in 11 of 54 patients with early prostatic carcinoma and normal lymphogram is not surprising, as the incidence of histologically confirmed lymph node metastases has been reported to be 7%–27% in low-grade prostatic cancer and 38%–82% in high-grade cancer [4, 5, 7, 8, 10, 12, 15, 24, 26, 28–42].

The sensitivity of lymphography in detecting metastases ranges from 37% to 75% [7, 15, 26, 30–32, 35–40, 43, 44], while the specificity ranges from 21% to 95%. In the Uro-Oncology Research Group (UORG) series of 114 patients with negative lymphograms, internal biopsy at staging surgery proved that 21 patients had nodal metastases [37].

The present series, in which 20% of lymphograms proved to be false negative by needle biopsy, coincides well with the UORG figure. Although about 17% of the biopsy material was not representative and usually only 10 lymph nodes per patient were biopsied, it is most likely that percutaneous biopsy is fairly sensitive in detecting patients with micrometastases. In the present series it was thought important to detect only positive lymph nodes by biopsy, not to find the number of afflicted lymph nodes. Even if attempts to puncture the nodes surrounding the obturator nerves and nodes immediately adjacent to them were made, micrometastases were found exclusively in higher pelvic lymph nodes in 5 out of the 11 positive patients. In one report [19] fine needle aspiration biopsy was performed in a mean of three nodes (always bilaterally in the obturator nodes) without detecting a single metastasis in 49 patients. The present series indicates that the nodes surrounding the obturator nerves do not always give positive material even if nodes cranial to them are afflicted. It has been reported that the nodes of the obturator nerve (the surgical obturator nodes) are the most common and frequently the first site of nodal spread in prostatic carcinoma [5, 34], but that does not mean that metastatic material can always be obtained there.

In the present series most biopsies were performed in the medial margin of the lymph nodes. It is a common opinion that metastases in lymph nodes begin at the margin and grow towards the centre [44]. After the initial pilot series, two to three biopsies of each node were performed and thereby the material became more representative.

No paralumbar nodes were punctured in the present series. Reports indicate that metastases to these nodes occur mainly when pelvic nodes are involved [15, 38, 40, 45]. Exceptions have been reported [46], though the reason may be that metastatic pelvic lymph nodes were not filled by contrast medium and thus were overlooked.

The findings in the present series coincide fairly well with an investigation [20] in which lymph node aspiration biopsy was compared with histological examination of lymphadenectomy specimens. By biopsy 15% false negative lymphograms were detected. However, it was discouraging that lymph node dissection revealed metastases in 15 of 50 patients, those 15 patients being negative at fine needle biopsy. This gives rise to the suspicion that the figure of 20% false negative lymphograms in the present series is too low, even if it coincides well with that of the UORG group [37].

Histological examination of 1223 lymph nodes in 62 patients with prostatic carcinoma established that in 40% of the lymph nodes only micrometastases were found [45]. The stages of the prostatic carcinomas are not given but the investigation demonstrates the high incidence of micrometastases impossible to detect at lymphography and indicates difficulties in finding all of them by biopsy. It is difficult to explain the negative experience of a group [47] stating that lymphangiography and fine needle aspiration biopsy are ineffective for staging early prostatic cancer. This report and another with one member of the same group as coauthor [19] probably indicate the use of a different technique when performing biopsy and also that too few (mean of three) nodes were biopsied.

The choice of needle varied widely in the present series. We consider that the type of needle is not important as long as it is a fine needle. The flexible Chiba needle has not proved to be very suitable when puncturing deep lesions, as it fairly often deviates from a straight course. Contamination from traversed organs has not been a problem in the cytological interpretation, and perusal of the literature indicates that this problem is rarely encountered. Complications of clinical significance are rare [17, 48].

Can computed tomography (CT) or ultrasound be used in staging early prostate neoplasm? Lymph nodes can only be interpreted as abnormal when enlarged and thus micrometastases are impossible to detect [e.g. 7, 49–51]. For detailed discussion see [52].

In conclusion, it may cautiously be stated that lymph node aspiration is an inexpensive and very safe procedure even if lymphography is somewhat time consuming. Positive cytological findings can spare several patients a staging lymphadenectomy and/or an unnecessary attempt at radical surgery. However, if lymphadenectomy is to be performed regardless of pelvic lymph node metastases (the lymphadenectomy being regarded as a treatment procedure), it is logical to biopsy para-aortic lymph nodes only, as the presence of metastases there would change the therapy. Negative cytological findings cannot be used as a reliable indicator for clinical management.

References

1. Catalona WJ, Stein AJ (1982) Accuracy of frozen section detection of lymph node metastases in prostatic carcinoma. J Urol 127:460–461

2. Fowler JE, Torgersson L, McLeod D, Stutz RE (1981) Radical prostatectomy with pelvic lymphadenectomy: observations on the accuracy of staging with lymph node frozen sections. J Urol 126:618–619
3. Gore RM, Moss AA (1983) Value of computed tomography in interstitial (^{125}I) brachytherapy of prostatic carcinoma. Radiology 146:453–458
4. McCollough DL, Prout GR, Daly JJ (1974) Carcinoma of the prostate and lymphatic metastases. J Urol 111:65–71
5. McLaughlin AP, Saltzstein SL, McCullough DL, Gittes RF (1976) Prostatic carcinoma: incidence and location of unsuspected lymphatic metastases. J Urol 115:89–94
6. Middleton AW (1981) Pelvic lymphadenectomy with modified radical retropubic prostatectomy as a single operation: technique used and results in 50 consecutive cases. J Urol 125:353–356
7. Paulson DF, Uro-Oncology Research Group (1979) The impact of current staging procedures in assessing disease extent of prostatic adenocarcinoma. J Urol 121:300–302
8. Prout GR, Heaney JA, Griffin PP, Daly JJ, Shipley WU (1980) Nodal involvement as a prognostic indicator in patients with prostatic carcinoma. J Urol 124:226–231
9. Sadlowski RW, Donahue DJ, Richman AV, Sharpe JR, Finney RF (1982) Accuracy of frozen section diagnosis in pelvic lymph node staging biopsies for adenocarcinoma of the prostate. J Urol 129:324–326
10. Zencke H, Farrow GM, Myers RP, Benson RC Jr, Furlow WL, Utz DC (1982) Relationship between grade and stage of adenocarcinoma of the prostate and regional pelvic lymph node metastases. J Urol 128:498–501
11. Babcock JR, Grayhack JT (1979) Morbidity of pelvic lymphadenectomy. Urology 13:483–486
12. Freiha FS, Pistenma DA, Bagshaw MA (1979) Pelvic lymphadenectomy for staging prostatic carcinoma: is it always necessary? J Urol 121:176–177
13. Correa RJ Jr, Kidd CR, Burnett GE, Gibbons RP, Cummings KB (1981) Percutaneous pelvic lymph node aspiration in carcinoma of prostate. J Urol 126:190–191
14. Johnson DE, von Eschenbach AC (1981) Roles of lymphangiography and pelvic lymphadenectomy in staging prostate cancer. Urology 17 (suppl):66–81
15. Spellman MC, Castellino RA, Ray GR, Pistenma DA, Bagshaw MA (1977) An evaluation of lymphangiography in localized carcinoma of the prostate. Radiology 125: 637:644
16. Dan SJ, Efremidis SC, Train JS, Cohen BA, Mitty HA (1982) Equivocal lymphogram and lymph node aspiration: importance in staging carcinoma of the prostate. Urol Radiol 4:215–219
17. Göthlin JH, Høiem L (1981) Percutaneous fine-needle biopsy of radiographically normal lymph nodes in the staging of prostatic carcinoma. Radiology 141:351–354
18. Khan O, Pearse E, Bowley N, Williams G, Krantz T (1983) Combined bipedal lymphangiography, CT scanning and transabdominal lymph node aspiration cytology for node staging in carcinoma of the prostate. Br J Urol 55:538–541
19. Kidd R, Correa R Jr (1984) Fine-needle aspiration biopsy of lymphangiographically normal lymph nodes: a negative view. AJR 141:1005–1006
20. Wajsman Z, Gamaro M, Park JJ, Beckley S, Pontes JE (1982) Transabdominal fine needle aspiration of retroperitoneal lymph nodes in staging of genitourinary tract cancer (correlation with lymphography and lymph node dissection findings). J Urol 128:1238–1240
21. Rothenberger K, Hofstetter A, Pfeifer KJ, Voeth C (1980) Feinnadelbiopsie retroperitonealer Lymphknoten beim Staging des Prostakarzinomas. Verh Dtsch Ges Urol 32:76–78
22. Piscioli F, Leonardi E, Reich A, Lucian L (1984) Percutaneous lymph node aspiration biopsy and tumor grade in staging of prostatic carcinoma. Prostate 5:459–468
23. Gleason DF (1977) Histologic grading and clinical staging of prostatic carcinoma. In: Tannenbaum J (ed) Urologic pathology. Lea and Febiger, Philadelphia, p 171
24. Donohue RE, Fauver HE, Whitesel JA, Pfister RR (1979) Staging prostatic cancer: a different distribution. J Urol 122:327–329
25. Gleason DF, Mellinger GT, UA Cooperative Urological Research Group (1974) Prediction of prognosis for prostatic adenocarcinoma by combined histologic grading and clinical staging. J Urol 111:58–64
26. Wilson CS, Dahl DS, Middleton RG (1977) Pelvic lymphadenectomy for the staging of apparently localized prostatic cancer. J Urol 117: 197–198
27. Flocks RH (1973) The treatment of stage C prostatic cancer with special reference to combined surgical and radiation therapy. J Urol 109:461–463
28. Arduino LJ, Glucksman MA (1962) Lymph node metastases in early carcinoma of the prostate. J Urol 88:91–93
29. Barzell W, Bean MA, Hilaris BS, Whitemore WF Jr (1977) Prostatic adenocarcinoma: relationship of grade and local extent to the pattern of metastases. J Urol 118:278–282
30. Bruce AW, O'Cleireachain F, Morales A, Awad SA (1977) Carcinoma of the prostate: a critical look at staging. J Urol 117:319–322
31. Castellino RA, Ray G, Blank M, Gowan D, Bagshaw M (1973) Lymphangiography in prostatic carcinoma. Preliminary observations. JAMA 223:877–881
32. Cerny JC, Farah R, Rian R, Weckstein ML (1975) An evaluation of lymphangiography in staging carcinoma of the prostate. J Urol 113:367–370
33. Esteve C (1978) Etude des adenopathies satellites du cancer prostatique. Indications du curage ganglionnaire. Thesis Université René Descartes, Paris
34. Flocks RH, Culp D, Porto R (1959) Lymphatic spread from prostatic cancer. J Urol 81:194–196
35. Hilaris BS, Whitmore WF Jr, Batata MA, Grabstald H (1974) Radiation therapy and pelvic node dissection in the management of cancer of the prostate. AJR 121:832–838
36. Kurth KH, Altwein JE, Hohenfellner R (1977) Die pelvine Lymphadenektomie als Staging-Operation des Prostatakarzinoms. Urologe 16:65–69
37. Liebner EJ, Stefani S, Uro-Oncology Research Group (1980) An evaluation of lymphography with nodal biopsy in localized carcinoma of the prostate. Cancer 45:728–734
38. Loening SA, Schmidt JD, Brown RC, Hawtrey CE, Fallon B, Culp DA (1977) A comparison between lymphangiography and pelvic lymph node dissection in the staging of prostatic cancer. J Urol 117:752–756
39. O'Donohue EPN, Shridar P, Sherwood T, Williams JP, Chisholm GD (1976) Lymphography and pelvic lymphadenectomy in carcinoma of the prostate. Br J Urol 48:689–696
40. Ray GR, Pistenma DA, Castellino RA, Kempson RL, Meares E, Bagshaw MA (1976) Operative staging of apparently localized adenocarcinoma of the prostate: results in fifty unselected patients. I. Experimental design and preliminary results. Cancer 38:73–83
41. Sadlowski RW (1978) Early stage prostatic cancer investigated by pelvic lymph node biopsy and bone marrow acid phosphatase. J Urol 119:89–93
42. Taddei L, Fuochi C, Menichelli E, Luciani L (1984) Accuracy of lymphography in staging of prostatic and bladder carcinoma: 88 cases with aspirative cytological and post lymphadenectomy histological verification. Diagn Imaging Clin Med 53:91–98
43. Hoekstra WJ, Shroeder FH (1981) The role of lym-

phangiography in the staging of prostatic cancer. Prostate 2: 433–440
44. Prando, A, Wallace S, von Eschenbach AC, Jing B-S, Rosengren J-E, Hussey DH (1979) Lymphangiography in staging of carcinoma of the prostate. Radiology 131:641–645
45. Schubert. J, Nitsche H, Heidl G, Gorski J (1981) Aussagen der Lymphographie beim Prostatakerzinom—vergleich mit der pelvinen En-bloc-Lymphadenektomie. Z Urol Nephrol 74:283–289 (Eng abstr)
46. Zingg EJ, Fuchs WA, Heritier P, Göthlin J (1974) Lymphography in carcinoma of the prostate. Br J Urol 46:549–554
47. Kidd R, Crane RD, Dail DH (1984) Lymphangiography and fine-needle aspiration biopsy: ineffective for staging early prostate cancer. AJR 141:1007–1012
48. Göthlin JH (1983) Radiologically guided biopsies. Eur J Radiol 3 (suppl):303–304
49. Emory TH, Reinke DB, Hill AL, Lange PH (1983) Use of CT to reduce understaging in prostatic cancer: comparison with conventional staging techniques. AJR 141:351–354
50. Golimbu M, Morales P, Al-Askari S, Shulman Y (1981) CAT scanning in staging of prostatic cancer. Urology 18:305–308
51. Weinerman PM, Arger PH, Pollack HM (1982) CT evaluation of bladder and prostate neoplasms. Urol Radiol 4:105–114
52. Masselot J, Ozanne F, Geoffray G, Seton G (1985) Cancer de la prostate: intérêt du scanner dans le bilan pre-radiothérapique. A propos de 49 cas revu à l'Institut Gustave-Roussy. Bull Cancer (Paris) 72:452–461

Chapter 18

Lymph Node Aspiration Biopsy and Lymph Node Dissection in Prostatic Cancer

Z. Wajsman

A significant number of patients with localised prostatic cancer have regional lymph node metastases. The incidence of pelvic lymph node involvement is closely related to the stage and grade of the tumour. Donohue et al [1] have reviewed the recent literature and reported that 23% of patients with stage A_2, 18% with stage B_1, 35% with stage B_2 and 46% with stage C disease have pathologically documented lymph node spread. Correlation of grading and lymph node metastasis was demonstrated by Kramer et al [2] when Gleason's grading system was used. No lymph node involvement was found in patients with Gleason's score 2, 3 or 4 in this study, nor in the study by Paulson et al [3] with the Urologic Oncology Group. However, the National Prostatic Cancer Project revealed a 30.5% incidence of lymph node invasion in low Gleason's scores [4]. On the other hand, most reports agree on the high incidence of lymph node metastases in patients with high-grade tumour.

In our last 18 months' experience at the University of Florida 24 radical prostatectomies were performed. Lymph node invasion was found in five patients. Three patients had poorly differentiated tumours, one patient had a moderately well-differentiated tumour and two patients had well-differentiated tumours. Of the five patients with positive lymph nodes, four were found to have seminal vesicle invasion and one had capsular involvement. None of these patients had an elevated acid phosphatase level. It is clear that a prior knowledge of lymph node status is critical in the treatment decision, since only 13% of patients with lymph node metastases survived tumour-free for 15 years [5]. One has to realise that, although good correlation seems to exist between tumour stage and grade and incidence of lymph node metastases, one cannot apply this to decisions in the treatment of an individual patient. Every effort should be made to document or rule out lymph node invasion based on histological diagnosis, either by fine needle aspiration or lymph node dissection.

Some urologists perform lymph node dissection separately and proceed with radical prostatectomy or radiation therapy only if there is no pathologically proved lymphatic invasion. Frozen section of lymph node at the time of definitive procedure seems to yield poor results in grossly normal glands. Catalona and Stein reported 8%–19% false negative frozen section in grossly normal nodes [6]. It appears then that the most logical way to proceed would be to perform a pelvic lymph node dissection and, if careful pathological examination did not reveal microscopic involvement, then radical surgery or radiation therapy could be applied. Recent information appears to suggest, however, that minimal lymph node involvement does not necessarily imply a hopeless prognosis [7, 8]. In addition, recent reports from the Mayo Clinic [9] suggest that radical prostatectomy with pelvic lymph node dissection will result in good survival when early hormonal therapy is applied in patients with stage D disease. Thus, prior knowledge of microscopic lymph node metastases does not necessarily rule out definitive procedures such as radical prostatectomy or radiation therapy.

Pelvic lymphadenectomy has a significant morbidity. The overall 20%–30% complication rate includes wound infection, lymphocele, pelvic abscess, leg and penile oedema, phlebitis and, occasionally, death from pulmonary embolism [10, 11]. More recent modifications of pelvic lymph node dissection which preserve the lymphatic tissue around the external iliac artery significantly reduce the leg and genital oedema without compromising the staging value of this procedure.

This was recently challenged by a few investigators, and it appears that limited node dissection may underestimate the number of lymph node metastases in 55%–80% of patients [12]. In view of the fact that even pelvic lymph node dissection may not detect lymphatic spread in a not insignificant number of patients, how accurate then is the fine needle aspiration of lymphangiographically visualised nodes? In 50 consecutive patients with prostatic and bladder cancer, lymphography and fine needle aspiration were performed, followed by pelvic lymph node dissection [13]. The accuracy of obtaining representative material from the lymph nodes was 83%. These 50 patients were part of a larger group of 100 patients with localised genitourinary malignancies. Seventy-four per cent had either localised prostate or bladder cancer. Of 43 patients with positive lymphography, only 14 patients had positive cytological findings. However, of the 40 patients with negative lymphographic results, six (15%) had positive cytological findings. All these patients also had negative computed tomography (CT) scan results.

In the group of 50 patients who underwent pelvic lymph node dissection, 23 patients had positive nodes by lymphadenectomy, but only 8 had positive aspiration. These results were discussed by Paulson [14], who questioned the validity of lymph node aspiration since the confirmation rate of positive lymphography was only 32%, while random aspirations of radiographically negative nodes provided only a 15% pick-up rate. He also pointed out that only 8 patients had positive aspirations of the 23 patients who had positive nodes at staging lymphadenectomy.

We have to analyse the value of fine needle aspiration in relation to its clinical application. Positive aspiration provides invaluable information and may spare the patients unnecessary major surgical procedures or unnecessary extensive radiation therapy. This method, however, is a tedious, time-consuming procedure which is not as inexpensive as it used to be and has its own complications. Negative fine needle aspiration is totally non-informative and cannot influence clinical decisions. This method also requires a specialised centre with an experienced radiologist and cytologist.

At present our staging procedures in evaluation of patients with prostatic cancer are as follows: patients with localised prostatic cancer who are candidates for either radical prostatectomy or radiation therapy are divided into high- and low-risk groups. A "high-risk" patient is defined as one with a significant chance of having metastatic nodal spread. The high-risk factor includes high-grade (Gleason's 7–10), large volume (stage C, palpable seminal vesicles), elevated acid phosphatase level and any black patient. In this group of patients "at risk", CT is performed and any suspicious finding aspirated. Negative aspiration or negative CT scan findings in these selected high risk patients will then result in submission of the patient to lymphography with fine needle aspiration only of suspicious nodes. The criteria are less rigid in patients who are borderline candidates for surgery, and lymphography plus random fine needle aspiration of lymphographically negative nodes is sometimes done. In patients with "no risk", radical prostatectomy with pelvic lymph node dissection is usually the treatment of choice for all B and A_2 lesions, and CT or lymphography is not performed. In our experience, lymphography and fine needle aspiration have limited value and are rarely used in our practice.

In our series of 25 patients treated by radical prostatectomy, 5 (20%) were found to have only microscopic disease. All had negative CT scan results and negative acid phosphatase levels. These patients would most likely have had a normal lymphogram and "only" one or two patients would have had a positive fine needle aspiration based on a 1/3 predicted pick-up. They would have been denied surgery, and probably even radiation therapy would not have been used in these patients with microscopically involved glands. The benefit of such decisions, in my opinion, is highly questionable.

In summary, fine needle aspiration remains an elegant and effective tool, when applied properly, for carefully selected patients, but has a relatively limited clinical application.

References

1. Donohue RE, Mani JH, Whitecel JA, Mohr S, Scanavino D, Auspurger RR, Biber RJ, Fauver HE, Wetlaufer JN, Pfister RR (1980) Pelvic lymph node dissection: Guide to patient

management in clinically locally confined adenocarcinoma of prostate. Urology 20:559–565
2. Kramer SA, Spahr J, Brendler C, Glenn J, Paulson D (1980) Experience with Gleason histopathological grading in prostate cancer. J Urol 124:223–225
3. Paulson DF, Piserchia P, Gardner W (1980) Predictors of lymphatic spread in prostatic adenocarcinoma, Urologic-Oncology Research Group Study. J Urol 123:697–699
4. Wajsman Z (1981) Lymph node evaluation in prostatic cancer: is pelvic lymph node dissection necessary? Urology 17 (Suppl) 80–82
5. Flocks RH (1973) The treatment of stage C prostatic cancer with special reference to combined surgical and radiation therapy. J Urol 109:461–463
6. Catalona WG, Stein AG (1982) Accuracy of frozen section detection of lymph node metastases in prostatic carcinoma. J Urol 127:460–461
7. Prout GR, Heaney JA, Griffin PP, Daly JJ, Shipley WU (1980) Nodal involvement as a prognostic indicator in patients with prostatic carcinoma. J Urol 124:226–231
8. Berzell W, Bean MA, Hilaris BS, Whitemore WF Jr (1980) Prostatic adenocarcinoma: relationship of grade and local extent to the pattern of metastases. J Urol 118:278–282
9. Myers RP, Zincke H, Fleming TR, Farrow GM, Furlow WL, Utz DC (1983) Hormonal treatment at time of radical retropubic prostatectomy for stage D_1 prostatic cancer. J Urol 130:99–101
10. Babcock YR, Grayhack YT (1979) Morbidity of pelvic lymphadenectomy. Urology 13:483–486
11. Freiha FS, Pistenma DA, Bagshaw MA (1979) Pelvic lymphadenectomy for staging of prostatic carcinoma: is it always necessary? J Urol 122:176–177
12. Catalona WG (ed) (1984) Prostate cancer. Grune and Stratton, Orlando, Florida
13. Wajsman Z, Gamarra M, Park JJ, Beckley S, Pontes JE (1982) Transabdominal fine needle aspiration of retroperitoneal lymph nodes in staging of genitourinary tract cancer (correlation with lymphography and lymph node dissection findings). J Urol 128:1238–1240
14. Paulson DF (1982) Editorial comments. J Urol 128:1240

Chapter 19

Lymph Node Aspiration Cytology: an Aid to the Staging of Prostatic Cancer

G. Williams

Introduction

Irrespective of the method of treatment used, be it radical surgery, radiotherapy or hormone therapy, lymph node spread from carcinoma of the prostate carries a bad prognosis [1]. Furthermore, there is little point in prescribing treatment aimed solely at the prostate if lymph node spread has already taken place.

A variety of techniques has been described for identifying lymph node involvement, including lymphangiography, computed tomography (CT) scanning and surgical node excision. Of these, only the latter provides definitive evidence of nodal involvement. Because of the inaccuracies involved in the former two procedures [2, 3] and the morbidity associated with surgery, many clinicians ignore the lymph nodes when staging the disease. This is obviously a reasonable procedure when metastases to other sites, e.g. bone, have been identified, but is unreasonable if radical, potentially curative therapy is to be offered, aimed solely at the prostate.

Controversy exists as to the value of lymphangiography. It does not identify the primary draining nodes nor does it identify microscopic metastases. An overall accuracy of 60% has been reported [4]. Abdominal CT scanning gives similar results [3], but neither technique provides tissue for confirmatory histological examination. Lymphadenectomy is undoubtedly the most accurate method of determining lymph node involvement. The primary and secondary drainage nodes can be removed and microscopic metastases identified. However, this procedure carries a significant morbidity in what is often an elderly or infirm population [5]. It has been recommended that routine pelvic lymph node dissection should not be advocated unless a very specific clinical decision can be made from the information gained [6]. To improve the accuracy of non-operative methods of diagnosis we have combined bipedal lymphangiography and pelvic CT scanning with percutaneous transabdominal lymph node aspiration cytology.

Patients and Methods

Fifty-seven patients with histologically proved carcinoma of the prostate have been studied. Twenty-seven patients were newly diagnosed or had never received prior hormonal therapy. The remaining 30 patients had been on treatment with oestrogens or had had an orchiectomy over a mean period of 20.3 months with a range of 2–70 months before study. These patients were studied to increase our experience with the technique and to assess any associated morbidity. The first 42 patients in this series have been reported elsewhere [7]. Our present practice is to study only those with T0 and T1, T2 disease [8] who show no other evidence of metastatic disease and in whom local radical therapy would be offered should no lymph node involvement be found. Lymphangiography and lymph node aspiration are carried out routinely in this group of patients. From our previous experience CT scanning is only performed when lymphangiography shows areas of non-opacification of nodes. In our total experience bipedal lymphangiography was successfully carried out in 55 patients. Abdominal and pelvic CT scans have been performed in 44 patients using a Siemens Somatom number 2 scanner 48 h after the lymphangiogram. The patients are prepared by oral administration of 800–1200 ml 4% Gastrografin to opacify the gastrointestinal tract. Alternate 8-mm slices are taken from the diaphragm to L-4, and contiguous slices from L-4 to the inferior pubic rami. The position and size of the nodes are recorded and nodes greater than 2 cm in diameter considered abnormal. Percutaneous lymph node aspiration is performed after light premedication with the patient in the supine position, using a 1% lignocaine local anaesthetic. An anterior transperitoneal approach is made using a 21-gauge Chiba needle, guided by fluoroscopy (Fig. 19.1). In patients in whom abnormal nodes had been reported on lymphangiography the nodes were aspirated, otherwise nodes were chosen randomly from a common iliac chain. In our initial studies at least one node was aspirated from each side. Our present practice is to aspirate as many nodes as possible from both sides. The aspirate is placed on albuminised glass slides, some of which were wet-fixed in alcohol and stained by the Papanicolaou method, the rest being air-dried and stained with the Giemsa stain.

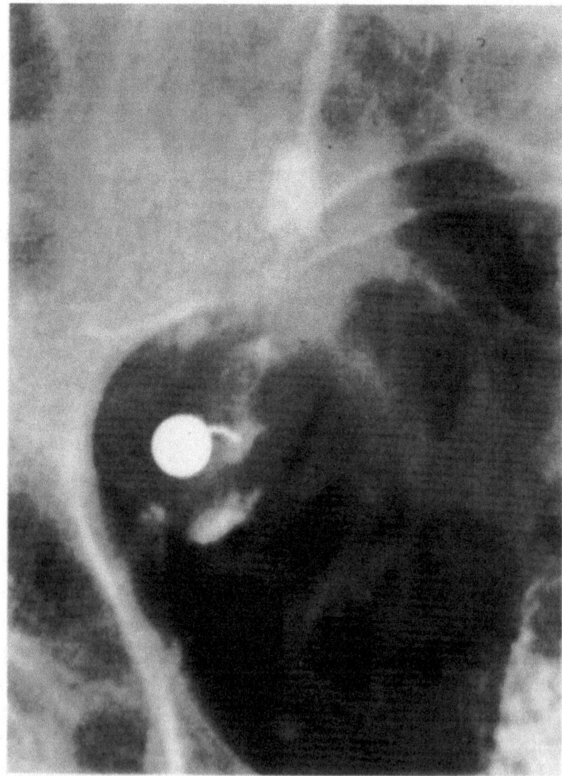

Fig. 19.1. A 21-gauge Chiba needle being guided by fluoroscopy into an abnormal lymph node identified on lymphangiography.

Results

Lymphangiography was carried out in 55 of the 57 patients, 2 patients being excluded because of severe chronic lung disease. A further three patients were not considered for this study because of this finding. Of the 26 newly diagnosed patients, 7 were abnormal, and of the 29 previously treated patients 10 were abnormal. In the untreated group cytological examination yielded positive results for malignant cells in four of seven with abnormal lymphangiograms and also identified a further two patients with normal lymphangiograms. In the treated group, cytological findings were positive in five patients, two with abnormal and three with normal lymphangiograms.

Repeat aspirations have been performed in three patients who had an initial negative cytological result for a node seen to be highly suspicious on lymphangiography. All three patients had negative cytological findings at the second aspiration.

Computed tomography scanning was performed in 44 patients and abnormal results were reported in 7. Aspiration cytology confirmed malignant cells in four of these. In the 11 patients with positive cytological findings, the acid phosphatase level was normal in 8.

Discussion

The lymphatic system is an important route of dissemination of prostatic cancer. Although the incidence of lymph node metastases can be predicted for each stage of the disease [9] or histological pattern of the primary tumour [10], this is not useful when advising treatment for the individual patient. There is evidence that the survival of untreated patients with focal disease is similar to that of age-matched controls [11]. It is therefore important that if no treatment is to be offered, then there should be good evidence that there is no nodal or metastatic spread. Because of the difficulties encountered with conservative methods of diagnosis and the morbidity associated with radical lymph node staging, the use of percutaneous transabdominal lymph node aspiration cytology offers considerable benefits. When positive, a definitive diagnosis of metastatic disease is obtained and staging surgery is unnecessary. When negative, providing the procedure has little or no morbidity, nothing has been lost, and if multiple nodes have been aspirated the patient can possibly be considered as having localised disease. In this study of 57 patients, 57 bilateral and 3 unilateral transperitoneal aspiration procedures were performed with no morbidity. Tenderness at the site of puncture was the only adverse report in this group of patients.

Although several techniques have been used in this study, only the presence of malignant cells in the lymph node aspirate has given unequivocal evidence of lymph node involvement. This was found in only 6 of 26 newly diagnosed or previously untreated patients, but affected the future management of all of them. Only 5 of 30 patients already on treatment had positive cytological findings, suggesting either an effective primary therapy or providing evidence for a high false negative rate of this technique.

This study again confirms the difficulties in interpreting lymphangiograms. The first 40 lymphangiograms were reviewed by two experienced radiologists, and in only 25 did they agree. Of the initial eight positive cytological specimens, lymphangiograms were reported normal in three of these by one radiologist and in six by the other, confirming the value of lymph node aspiration in the diagnosis of lymph node metastases. CT scanning was of most value in identifying lymph nodes not opacified on lymphangiography, and in this situation four of six patients had positive cytological findings. In view of the cost and high proportion of false negatives, CT scanning cannot be recommended as a routine staging procedure.

As a result of our experience with aspiration cytology of lymph nodes with its low morbidity we have gradually altered our management of patients who are found to have an apparently localised low-stage prostatic cancer. Provided that knowledge of the lymph node status will result in a specific management decision, we feel that in such patients it is reasonable to perform lymphangiography and multiple bilateral lymph node aspirations for lymph node staging. When no opacified areas are found, CT scanning is indicated. If cytological findings are negative after multiple aspirations, then further more radical attempts at lymph node staging may be indicated.

Acknowledgements. I thank my colleagues in the Departments of Radiology and Cytology for their invaluable assistance with this work and Miss Anne Cowley for typing the manuscript.

References

1. Paulson DF (1980) The prognostic role of lymphadenectomy in adenocarcinoma of the prostate. Urol Clin North Am 7:615–622
2. Hoekstera T, Schroeder FH (1981) The role of lymphangiography in the staging of prostate cancer. Prostate 2:433–435
3. Golimbu M, Morales P, Al-Askari S, Shulman Y (1981) CT scanning in staging of prostate cancer. Urology 18:305–308
4. Sherwood T, O'Donohue EPN (1981) Lymphograms in prostatic cancer. False positive and false negative assessments in radiology. Br J Radiol 54:15–17
5. Babcock JR, Grayhack JT (1979) Morbidity of pelvic lymphadenectomy. Urology 13:483–486
6. Chisholm GD, Habib FK (1981) Prostatic cancer: experimental and clinical advances. In: Hendry WF (ed) Recent advances in urology/andrology, No 3. Churchill Livingstone, Edinburgh, pp 221–233
7. Khan O, Pearce E, Bowley N, Williams G, Krausz T (1983) Combined bipedal lymphangiography, CT scanning and transabdominal lymph rate aspiration cytology for node staging in carcinoma of the prostate. Br J Urol 55:538–541
8. UICC: Union Internationale Contre le Cancer (1978) TNM. Classification of malignant tumours, 3rd edn. International Union Against Cancer, Geneva
9. Chisholm GD (ed) (1980) Urological malignancy: prostate. In: Tutorials in postgraduate medicine: urology. Heinemann, London, pp 223–246
10. Kramer SA, Spahr J, Brendler CB, Glenn JF, Paulson DF (1980) Experience with Gleason's histopathological grading in prostate cancer. J Urol 124:223–225
11. Byar DP (1977) VACURG studies on prostatic cancer and its treatment. In: Tannenbaum M (ed) Urologic pathology: the prostate. Lea and Febiger, Philadelphia, pp 241–267

Chapter 20

Lymph Node Aspiration Biopsy in Patients with Prostatic Cancer

J. Zornoza

Fine needle aspiration biopsy has been practised for many years and has been used to assess both neoplastic and inflammatory disease processes in a number of organs (e.g. breast, lymph nodes, thyroid, salivary glands and lungs). During the 1960s several Swedish authors presented reports that substantiated the diagnostic possibilities of this method.

Over the last several years radiological localisation of lymph nodes has grown more precise because of the new imaging modalities of ultrasound and computed tomography (CT). Although the greater accuracy in diagnosis allows better determination of the extent of disease, some lesions remain dilemmas. Percutaneous aspiration of lymph nodes for diagnostic or therapeutic purposes is useful in many clinical situations.

Since lymphangiography was introduced by Kinmoth in 1952, this procedure has experienced a varying degree of popularity. At present, lymphangiography is an accepted procedure for the pretreatment evaluation of the extent of involvement in patients with lymphomatous diseases and in staging patients with epithelial tumours (carcinoma of the cervix, ovary, testicle, bladder and prostate).

The importance of assessing possible metastasis to lymph nodes in patients with prostatic cancer has been recently emphasised [1]. Staging pelvic lymphadenectomies have revealed lymph node involvement in approximately 24% of patients with stage A_2 disease, 14% of those with stage B_1, 40% of those with stage B_2 and 50% of those with stage C disease. However, although by staging pelvic lymphadenectomy with careful histological examination one can accurately assess the extent of disease, this procedure is not without its drawbacks. Pelvic lymphadenectomy has been associated with a 20%–34% incidence of morbidity, including principally wound complications but also atelectasis, ileus, sepsis, pulmonary embolus, hepatitis, ureteral injury, renal failure, thrombophlebitis, lymphocele and penile or lower extremity oedema. Oedema is particularly likely to develop when postoperative definitive radiation therapy is delivered to the prostate. Moreover, when nodal disease is not encountered or when the nodal disease is extensive, pelvic lymphadenectomy does not influence the outcome for the patient. Thus, there is no evidence to support a therapeutic role for lymphadenectomy. Obviously any other staging procedure by which one could accurately assess the status of pelvic lymph nodes without subjecting patients to an open surgical procedure and anaesthetic risk would be preferable.

Although lymphangiography has long been employed to assess nodal metastasis, it has not been widely accepted because of concern about a high incidence of false positive and false negative interpretations. Some believe that the bipedal lymphangiogram does not visualise the nodes that primarily drain the prostate, although Merrin and colleagues have demonstrated that the obturator group of nodes is consistently visualised on bipedal lymphangiograms [2]. True, the hypogastric and presacral nodes are not consistently visualised by this procedure, but neither are they usually removed at the time of lymphadenectomy.

Recognition of the replacement of lymph nodes by a metastatic tumour is sometimes a problem, making the interpretation of lymphangiograms difficult. Strict diagnostic criteria should be used to prevent a high incidence of false positive lymphangiograms. However, proper criteria do not always insure accuracy because metastasis is not the only possible cause of a filling defect in a lymph node. Similar changes may be seen in instances of caseous fibrosis, fatty replacement and conglomerate lymph nodes. all these factors lower the clinical value of the examination.

Computed tomography allows only gross morphological evaluation of the nodes and can only define the size of the lymph node, not the internal architecture. However, it is a very precise method of guidance for percutaneous biopsy.

Technique

Once the appropriate radiological examination has been performed to localise the lesion, the patient is placed on the examining table for the procedure. Biopsies are performed on out-patients and inpatients and no fasting is required. The procedure rarely requires premedication. If a patient is anxious, then 10 mg Valium (diazepam) are given intramuscularly. An anterior transabdominal approach is used for almost all biopsies performed.

When image-intensifier fluoroscopy is used, a radiopaque object is positioned on the skin directly over the lesion from which a specimen is to be taken. If a large tumour is present, then the indicator should be located at the edge of the lesion because the centre is frequently necrotic and, although a large amount of aspirated material can be obtained, it usually contains no viable cells.

The skin and subcutaneous tissues down to the peritoneum are anaesthetised locally with 2% Xylocaine (lignocaine). A small incision with a surgical blade and separation of the tissues with a haemostat facilitate the entry of the needle. At this point in the procedure it is very important to maintain the needle on a straight path. A minimal deviation from the desired course will ultimately misplace the needle, which cannot be realigned by bending the hub portion. In such circumstances it is necessary to withdraw the needle completely and readvance it. The position of the needle should be checked frequently under fluoroscopy to be sure of the desired vertical course. If a 15- or 20-cm needle is used, then it is recommended that one hand advance the needle while the other stabilises it at the skin. This should be done in small increments and, after each advance, fluoroscopy should be performed. Manual control of the needle is preferable to the use of rubber-tipped haemostats in maintaining stability (Fig. 20.1).

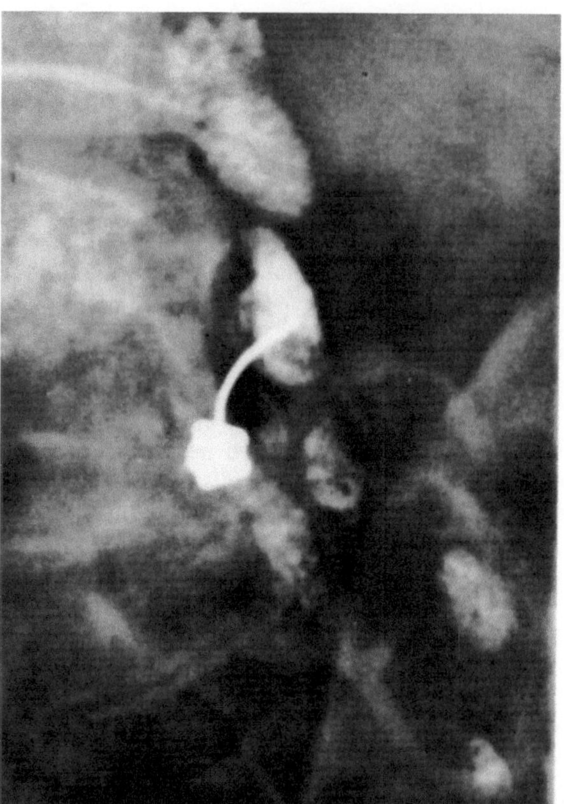
a

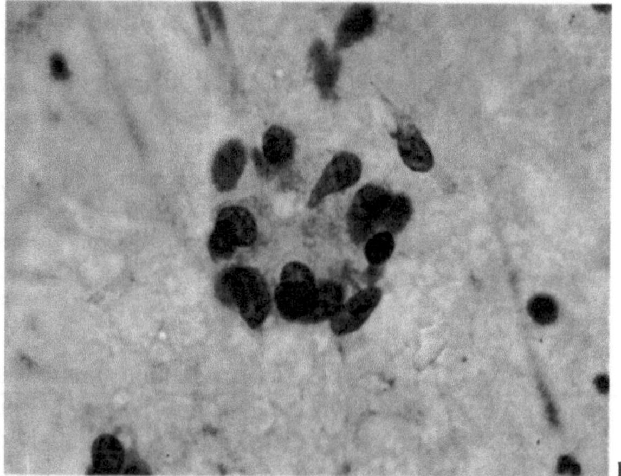
b

Fig. 20.1 a Anteroposterior radiograph of the low lumbar region shows a needle in a left common iliac lymph node. **b** Metastatic adenocarcinoma. (Papanicolaou stain, × 73)

Once the abdominal cavity has been entered, no resistance is met until the tip of the needle reaches the posterior parietal peritoneum. At this time, the patient may experience pain, which can be alleviated by the injection of a small amount of Xylocaine (lignocaine). Small-gauge needles (22–23 gauge) tend to bend at this point if they are advanced with a quick motion. Slight, continuous pressure allows an easier passage of the needle through the anterior retroperitoneal fascia.

The introduction of the needle is time consuming. Because it is difficult for a patient to achieve a consistent degree of respiration and any abrupt respiratory movement is undesirable while the needle is in the abdominal cavity, the patient is allowed to breathe normally during the procedure. Often, an increased resistance to the needle indicated that the capsule of the lymph node has been reached. The desired point from which the specimen is to be taken can also be verified by oblique projections and biplane fluoroscopy when the biopsy is performed under fluoroscopy, and by CT scan when the biopsy is performed under computed tomography.

When the biopsy is performed under fluoroscopy, the patient is rotated slightly to both sides to determine the relative depth of the needle (Fig. 20.2). When the node is punctured, it will move in concert with the needle tip, indicating an accurate placement. Because the lesions are usually small (1–2 cm), the needle should be moved very gently to detach material. Large excursions would produce an undesirable amount of blood and extranodal

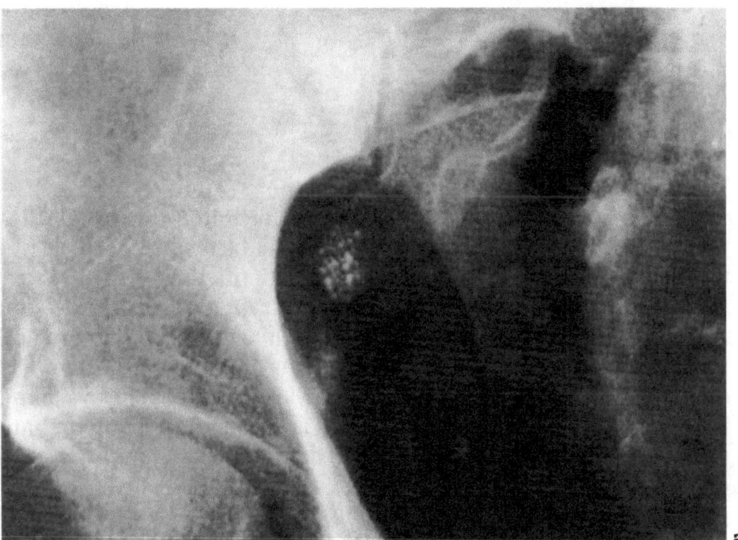

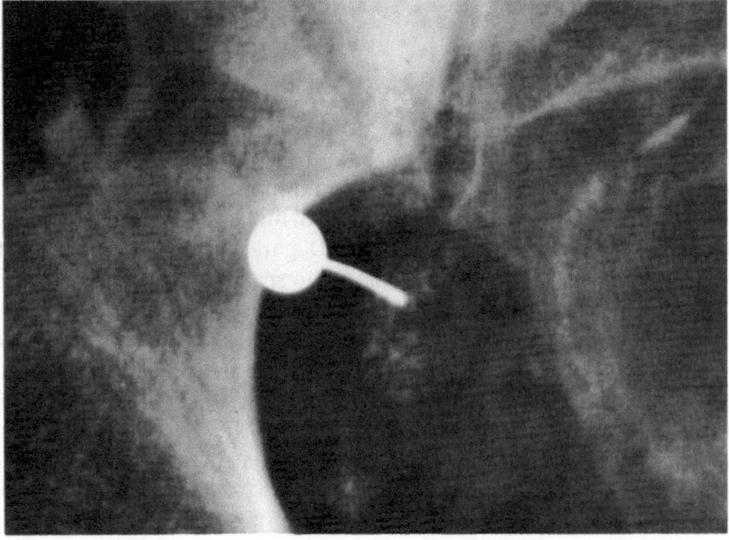

Fig. 20.2a,b A 64-year-old man with the diagnosis of adenocarcinoma of the prostate was treated with radiation therapy. Two years later, a follow-up radiograph of the pelvis revealed enlargement of an external iliac node. **a** Normal node at the time of staging (1976); **b** 2 years later enlargement of the lymph node is seen. Metastatic adenocarcinoma.

tissue. The presence of oil droplets in the aspirate demonstrates that the node previously opacified during lymphangiography has been punctured.

At the time of biopsy, the needle is rotated clockwise and counter-clockwise around its longitudinal axis. This manoeuvre detaches cellular material near the needle point. After the needle has been rotated a few times, the stylet is removed and a disposable 12-ml plastic syringe is attached to the hub. Suction is applied while the needle is gently moved up and down several times with approximately a 1-cm excursion. Suction is relieved when the aspiration has been completed and the pressure in the syringe is allowed to equalise before the needle is withdrawn from the lesion. This manoeuvre retains the tissue in the needle, from which it can be ejected easily onto glass slides or into a container of preservative solution [3].

Materials and Methods

Thirty-five patients with carcinoma of the prostate whose lymphangiogram findings were "definitely positive" or appeared "suspicious" were subjected to fine needle aspiration biopsy. In 20 patients the initial interpretation of the lymphangiogram was definitely positive, while in 15 patients the studies were considered abnormal or suspicious but not diagnostic of metastasis. The criterion for a definitely positive lymphangiogram finding is steadfastly maintained to be a concentric filling defect present on the nodal phase that is not traversed by lymphatics on the flow phase.

Results

In 10 of the 20 patients (50%) with definitely positive lymphangiogram findings, percutaneous fine needle aspiration and cytological examination of the specimens confirmed the diagnosis of metastatic focus of adenocarcinoma. Of the 10 patients in whom metastasis was unconfirmed, the specimens from 3 were considered inadequate for diagnosis. Of these 10 patients 5 subsequently underwent open node biopsy, which determined that 4 were histologically positive and 1 histologically negative for metastatic carcinoma. Therefore, for 15 of the 20 patients who had definitely positive lymphangiogram findings, histological confirmation was available; 14 of the 15 were truly positive, yielding a true-positive rate of 93%. The percutaneous needle biopsy spared 50% of these 20 patients an open procedure.

Of the 15 patients with suspicious lymphangiograms, fine needle aspiration biopsy findings were positive in 9 and negative in 6. None of the six patients with negative biopsy findings had subsequent open lymph node biopsy. In this group of 15 patients with suspicious lymphangiograms, therefore, 60% were spared further surgical intervention, since fine needle aspiration biopsy confirmed the diagnosis of metastatic carcinoma.

Overall, in 19 of the 35 patients (54%) who underwent aspiration biopsy, the lymphangiographical diagnosis was cytologically confirmed; these 19 patients were spared any further interventional assessment.

No complications were encountered in this group of patients except for mild to moderate pain or discomfort at the time the biopsy was performed. This was easily controlled by local anaesthetic. All procedures were performed without hospitalisation.

Discussion

Accurate diagnosis of metastatic deposits by bipedal lymphangiography depends on adherence to a strict criterion. Frequently, however, the essential requisite of a filling defect in the lymph node not traversed by lymphatics is absent, although other abnormalities suggest the possibility of metastatic disease. These include enlarged nodes, distorted or obstructed lymphatics, collateralised lymphatic channels and lymphatic-venous communications. Under these circumstances, further efforts to confirm the presence or absence of metastatic disease are warranted. Further evaluation is important because nodal metastases may be equated with systemic disease, which renders attempts at local control ineffective.

In order to spare patients the morbidity of pelvic lymphadenectomy, other methods for diagnosing lymph node metastasis are desirable. Percutaneous needle aspiration can be a valuable adjunct to lymphangiography and can spare over 50% of the patients open pelvic surgery for histological confirmation. Both the techniques and safety of this biopsy procedure for pelvic and para-aortic nodes are well established [4]. Furthermore, as more sophisticated, non-invasive radiological tools such as CT are used increasingly, percutaneous fine needle aspiration biopsy can also be employed as an adjunct to these modalities.

References

1. Johnson DE, von Eschenbach AC (1981) Roles of lymphangiography and pelvic lymphadenectomy in staging prostatic cancer. Urology 17:66–81
2. Merrin C, Wajsman Z, Baumgartner G, Jennings E (1977) The clinical value of lymphangiography: are the nodes surrounding the obturator nerve visualized? J Urol 117:762–764
3. Zornoza J, Wallace S, Goldstein HM, Lukeman M, Jing BS (1977) Transperitoneal percutaneous retroperitoneal lymph node aspiration biopsy. Radiology 122:111–115
4. Zornoza J (ed) (1981) Abdomen. In: Percutaneous needle biopsy. Williams and Wilkins, Baltimore, pp 102–140

Chapter 21

Combined Lymph Node Aspiration Cytology and Tumour Grade in Staging of Prostatic Cancer

L. Luciani, P. Scappini, G. L. Failoni and F. Piscioli

Introduction

The prognosis and treatment of prostatic cancer are strictly related to the status of the draining nodes. Imaging methods commonly employed in the study of the pelvic nodes do not allow exact knowledge of the presence or absence of metastatic deposits and therefore false positive and false negative results are often given. Histological examination of the nodes should be the procedure of choice, but the role of staging pelvic lymphadenectomy in the management of prostatic cancer is still controversial [1, 2].

Aspiration biopsy cytology (ABC) has been proposed as an alternative to surgical staging in the attempt to avoid morbidity and mortality in those patients with disease-free nodes. Negative cytological findings cannot be considered diagnostic because of the possibility of micrometastasis or errors in sampling. Combined ABC and grading systems can contribute useful information in evaluating the metastatic potential of the disease.

In this chapter we evaluate the reliability of three grading systems in predicting the presence of nodal metastases and compare the results of grading systems with those of cytological and histological examinations in the staging of patients with prostatic cancer.

Materials and Methods

A total of 91 cases of biopsy-proven prostatic adenocarcinoma, which had undergone staging lymphadenectomy, were classified according to the Gaeta [3, 4], Gleason [5–7] and the M. D. Anderson Hospital (MDAH) [8] grading systems. All patients were evaluated to exclude the presence of metastasis other than those eventually involving the lymph nodes. Of these patients, 55 also had ABC of the nodes before surgery. The needle used was a modified Chiba needle[1] of 22–24 gauge in diameter and 18–21 cm in length.

[1] Cyto-Aspir (patent pending). Cook Urological Inc., Spencer, Ind., USA.

Results

No nodal metastases were found in 21 patients with a Gleason score of 2–4 (Table 21.1). By contrast, nodal involvement was present in all patient groups graded according to the Gaeta and MDAH systems in percentages varying from 14%–47% for grades I and II to 48%–66% for grades III and IV respectively (Tables 21.2, 21.3).

Table 21.1. Nodal metastases and patients' distribution according to the Gleason grading system

Grade	Total patients	Patients with N+	%
2	4	0	–
3	5	0	–
4	12	0	–
5	7	1	14
6	12	2	16
7	16	7	44
8	14	12	86
9	9	6	66
10	12	10	83
Total	91	37	40

N+, nodal metastases.

Table 21.2. Nodal metastases and patients' distribution according to the MDAH grading system [8]

Grade	Total patients	Patients with N+	%
I	28	5	18
II	32	15	47
III	13	8	61
IV	18	11	61
Total	91	37	40

N+, nodal metastases.

Table 21.3. Nodal metastases and patients' distribution according to the Gaeta grading system [3, 4]

Grade	Total patients	Patients with N+	%
I	11	2	18
II	28	4	14
III	29	14	48
IV	23	17	74
Total	91	37	40

Complete agreement was found between histological and cytological findings in patients with a Gleason sum <5. There were 4 false negative results in the remaining cases (Table 21.4). As regards the other grading systems, there was no significant relationship between patients' stratification and histological and cytological findings. No false positive cytological findings were found.

Discussion

Grading systems have been elaborated in an attempt to predict the biological and clinical behaviour of prostatic carcinoma. Gleason's grading system has been tested the most and has shown such good correlation between the grade of the tumour and the incidence of lymph node metastases found in staging lymphadenectomies that several authors [9, 10, 11] have even suggested avoiding lymphadenectomy in patients with the highest and lowest histological scores.

Sagalowsky et al. [12], however, although they did not identify nodal metastases in tumour grades 2–4, believe that Gleason grading alone does not always reflect the status of the pelvic lymph nodes and, therefore, pathological staging should always be performed before a radical prostatectomy for cure.

Even Zincke et al. [13] reported that 6%–35% of 68 patients with pelvic nodal disease had a Gleason sum of 4 or $\leqslant 5$ respectively. Nevertheless questions were put by Paulson [14] about the criteria on which this study was based, especially as regards the determination of the Gleason sum on the entire surgical prostatic specimen, since in such cases grades should be increased by 1 or 2 numbers to be comparable with those obtained by needle biopsy or by review of excised prostatic chips.

Correlation between grade of differentiation and tumour progression has been proposed by Parfitt et al. [15], who reported the presence of diffuse tumour on repeat transurethral resection in 4 out of 16 patients with previous A_1 prostatic carcinoma. However, the number of patients is too small to draw significant conclusions on this matter.

Guinan et al. [16] evaluated the progression of the disease during a follow-up period of 6 months to 8 years in 111 patients with prostatic carcinoma who had undergone radical surgery and had been classified according to the Gleason and Broders [17] grading systems and to the Whitmore [18] staging procedure.

Of 72 patients with no progression of the disease 44% had a Gleason score of $\leqslant 5$ and of 39 patients with disease progression 87% had a sum $\geqslant 6$.

These results have been questioned by Byar [19], who stressed the importance of more homogeneous

Table 21.4. Correlation between histological and aspiration cytological findings in the lymph nodes and tumour grade

	Grade	Total patients	No. of patients with histologically positive nodes	No. of patients with cytologically positive nodes
Gleason grading system	2	3	—	—
	3	3	—	—
	4	5	—	—
	5	4	—	—
	6	8	2	1
	7	9	4	4
	8	9	8	7
	9	7	5	5
	10	7	6	4
Gaeta grading system	I	5	1	1
	II	17	3	2
	III	17	9	8
	IV	16	12	10
MDAH grading system	I	18	4	2
	II	15	7	7
	III	11	7	6
	IV	11	7	6
Total		55	25	21

follow-up periods and the possibility of obtaining higher accuracy with the Gleason system because of the intermediate grades 2 and 6 and 7 and 10. Nonetheless, the reliability of the Gleason sum could be damaged by this continuous changing in the intermediate grades.

Thomas et al. [20] compared the results of the Gaeta and Gleason grading systems in 130 patients with prostatic cancer. They found that 8 out of 57 patients with Gaeta's grade II tumours had stage C or D disease, but that none of the 18 patients with a Gleason sum $\leqslant$ 5 had stage C or D cancer, demonstrating a better correlation between the stage and the grade of the disease.

By contrast, Gaeta et al. [2] reviewed 169 cases of prostatic adenocarcinoma and found that the deaths per year of follow-up varied from 0.036 for grade I to 0.349 for grade IV patients and that tumour stage was worse the higher the tumour grade. This might mean that there had been a deterioration of the histological type of the primary neoplasm or the metastatic deposits with the progress of the disease, but such an event has never been clearly demonstrated, so that we may assume that a low-grade tumour is related to a less malignant disease.

In a group of patients classified according to both the Gaeta and Gleason grading systems 28% of 32 patients with stage D disease were grade I or II (Gaeta) but 34% of the same patients had a pattern score of 4 or lower [4].

As regards other grading systems, Saltzstein et al. [21] reported 22% of nodal involvement in 18 patients with grade I anaplasia according to the Mostofi and Price histological classification and stage B_1 and B_2 or C prostatic adenocarcinoma.

Brawn et al. [8] compared the predictive value of the MDAH grading system with those of Mostofi [22] and Gleason in determining the survival of 182 patients with stage C prostatic carcinoma and found that the MDAH grading system was the most successful in determining the biological behaviour of the tumour as regards the patient's life expectancy.

Myers et al. [23] graded the nuclear features of the cancer cells into three main categories: "not prominent" 1, "intermediate" 2 and "prominent" 3. They associated this nuclear grading to each Gleason pattern and then proceeded to sum up, obtaining two numbers: the Gleason score and the nuclear sum. These authors found that 9 out of 13 patients with stage C prostatic carcinoma had "prominent" or "intermediate" nucleoli in the primary Gleason pattern and that the mean interval to progression was shorter in this group than in the 4 patients whose nucleoli were judged to be "not prominent". However, these findings need corroboration from other studies.

The nuclear roundness factor (NRF) was elaborated by Epstein et al. [24] with regard to how closely the nuclear feature of the cancer cells in a histological section approximated to a perfect circle. This parameter seems to be a more accurate predictor of the metastatic potential of prostatic cancer than conventional grading systems. Major limitations to this method are that it cannot be applied to needle biopsy and it is adversely affected by the presence of multiple areas of malignancy which may not be accurately sampled. However, only series with few patients have been reported in the literature, and this method also needs to be tested further.

From a review of the literature it becomes evident that the reliability of prostatic grading systems are strictly related to the ability of the testing pathologists. The overall impression is that the best results were obtained in the institutions where these systems were elaborated.

In fact, one of the major impediments to the diffusion of the grading systems has been their variable reproducibility [25, 26]. In this connection the Gleason's score has proved to be highly accurate in different series from many authors. Even in our experience the distribution pattern of the patients according to the Gleason grading system has shown the possibility of distinguishing patients with a low or absent risk of nodal metastases from those with a high incidence of metastatic disease. Nevertheless, to ensure adequate clinical evaluation of patients with prostatic cancer, the results of grading systems need to be interpreted and integrated together with those of other diagnostic methods.

Post-lymphographic pelvic node aspiration cytology has been used as a means of improving the reliability of lymphography in staging prostatic carcinoma. Excluding the case of a positive aspiration cytological finding, which is always conclusive of a stage D disease, the remaining findings are inconsistent as regards their diagnostic accuracy and are of no clinical use, requiring histological verification. The combination of aspiration cytology and the Gleason grading system can constitute a further improvement in the staging of clinically localised prostatic carcinoma. This possibility has already been proposed, but data from specific series have been reported in very few studies [27–30].

A negative nodal aspiration finding can have clinical significance if the Gleason score is lower than 5. In our experience, not one of the 21 patients with a Gleason sum $\leqslant 4$ had nodal metastases on histological control; therefore, we may assume that in such cases negative cytological findings are truly negative and that no surgical staging procedures are required.

Patients with a Gleason grade higher than 7 should be considered at greater risk of nodal metastases; however, negative aspiration results can never be accepted as conclusive, and nodal dissection remains mandatory.

In the case of intermediate categories, i.e. from 5 to 7, a negative aspiration finding may be considered diagnostic when representative material is obtained from most of the regional nodes. However, considering that such an accurate procedure is not easy to accomplish, these patients should undergo surgical staging for treatment selection.

In conclusion, the combined use of aspiration cytology and the Gleason grading system can constitute an effective method in staging prostatic carcinoma, permitting selection of patients who could be spared an unnecessary lymphadenectomy. The possibility of false negative findings in the group of patients at lower risk of metastatic diffusion should always be kept in mind, but such an event may become of secondary importance if one takes into account the complications of the node dissection and also the possible failure of accurate serial histological examination in detecting microscopic nodal metastases.

References

1. Olsson CA (1985) Staging lymphadenectomy should be an antecedent to treatment in localized prostatic cancer. Urology 25 (Suppl):4–6
2. Paulson DF (1985) Staging lymphadenectomy should not be an antecedent to treatment in localized prostatic cancer. Urology 25 (Suppl):7–14
3. Gaeta JF, Asirwathman JE, Miller G, Murphy GP (1980) Histologic grading of primary prostatic cancer: a new approach to an old problem. J Urol 123:689–693
4. Gaeta JF (1981) Glandular profiles and cellular patterns in prostatic cancer grading. National Prostatic Cancer Project System. Urology 17 (Suppl):33–37
5. Gleason DF (1966) Classification of prostatic carcinomas. Cancer Chemother Rep 50:125–128
6. Gleason DF, Mellinger GT (1974) Veterans Administration Cooperative Urologic Research Group: prediction of prognosis for prostatic adenocarcinoma by combined histologic grading and clinical staging. J Urol 111:58–64
7. Gleason DF (1977) Histologic grading and clinical staging of prostatic carcinoma. In: Tannenbaum M (ed) Urologic pathology: the prostate. Lea and Febiger, Philadelphia, pp 171–197
8. Brawn PN, Ayala AG, Von Eschenbach AC, Hussey DM, Johnson DE (1982) Histologic grading study of prostate adenocarcinoma: the development of a new system and comparison to other methods. A preliminary study. Cancer 49:525–532
9. Barzell W, Bean MA, Hilaris BS, Whitmore WF Jr (1977) Prostatic adenocarcinoma: relationship of grade and local extent to the pattern of metastases. J Urol 118:278–282
10. Kramer SA, Spahr J, Brendler CB, Glenn JF, Paulson DF (1980) Experience with Gleason's histopathologic grading in prostatic cancer. J Urol 124:223–225
11. Paulson DF, Piserchia PV, Gardner W (1980) Predictors of lymphatic spread in prostatic adenocarcinoma: Uro-Oncology Research Group Study. J Urol 123:697–699

12. Sagalowsky AI, Milam H, Revely LR, Silva FG (1982) Prediction of lymphatic metastases by Gleason histologic grading in prostatic cancer. J Urol 128:951–952
13. Zincke H, Farrow GM, Myers RP, Benson RC Jr, Furlow WL, Utz DC (1982) Relationship between grade and stage of adenocarcinoma of the prostate and regional pelvic lymph node metastases. J Urol 128:498–501
14. Paulson DF (1982) Editorial comment. J Urol 128:501
15. Parfitt HE Jr, Smith JA Jr, Gliedman JB, Middleton RG (1983) Accuracy of staging A_1 carcinoma of the prostate. Cancer 51:2346–2350
16. Guinan P, Talluri K, Nagubadi S, Sharifi R, Ray V, Shaw M (1983) Evaluation of Gleason classification system in prostate cancer. Urology 21:458–460
17. Broders AC (1922) Epithelioma of the genito-urinary organs. Ann Surg 75:570–575
18. Whitmore WF (1973) The natural history of prostatic cancer. Cancer 32:1104–1112
19. Byar DP (1983) Grading prostate cancer. Urology 22: 562 (letter)
20. Thomas R, Lewis RW, Sarma DP, Coker GB, Rao MK, Roberts JA (1982) Aid to accurate clinical staging. Histopathologic grading in prostatic cancer. J Urol 128:726–728
21. Saltzstein SL, McLaughlin AP (1977) Clinico-pathologic features of unsuspected regional node metastases in prostatic adenocarcinoma. Cancer 40:1212–1215
22. Mostofi FL, Price EB Jr (1973) Tumors of the male genital system. In: Atlas of tumor pathology. Second Series, Fascicle 8. Armed Forces Institute of Pathology, Washington DC, 218–223
23. Myers RP, Neves RJ, Farrow GM, Utz DC (1982) Nuclear grading of prostatic adenocarcinoma: light microscopic correlation with disease progression. Prostate 3:423–432
24. Epstein JJ, Berry SJ, Eggleston JC (1984) Nuclear roundness factor: a predictor of progression in untreated stage A2 prostate cancer. Cancer 54:666–671
25. ten Kate FJW, Gallee MPW, Schmitz PIM, Joebsis AC, van der Heul RO, Prins MEF, Blom JHM (1986) Problems in grading of prostatic carcinoma: interobserver reproducibility of five different grading systems. World J Urol 4:147–152
26. Rousselet MC, Saint-André JP, Six P, Soret JY (1986) Reproducibilité et valeur pronostique des grades histologiques de Gleason et de Gaeta dans les carcinomes de la prostate. Ann Urol 20:317–332
27. Wajsman Z (1981) Lymph node evolution in prostatic cancer. Is pelvic lymph node dissection necessary? Urology 17 (Suppl):80–82
28. Piscioli F, Leonardi E, Reich A, Luciani L (1984) Percutaneous lymph node aspiration biopsy and tumor grade in staging of prostatic carcinoma. Prostate 5:459–468
29. Luciani L, Scappini P, Pusiol T, Piscioli F (1985) Comparative study of lymphography and aspiration cytology in the staging of prostatic carcinoma. Urol Int 40:181–189
30. Piscioli F, Scappini P, Luciani L (1985) Aspiration cytology in the staging of urologic cancer. Cancer 56:1173–1180

Chapter 22

Cytological Findings in Conservative Treatment of Prostatic Cancer

F. Pagano, G. Costantin and F. Zattoni

Treatment response in the case of prostatic cancer may sometimes be difficult to assess. It should be emphasised that response criteria must be objective, quickly carried out, easily repeatable and with a favourable cost-benefit ratio. On the other hand, a precise evaluation of cancer patients on conservative treatment is well-recognised to be important in order to decide in good time whether current therapy should be continued or modified.

Several authors [1–3] suggested the assessment of morphological changes in tumour cells as a valid procedure in the follow-up of patients with prostatic cancer. From this point of view, cytological examination by aspiration biopsy appears to be ideal because it is highly reliable [4] and at the same time a technically simple method causing the patient only slight trauma.

Modifications induced by conservative treatment in prostatic tumour cells are well described. Hormone therapy produces characteristic changes of cell nucleus and/or cytoplasm: the nucleus may present pyknosis and a very coarse granular or irregularly distributed chromatin; the nucleolus may be difficult to recognise and cytoplasm is often vacuolised. Squamous cell metaplasia, as a result of hormone therapy, is often found in cytological smears. Radiation-treated carcinoma cells show highly nuclear polymorphism and a great variation in size. (See Figs. 22.1–22.5 for examples.)

Good correlation was achieved by comparing the clinical course of disease, cytological changes induced by conservative therapy and biological behaviour of prostatic carcinoma [5–8].

In our institute, aspiration biopsy of the prostate gland is a reliable tool for diagnosis of prostatic cancer and an aid to assessing the tumour response to therapy. The following work deals with the problems of aspiration biopsy of conservatively treated carcinomas of the prostate with respect to morphological parameters and clinical implications.

Material and methods

Between 1978 and 1983 we examined 127 consecutive patients treated for prostatic cancer. They were followed-up over a period of 12–72 months (Tables 22.1–22.3). Cytological control biopsies were performed according to Franzen's technique in all the patients at intervals of 3–12 months after the beginning of therapy. At the time of aspiration biopsy, multiple specimens were taken from different parts of the gland in order to achieve a "map" of the prostate. Specimens were judged unsatisfactory for cytological examination in 5.4% of the cases. Cytological response to treatment was defined

Table 22.1. Follow-up of 127 patients conservatively treated for prostatic cancer

No. of aspiration biopsies performed	336 (2.64 times/patient) (5.4% unsatisfactory specimens)
Age at study entry	Mean age 67 years Range 37–83 years 9% younger than 55 years

Table 22.2. Treatment of 127 patients with prostatic cancer

Oestrogens	50
Orchiectomy and/or anti-androgens	19
Oestramustine phosphate	13
Radiation therapy	23
Various treatments	14
No treatment	8
Total	127

Table 22.3. Study entry and years of follow-up of 127 patients conservatively treated for prostatic carcinoma

Year	Total per year	No. of patients		
		Alive	Dead	% Dead
6	12	12	—	—
5	22	10	2	9.0
4	39	17	5(1)[a]	12.8
3	53	14	10	18.8
2	73	20(1)[b]	7	9.5
1	98	25(2)[b]	5(1)[a]	5.1
	Total	98	29	

[a] Death due to other causes.
[b] Underwent radical prostatectomy.

according to the presence of cytomorphological regression and recognisable cancer cells.

Results

In Table 22.3 the time distribution of the follow-up of 127 patients is presented. Out of seven untreated patients only one had clinical progression 3 years after cancer diagnosis. In Tables 22.4 and 22.5 correlation between cytological findings and clinical response in 120 conservatively treated patients is shown. Cytological findings showing little or no regression under therapy or showing cancer dedifferentiation are found in 51 out of 59 patients with clinical disease progression (86.4%). Cytological dedifferentiation (Table 22.6) was discovered in 11 patients followed from 2 to 5 years. Up to the time of writing, 29 patients have died, 27 as the result of cancer progression. Frequency of death with regard to differentiation grading is higher for poorly differentiated carcinomas unconnected with treatment modalities. A higher death incidence resulted in the 3rd year of follow-up (18.8%).

Table 22.4. Correlation between cytological findings and clinical response in 97 patients after contrasexual therapy

	No. of patients	%	
Clinical stability/cytological regression	43	44.4	
Clinical progression/cytological regression	6	6.2	
Clinical progression/no cytological regression	34	35.1	40.2
Clinical progression/cytological dedifferentiation	5	5.1	
Clinical stability/no cytological regression	8	8.2	9.2
Clinical stability/cytological dedifferentiation	1	1.0	
Correlation		84.53%	

Table 22.5. Correlation between cytological findings and clinical response in 23 patients with radiotherapy

	No. of patients	%	
Clinical stability/cytological regression	6	26.1	
Clinical progression/cytological regression	2	8.7	
Clinical progression/no cytological regression	8	34.8	52.2
Clinical progression/cytological dedifferentiation	4	17.4	
Clinical stability/no cytological regression	2	8.7	13.0
Clinical stability/cytological dedifferentiation	1	4.3	
Correlation		78.2%	

Table 22.6. Cytological dedifferentiation grades versus clinical response to treatment in 11 patients with prostatic cancer

	Clinical stability		Clinical progression	
	G1–G2	G2–G3	G1–G2	G2–G3
Radiotherapy	1	—	2	2
Contrasexual therapy	1	—	3	2

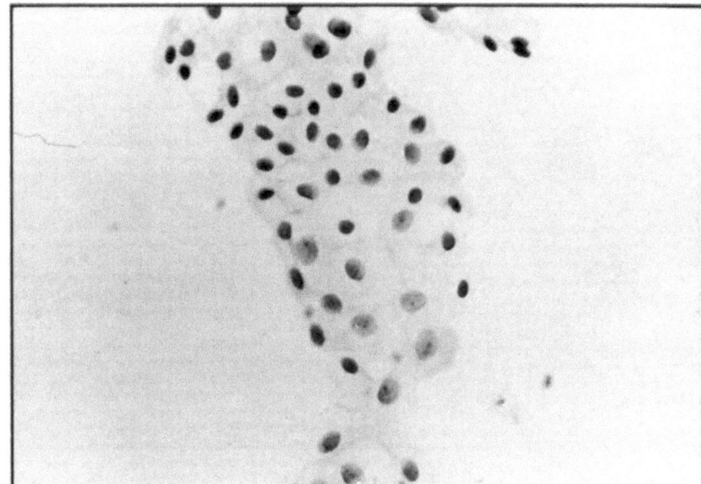

Fig. 22.1. Change of prostatic grandular epithelium into metaplastic squamous epithelium. (Papanicolaou stain, × 30)

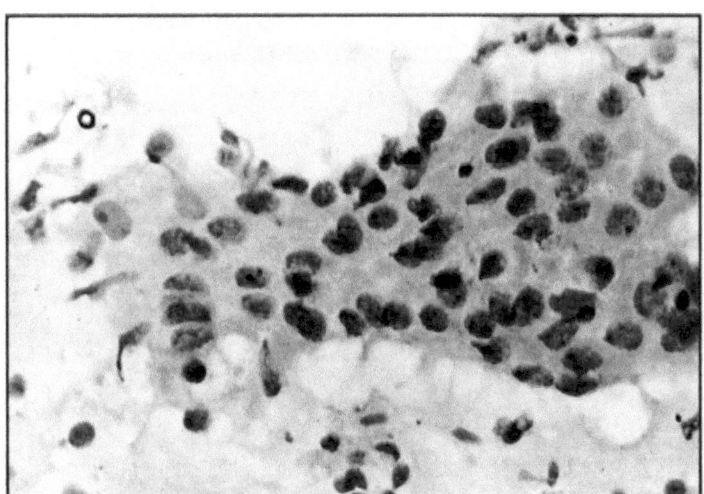

Fig. 22.2. Moderately differentiated prostatic carcinoma before therapy. (Papanicolaou stain, × 30)

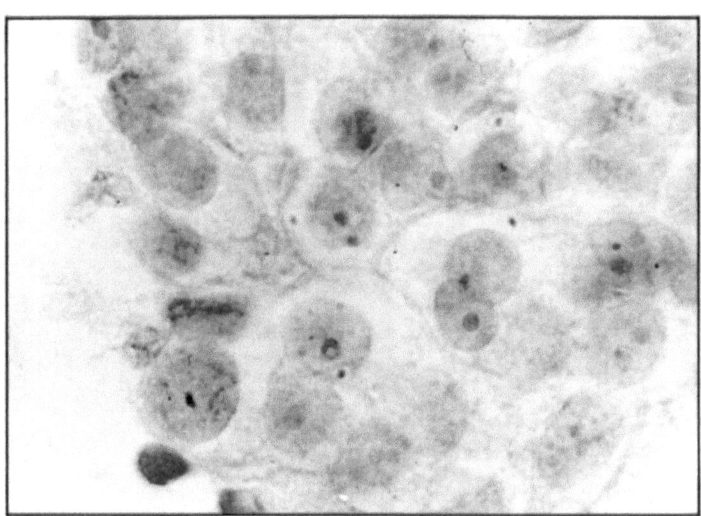

Fig. 22.3. Moderately differentiated prostatic carcinoma after hormone therapy lasting 8 months: cytoplasmatic vacuolisation and wasting of nucleoli. (Papanicolaou stain, × 66)

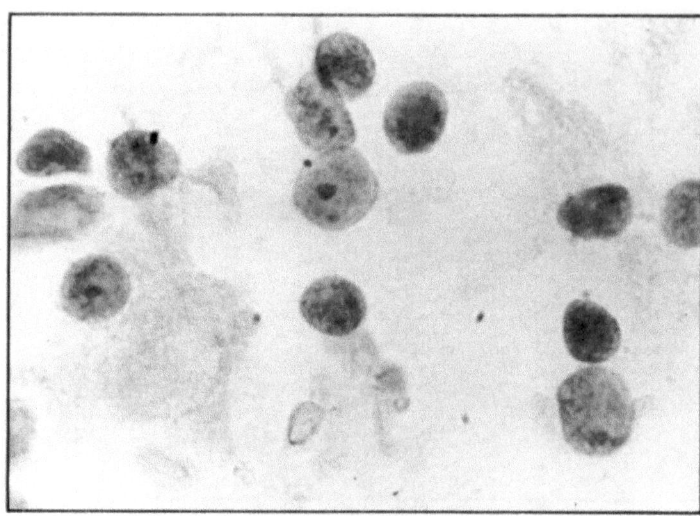

Fig. 22.4. Poorly differentiated prostatic carcinoma before radiotherapy. (Papanicolaou stain, × 66)

a

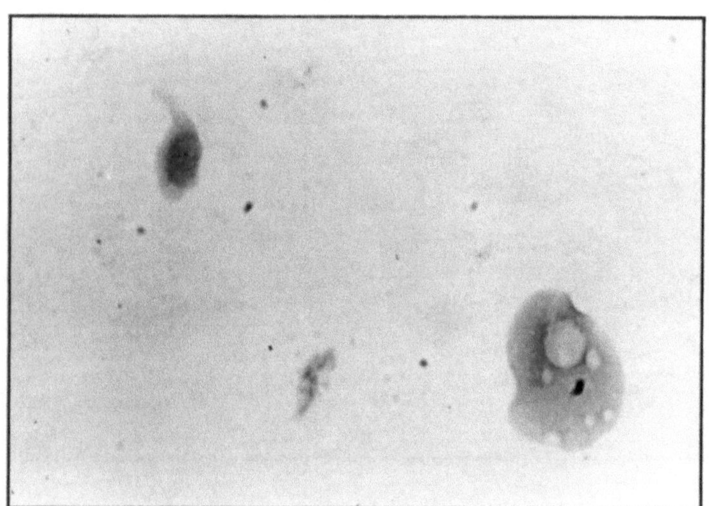

b

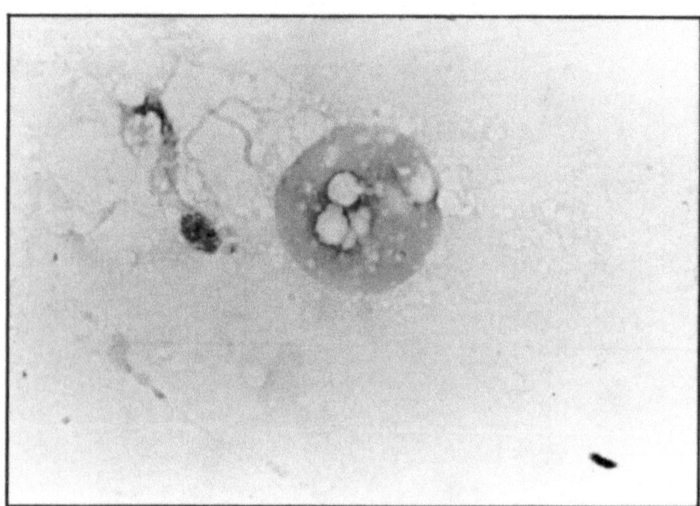

Fig. 22.5a, b. Poorly differentiated prostatic carcinoma. Same case as in Fig. 22.4, 4 months after radiotherapy; marked nuclear regression. (Papanicolaou stain, × 66).

Discussion

Some interesting data can be pointed out from our experience. As far as radiation-treated patients are concerned, we agree that no valid conclusions can be made on the basis of rectal findings because of the great number of false positive and false negative results [9]. Even the evaluation of therapy response by means of morphological data is still open to question [10, 11].

Correlation between cytological findings and clinical events (78.2% of the cases in this study) proved that aspiration biopsy can provide useful information on tumour behaviour. Patients who have positive cytological findings after treatment often demonstrate clinical tumour progression too [12]. The time of eventual residual tumour disappearance is controversial. Some authors stress that the presence of neoplastic cells within 12–18 months after radiotherapy does not necessarily mean disease progression [13]. Persistence of neoplastic cells after that is universally considered an indication of residual tumour or relapse [12].

Patients with positive biopsy findings after radiation therapy have more chance of disease progression than patients with negative findings [8, 9, 12]. Cytological findings can adequately reflect the prognosis of disease. Data confirmed that negative findings do not indicate absence of residual tumour, but a cytologically positive response must be evaluated as a failure of treatment and the current therapy should be changed.

In 49% of biopsies from radio-treated patients Kiesling reported changes in tumour grading towards more anaplastic lesions in comparison with the pretreatment findings [14]. Perhaps well-differentiated cells are more susceptible to radiation effects and consequently disappear. This does not happen to poorly differentiated cells, which persist because of the greater radioresistance.

Dedifferentiation of prostatic carcinoma is an unfavourable sign—disease progression was observed in 80% of such cases. Since fibrosis is quite prominent after radiation, samples for cytological examination obtained by aspiration biopsy can sometimes be unsatisfactory [2].

"Empty" smears can indirectly signify excellent sensitivity to the treatment. Regression was the same in all hormone treatments of carcinomas: oestrogens, orchiectomy and/or antiandrogens, and oestramustine phosphate therapy produced the same cellular degeneration. Cytological regression was observed 4 months after the beginning of the therapy [1]. In the present study significant cell modifications appeared only 6–8 months later. Comparing morphological regression and clinical follow-up, our results provide a more favourable correlation than those of other authors [8–15].

Conclusion

Evaluation of cancer treatment has to be based on the most objective possible criteria. With this aim, cytological findings must be successfully combined with other traditional parameters. In this way the final evaluation of the treatment results is even supported by morphological data which are able to visualise cell modifications during treatment.

References

1. Esposti PL (1971) Cytologic malignancy grading of prostatic carcinoma by transrectal aspiration biopsy. Scand J Urol Nephrol 5:199–209
2. Spieler P, Gloor F, Egle N, Bandhauer K (1976) Cytological findings in transrectal aspiration biopsy on hormone- and radio-treated carcinoma of the prostate. Virchows Arch [Pathol Anat] 372:149–159
3. Faul P (1979) Diagnostische Möglichkeiten der Prostata-Zytologie unter besonderer Berücksichtigung des Einflussess des zytologischen Differenzierungsgrades beim Prostatakarzinom auf die 5-Jahres-Überlebenszeit. Arch Geschwulstforsch 49(1):18–28
4. Zattoni F, Pagano F, Rebuffi A, Costantin G (1983) Transrectal thin-needle aspiration biopsy of prostate: four years experience. Urology 22:69–72
5. Leistenschneider W, Nagel R (1980) Zytologisches Regressions grading und seine prognostiche Bedeutung beim konservativ behandelten Prostata-Karzinom. Aktuel Urol 11:263–275
6. Leistenschneider W, Nagel R (1984) Control of response to extramustine phosphate therapy through cytology and DNA analysis of cell nuclei in prospective study. Eur Urol 23(6):81–88
7. Dohm G, Honbach C (1984) Pathology and classification of prostate malignancies: experience of the German Prostate Cancer Registry. In: Jacobi GH, Hoenfellner R (eds) Prostate cancer. International perspectives in urology, vol 3. Williams and Wilkins, Baltimore, pp 95–113
8. Bandhauer K (1983) Value of cytology for follow-up of prostatic cancer after contrasexual and irradiation therapy. Prostate 4 (4):428
9. Frankel JM, Scardino PT, Carlton EC (1983) The accurate predictive value of the late prostate exam and biopsy in prostatic cancer treated with radiotherapy. Abstract 299, AUA Meeting, Las Vegas, 17–21 April, 1983
10. Cupps RE, Utz DC, Fleming TR, Carson Culley C, Zincke H, Myerns RP (1980) Definitive radiation therapy for prostatic carcinoma: Mayo Clinic Experience. J Urol 124:855–859
11. Cox JD, Kline RW (1983) Do prostatic biopsies 12 mths or more after external irradiation for adenocarcinoma, stage III, predict long-term survival? Int J Radiat Oncol Biol Phys 9:299–303

12. Scardino PT, Wheler TM (1985) Prostatic biopsy after irradiation therapy for prostatic cancer. Urology 25 (Suppl) (2) 39–46
13. Ackerman LV, Del Regato JA (1970) Radiation therapy in the treatment of carcinoma of the prostate. In: Cancer, 4th edn. Mosby, St Louis, p 79
14. Kiesling VJ, McAninch JW, Goebel JL, Agee RE (1980) External beam radiotherapy for adenocarcinoma of the prostate: a clinical follow-up. J Urol 124:851–853
15. Schmeller JD, Laible V, Leick H (1984) Aspiration cytology of prostatic carcinoma following endocrine therapy. Abstract 551, AUA Meeting, New Orleans, 6–10 May, 1984

Section IV

Bladder Cancer

Chapter 23

The Histopathology of Human Bladder Cancer

P. Dalla Palma and M. Piazza

Epidemiology

In relation to the increase in oncogenic agents in the environment in which we live [1–5], wide variations have been noted in the incidence of carcinoma of the bladder throughout the world. Carcinoma of the bladder is more frequent in industrialised areas rather than undeveloped ones and in urban areas rather than rural ones.

Since 1895, when Rehn in Germany [6] first demonstrated the risk of bladder cancer in workers exposed to aniline, many other occupational factors have been indicated which could provoke or promote the development of the carcinoma. Actually, the categories of workers considered at risk are those who handle dye, rubber, leather and leather products, paint, and organic chemical compounds [5, 7–11].

Of equal importance are daily habits: smokers seem to be subject to bladder cancer twice as often as non-smokers [5, 12]. Still under study are other factors such as saccharine and cyclamates. In addition, other factors have been implicated: coffee abuse [9, 13, 14], phenacetine abuse [15], opium [16], INH [17], schistosomiasis [18], and pteridium aquilinum infections [19]. Rats fed with *N*-methyl-*N*-nitrosurea (MNU) or with *N*(4(5-nitro-2-furanyl)-2-tiazolil) formamimide (FANFT) after a few months develop bladder carcinomas morphologically analogous to the carcinomas seen in man [20, 21].

Factors related to the host organism [22] determine susceptibility or resistance, i.e. the capacity to produce an adequate immunological response to oncogenic agents varies greatly from one person to another, often rendering unpredictable the prognosis of the single patient affected with bladder neoplasia [2, 23–28].

Since the inside of the bladder was observed for the first time with the cystoscope, before a steady increase in the incidence, the mortality for bladder carcinoma has remained substantially unchanged. Each year, independently of its treatment, one in 4–5 males and one in 3–4 females with bladder carcinoma die from their disease. The male/female ratio is 3.5:1. In the case records of more than 4000 bladder neoplasias observed in the Institute of Pathological Anatomy at the University of Padua, from 1965 until 1985, that relationship is approximately 9:1. Carcinoma of the bladder is a typical neoplasia of the adult, and, like the majority of other neoplasias of the urinary tract, its incidence increases with age with a maximum incidence between 60 and 70 years of age.

Histology

More than 90% of the malignant tumours originate from the urothelium. The remaining neoplasias include rhabdomyosarcomas [29], leiomyosarcomas [30], pheochromocytomas [31], malignant lymphomas [32], mixed mesodermal tumours [33], melanomas [34], and primitive carcinoid tumours [35], all of which are very rare.

The sarcomas tend to create voluminous masses, as large as 10–15 cm in diameter, which consequently protrude into the bladder lumen. Their sarcomatous nature is suggested by their macroscopic appearance, their grey-white fish flesh colour and their soft consistency.

Epithelial neoplasias are often multiple and represent the local manifestation of a diffuse "abnormality" of the urothelium which has a variable presentation with regard to time and position [36–39]. However, cystoscopically, the majority of infiltrating tumours are unifocal, and, according to Melicow [22], posterior and lateral wall tumours (about 70%) are generally papillary and slightly infiltrating. Tumours of the dome (approximately 20%) are generally silent and are discovered only after they have been present for a long time, i.e. when they are already infiltrating. Tumours of the trigone and neck, 20%, invade the underlying structures rapidly and massively and tend to give early metastasis, like those tumours which develop in diverticulas.

In the last 60 years many histological classifications have been proposed for bladder carcinomas, not all of which can be easily compared [40]. The most widely used classification is that proposed by the World Health Organisation [41]. We have essentially followed that classification, differing only slightly because of recent acquisitions and because of our personal experience.

Papilloma

Only 2.3% of all papillary tumours are papillomas [40]. They are formed of a thin connective tissue and vascular core, covered by normal urothelium which is disposed in no more than seven layers. Mitoses are rare and, when present, are located in the basal layer (Fig. 23.1).

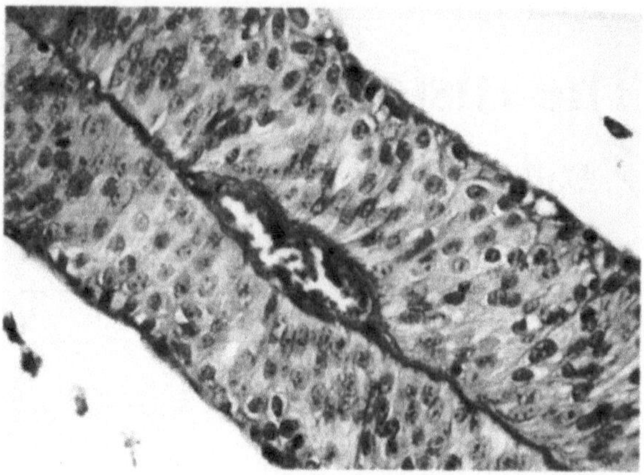

Fig. 23.1. Papilloma. Normal urothelium covers thin vascular core. (H&E, × 270)

Inverted papilloma

The inverted papilloma is an infrequent polyploid outpouching formed by a proliferation of solid cystic or urothelial elements without evidence of atypical cells. The basal membrane is intact and covered by a normal urothelium. This form was first described by Potts and Hirst in 1963 [42], who labelled it inverted papilloma because of its similarity to the better known pathological condition of the nasal cavity. Inverted papilloma is probably an "adenomatous" hyperplasia of the nests of von Brunn [43] (Fig. 23.2). Rarely, these papillomas undergo a malignant degeneration.

Squamous Cell Papilloma

The squamous cell papilloma is an extremely rare variety seen exclusively in countries, such as Egypt, where schistosomiasis is endemic [44]. Some cases reported in the literature correspond to viral condylomatous lesions.

Papilloma of Infancy

Bladder tumours which occur in infancy are very rare and do not progress toward more aggressive

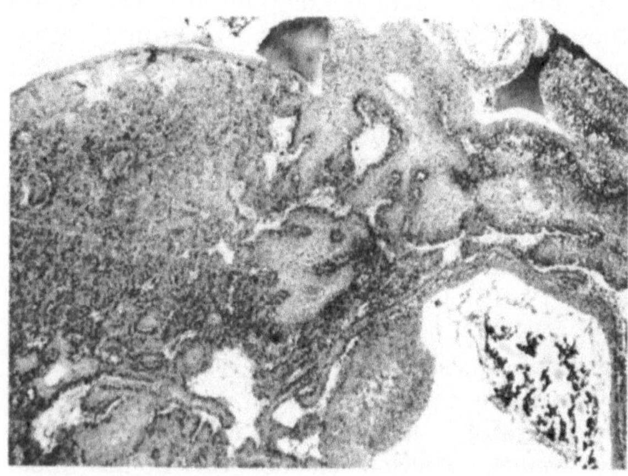

Fig. 23.2. Inverted papilloma. Solid epithelial nests covered by normal urothelium. (H&E, × 40)

forms. They, like other papillomas, may represent hyperplastic processes rather than the neoplasias.

Urethral Polyp

In the bladder neck small polyploid outpouchings consisting of ectopic prostatic tissue are seen [45]. This condition is a true hamartoma.

Nephrogenic Adenoma

The nephrogenic adenoma is a tumoural or pseudotumoural lesion characterised by closely packed tubular formations which consist of a single layer of uniform, and cuboidal cells with a clear cytoplasm. It is a rare form, probably an adenomatous metaplasia of the urothelium reacting to an inflammation [46] (Fig. 23.3).

Carcinoma of the Bladder

Three principal types of carcinomas are recognised histologically: transitional cell (90%), squamous cell (5%) and adenocarcinoma (1%). The remaining 4%–5% is composed of various combinations of the above-mentioned forms (mixed cell carcinoma). In our recent review of 178 cystectomies, performed for infiltrating carcinoma, 160 were transitional carcinoma (89.9%), 5 were epidermoid (2.8%), only 1 was adenocarcinoma (0.6%) and 12 mixed carcinomas (6.7%). Mixed forms of carcinoma are more frequent in surgical specimens than in biopsy fragments obtained endoscopically where the limited size of the sample does not always represent a type of whole tumour.

With regard to the macroscopic and microscopic features, carcinoma of the bladder is classified as follows:

1. Papillary (exophytic), non-invasive.
2. Papillary (exophytic), invasive.
3. Non-papillary (planophytic), non-invasive, i.e. in situ.
4. Non-papillary (endophytic), invasive.

Papillary Carcinoma

Papillary tumours occur very frequently and are often multiple, with a tendency to recur either in the same site or elsewhere. Especially in the forms which demonstrate a greater cellular anaplasia, there can frequently be neoplastic infiltration of the vascular connective tissue. Invasion of the stroma is not associated with a worsening of the prognosis [47]. It is therefore justified to classify all these tumours in a single pathological grade, the TA, or more precisely T1A, in that the stromal core is a part of the underlying lamina propria (Fig. 23.4).

Papillary tumours which invade the implantation base have a distinctly worse prognosis, similar to that for malignant transformation of polyps of the large intestine. These tumours are classified as T1 or T1B, as opposed to TA, in that they infiltrate the subepithelial connective tissue. i.e. the lamina propria of the bladder. Such a distinction is not always easy to make, and a separate examination of the exophytic component and of the implantation base is very important. Frequently, pseudoinvasive aspects are seen which are due to the fact that

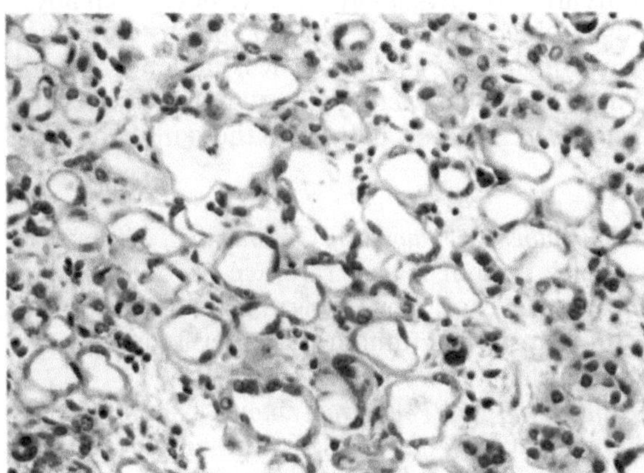

Fig. 23.3. Nephrogenic adenoma composed by packed tubular monolayered formation. (H & E, × 27)

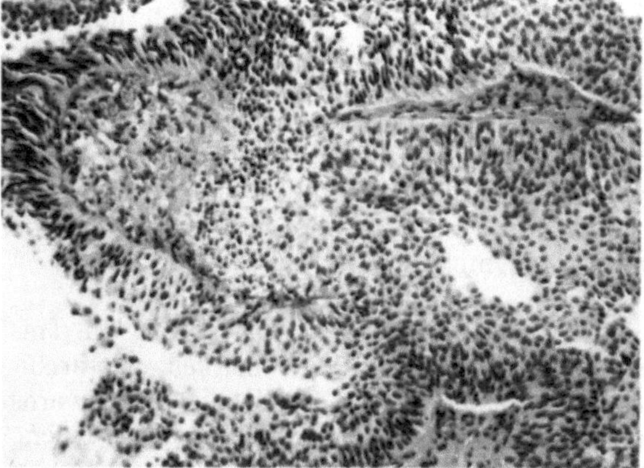

Fig. 23.4. Neoplastic invasion of vascular connective tissue by transitional cell carcinoma. (H & E, × 85)

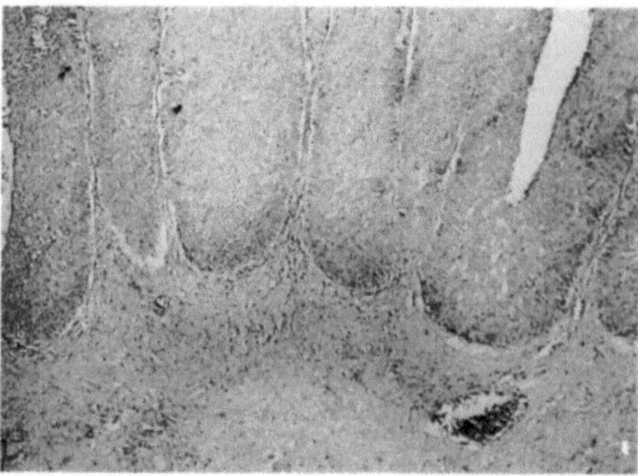

Fig. 23.5. "Pushing borders" of a papillary carcinoma. (H & E, × 270)

papillary tumours can present with their inferior margins irregularly formed along a wide tract, the so-called "pushing borders" of Brawn [48], without passing beyond the basal membrane (Fig. 23.5).

Transitional (Urothelial) Carcinoma Grade I. This is a papillary tumour with a thin vascular connective tissue stalk covered by urothelium disposed in seven to ten layers with scarce anaplasia. There is a low nuclear pleomorphism, slight increase in the nuclear/cytoplasmic ratio, dispersed chromatin, slight irregularity in the basal and superficial cell layer maturation, and rare mitotic figures. Generally this form does not infiltrate its implantation base.

Transitional (Urothelial) Carcinoma Grade II. This tumour consists of short and scarcely ramified papillae which are covered by transitional elements which show slight anaplasia in their cytoarchitecture. There is also a moderate nuclear polymorphism, and the nuclear/cytoplasmic ratio is significantly increased; large nucleoli and few mitotic figures are demonstrated. Maturation of the superficial cell layers is irregular, and the "umbrella" cells disappear. This type of papillary carcinoma can be non-invasive, with "pushing borders", or can infiltrate the implantation base, with a tentacular type of growth and with invasion of the superficial lymphatics.

Transitional (Urothelial) Carcinoma Grade III. This is a papillary tumour with marked cyto-architectural anaplasia; there is no tendency towards maturation from the basal to the superficial layers. The papillae are very short and often fused together. There is a marked polymorphism, often with giant, frankly atypical, elements (Fig. 23.6)

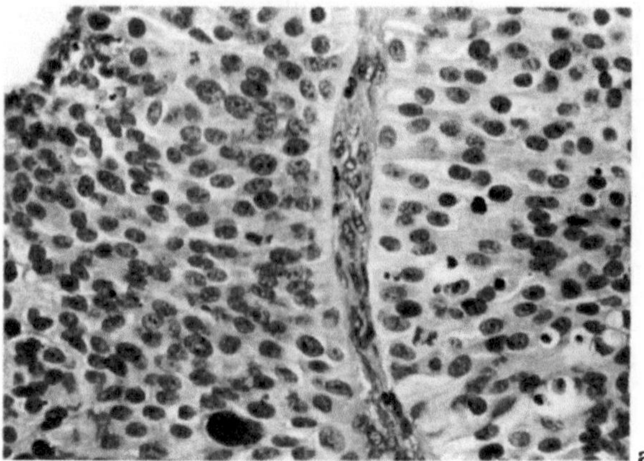

a

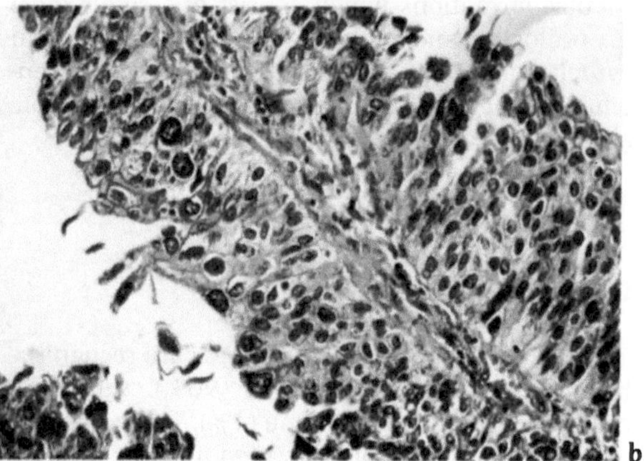

b

Fig. 23.6a, b. Two examples of papillary G2–3 transitional cell carcinoma. (H & E, × 270)

The nuclear/cytoplasmic ratio is significantly increased. There are also frequent karyokinetic figures. The nuclear membrane is irregularly thickened and the cromatin is clumped. Non-invasive forms are rare (personally we have never seen any); those which infiltrate the lamina propria are frequent. In these later cases it is no longer useful to distinguish between papillary and non-papillary tumours in that they both have the same prognosis [25].

Squamous Cell Papillary Carcinoma. This form of papillary tumour is very rare; we have never seen any. Instead the discovery of small metaplastic nests in the transitional tumours is rather frequent.

Cylindrical Cell Papillary Tumour. This tumour is also very rare (Fig. 23.7). In the archives of the Institute of Anatomic Pathology there is only one case of metaplastic glandular tumour of exclusively exophytic character of both pelvic, ureteral and

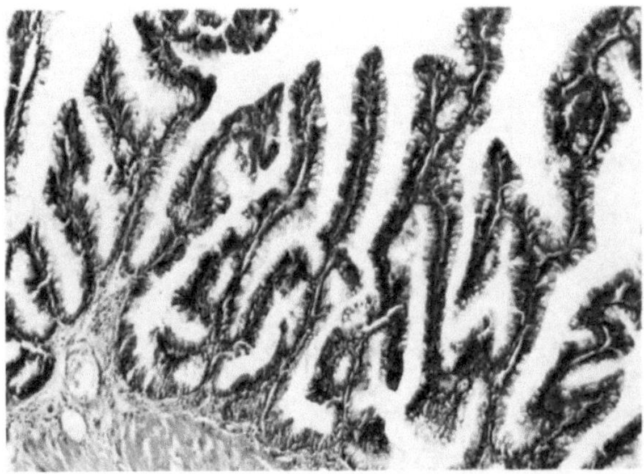
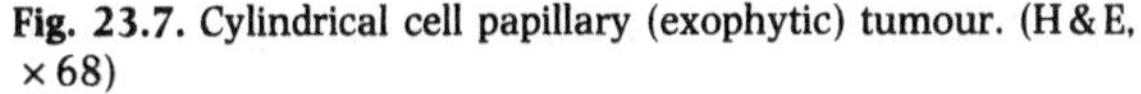

Fig. 23.7. Cylindrical cell papillary (exophytic) tumour. (H & E, × 68)

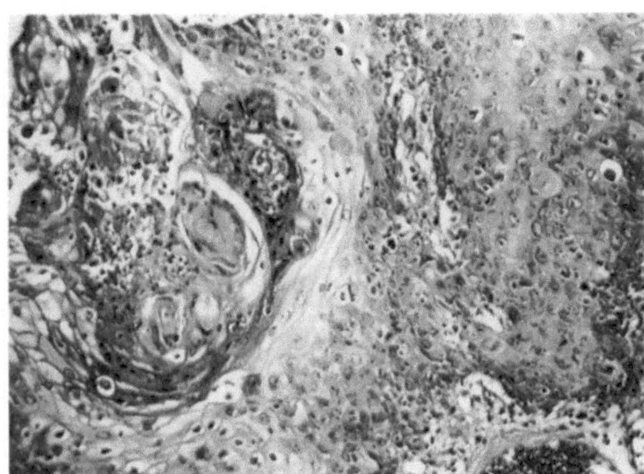

Fig. 23.8. Squamous cell carcinoma. (H & E, × 170)

bladder urothelium. This female patient had mucus in her urine for over 20 years.

Non-papillary Carcinomas

Non-papillary carcinomas, by definition, should not project into the bladder lumen and should, therefore, appear as solid masses. Even if some of these forms derive from a previous papillary tumour in which the exophytic portion has disappeared, the majority derive from an area of flat carcinoma in situ.

Transitional Carcinoma. This is the most common variety of non-papillary tumour and presents the same cytohistological grades as the corresponding papillary forms, even if there is a clear prevalence of the less differentiated forms. In fact, an infiltrating non-papillary G1 carcinoma, if it even exists, has still not been reported in the literature; the majority of these forms are, therefore, G2 or G3 carcinomas. In our most recent case reports of 178 infiltrating carcinomas, the G2 cancers accounted for 29.8% and the G3 cancer 70.2% of the cases. Generally one deals with ulcerated tumours, composed of irregular trabeculae of transitional elements that often fill the lumen of lymphatic vessels and, early in its course, infiltrate the muscular wall. The stroma is rich in collagenous fibres, and a variable number of lymphoplasmocytic inflammatory elements are present, often aggregated in follicles but sometimes irregularly dispersed.

Epidermoid Carcinoma. This neoplasia is formed by squamous cells with intercellular bridges, pearls and keratohyaline granules. It can present three histological grades: (1) well differentiated, (2) moderately differentiated and (3) scarcely differentiated (Fig. 23.8). The stroma can present pseudosarcomatous aspects with spindle cells. Flat epidermoid carcinoma "in situ" does not differ structurally from analogous tumours of the uterine cervix, oesophagus etc. True squamous carcinoma can represent 50% of all bladder carcinomas in countries where schistosomiasis is endemic.

Adenocarcinoma. This is a rare neoplasia formed by glandular and tubular structures having various grades of differentiation. Forms derived from the urachus and allontois are differentiated from metaplastic ones in subjects with cystic and glandular cystitis or with bladder extrophy. Histologically they are similar to intestinal adenocarcinomas with frequent mucous differentiation, and thus it is very

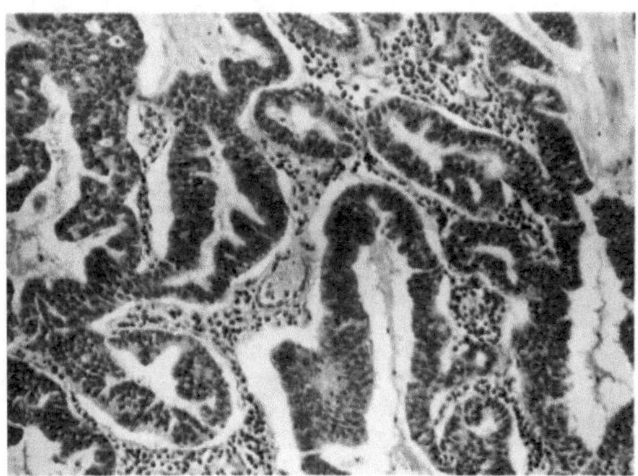

Fig. 23.9. Bladder adenocarcinoma. (H & E, × 170)

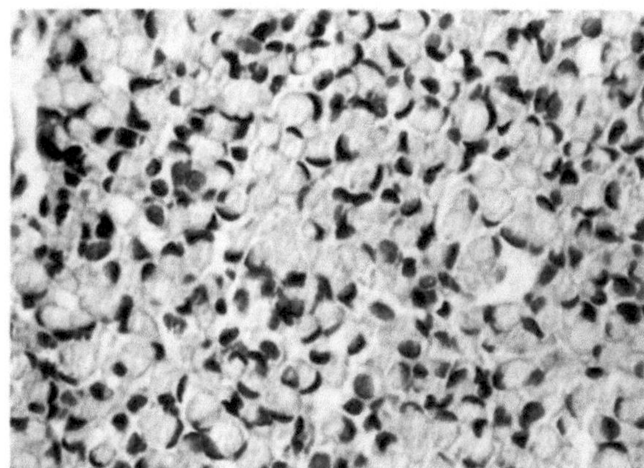

Fig. 23.10. "Signet ring cell" carcinoma. (H & E, ×272)

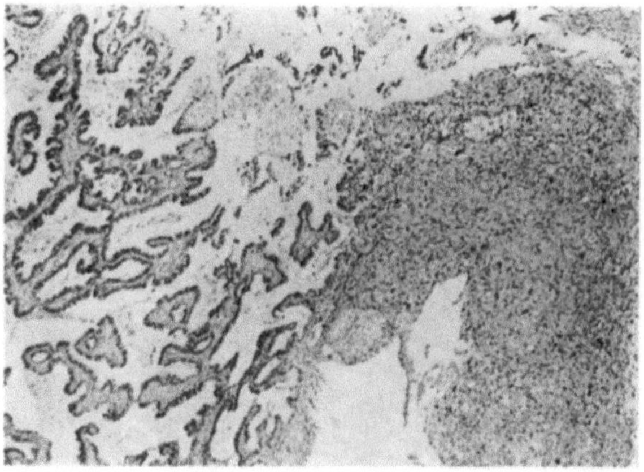

Fig. 23.12. Mixed (transitional and adeno-) carcinoma. (H & E, ×272)

difficult to distinguish them from metastatic tumours to the bladder which are also of cloacal derivation. Histochemical study for various types of mucin and CEA are not helpful in clarifying their origin (Fig. 23.9). Recently, signet ring neoplasia [49] (Fig. 23.10) and mesonephric adenocarcinomas (Fig. 23.11) have been described. The latter tumours are the malignant form of nephrogenic adenomas and characterised by clear or hobnail cells [50].

Mixed Carcinomas. Foci of metaplastic squamous cells are frequently present in transitional carcinoma. Our most recent cases present metaplastic squamous foci in 25 cases, less than 10%. Thus, it is a rather frequent occurrence which does not affect the prognosis. When the squamous metaplastic form is more extended, it is described as a mixed carcinoma, and in this latter case, with similar histological grading, the prognosis is intermediate between transitional and squamous pure carcinomas.

A grandular component is less frequent or at least less easily identified (Fig. 23.12). Sometimes histochemical staining is necessary for mucin to identify single metaplastic elements. Even in this case it is the extension of the metaplastic component which can condition a worse prognosis.

Undifferentiated Carcinoma. This was an extremely rare neoplasia in our series. If numerous samples of the neoplasia are taken, generally it is possible to identify small foci of transitional cell carcinoma (Fig. 23.13). On the other hand the differential diagnosis must be considered, especially with lymphomas. In the literature there are reports of giant and spindle cell carcinoma among the undifferentiated carcinomas. The prognosis is very poor since at the

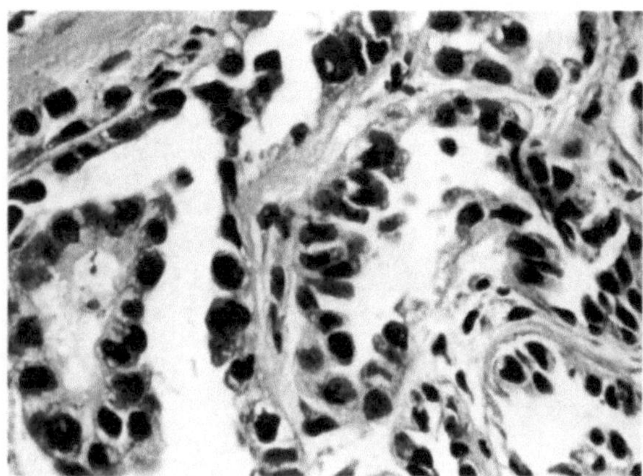

Fig. 23.11. Mesonephric adenocarcinoma of the trigone. (H & E, ×272)

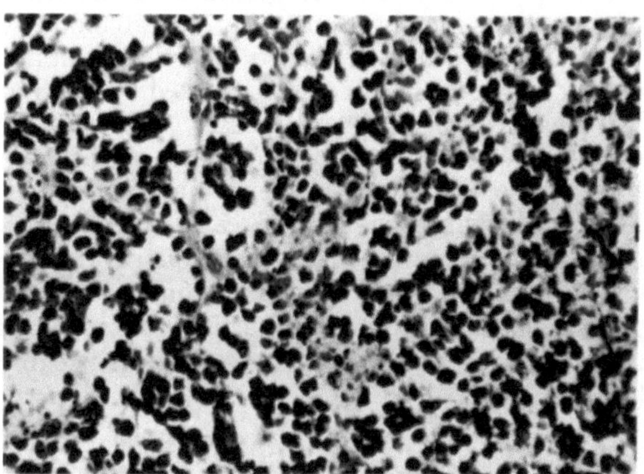

Fig. 23.13. Undifferentiated carcinoma. (H & E, ×85)

time of the diagnosis the stage is generally advanced.

Spread

No matter which classification is followed, the most important parameters are pathological grade and stage [51]. The latter parameter cannot be determined with endoscopic biopsy where the infiltration of the muscularis can be rarely seen. Muscular invasion divides neoplasia into superficial (with a fair prognosis) and deep forms. In the 178 carcinomas we reported, an infiltration of the muscularis was present in 50% of the cases (24.2% of T2 and 25.8% of T3A); an invasion of the perivesical fat, T3B, was detected in 11.2% of the cases. At the time of the diagnosis, the neoplasias were generally still limited to the urogenital tract, and involvement of the prostate, seminal vesicles and female pelvis occurred by a process of continuity of the pelvic organs in 7.3% of cases. The prostate can also be invaded (Fig. 23.14), with pagetoid modality, at first from the urethra and then from the periurethral prostatic ducts.

It is also important to evaluate the possible invasion of lymphatics. In our most recent series, lymphatic invasion was present in 79.2% of cases (44.4% in superficial lymphatics and 34.8% in deep lymphatics). Lymphatic invasion does not necessarily involve lymph node metastasis but it certainly requires accurate study. We observed metastasis of local lymph nodes in only 13.3% of cases. We do not have reliable data on the real incidence of distant metastases which are usually tardive and involve especially the liver, lungs and spinal cord.

However, each tumour has early stages when it is possible to be treated. The pathologist can have an important role in secondary prevention in at-risk patients. In 70 cases among those recently re-examined, the urothelial mucosa distant from the neoplasia showed patterns of severe dysplasia or carcinoma in situ associated in 25% of G2 and in 55.3% of G3 carcinomas.

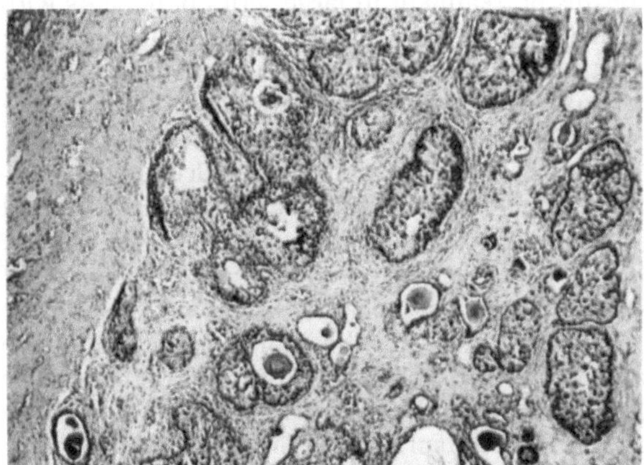

Fig. 23.14. "Intraductal" invasion of the prostate. (H&E, ×40)

Concept of Intraurothelial Neoplasia

In the bladder the distinction of preneoplastic alterations into two distinct categories (dysplasia and carcinoma in situ) does not correspond to the natural course of the tumour. It deals with a real problem of practical interest since the clinician considers "dysplasia" an alteration which is not necessarily progressive and therefore long lasting, whereas "carcinoma in situ" represents a true carcinoma even if it is not yet invasive. Indeed, some dysplasias of high degree have a morphological and biological potential similar to carcinoma in situ and represent different stages of the same process [52–59]. For this reason, we have introduced the concept of intraurothelial neoplasia (IUN) to indicate the continuous spectrum of alterations preceding the appearance of infiltrating carcinoma [60, 61]. The observation of infiltrating carcinomas in the same site of intraepithelial lesions which are still well differentiated, that is, dysplasias of various degrees, implicates two different possibilities: the passage dysplasia-carcinoma is not mandatory, or the carcinoma in situ is so rapid that it escapes the most careful observation. We observed a carcinoma in situ which evolved, at the same site, into an infiltrating form in only 3 months.

Pathogenesis

The study of neoplastic alterations of the bladder permits us to consider the pathogenesis of bladder carcinoma. The importance of papillary neoplasia in the pathogenesis of ulcerative and infiltrating forms, stressed by Melicow [22], is limited by the demonstration that these forms generally take origin from the contiguous flat carcinoma in situ. We believe that, at first, normal urothelium becomes hyperplastic without significant atypical elements and can be expressed with papillary growths which are easily seen with the endoscope, but having little biological meaning. In some cases the process stops at this stage, in other cases it progresses, with resulting hyperplasia plus anaplasia.

The neoplasia can maintain exophytic papillary aspect or invade the basal membrane, thus becoming infiltrating. A third possibility is that lesions remain "flat" and, through modifications characteristic of intraurothelial neoplasia, are transformed into a carcinoma which is almost always infiltrating and non-papillary. While periodic endoscopic control of primary alterations is necessary to identify their possible progression in the same site, the simultaneous evaluation of the remaining mucosa allows us to predict recurrences in different sites.

With classic histology it is not always possible to visualise the early morphology of intraurothelial neoplasia. Cytology, which is so useful in the diagnosis of neoplasia characterised by anaplasia, does not allow us to recognise lesions with slight cytonuclear atypia. Morphometry with evaluation of the nuclear area, round factor of the nucleus and slope angle of non-round nuclei lets us mathematically differentiate cells of various degrees. Unfortunately, we have no personal experience with morphometry. Immunohistochemical research of tumoural markers, such as CEA, gave us results which are scarcely satisfactory and very poorly reproducible, depending on the type of antiserum used [62].

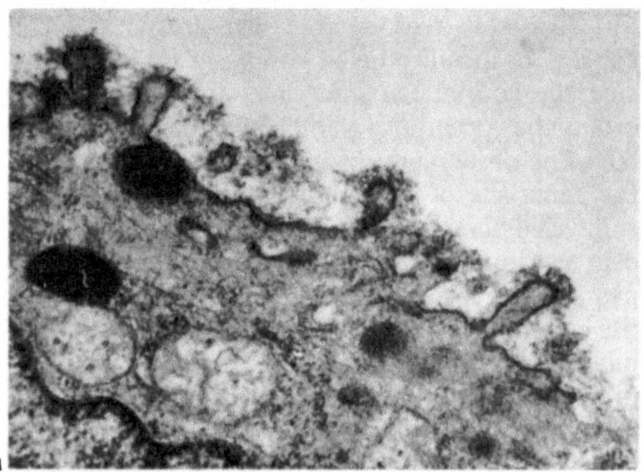

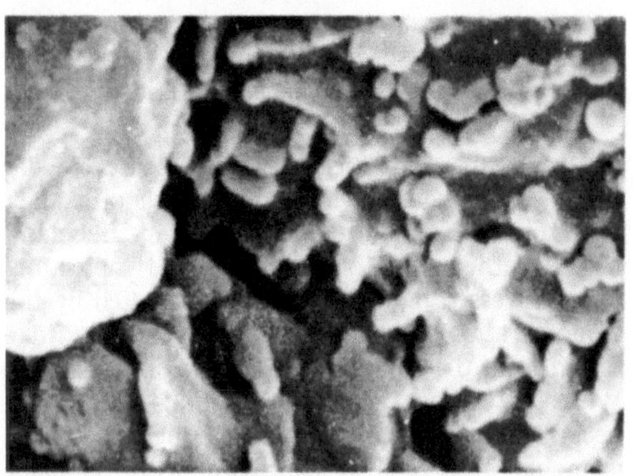

Fig. 23.15a, b. Pleomorphic microvilli detected with TEM (**a**) and SEM (**b**).

A more useful prognostic guide seems to be the research of group-specific ABO antigens in which we performed the specific erythrocytic adherence tests and immunohistochemical method. These antigens are in fact lost precociously in carcinomas as the cytological grade increases, and thus their presence is associated with a better prognosis. However, there are some invasive carcinomas that retain the group-specific antigens [57, 63, 64].

Ultrastructural study seems to offer greater potential for detecting the initial phases of bladder carcinoma [20, 65–67]. Carcinogenic stimulus leads to proliferation of urothelium, and the increased turnover seems to follow a lesion of the Golgi apparatus involved in the synthesis of AUM. The epiphenomenon of the lesion is the alteration of its typically asymmetric structure with appearance of microvilli which are uniform at first and then pleomorphic and which are well detected both with transmission electron microscopy (TEM) and scanning electron microscopy (SEM). (Fig. 23.15). Successively, there is an altered distribution of cytoplasmic organelles, loosening of tight junctions and modifications of the basement membrane with constant presence of neoangiogenesis [68].

Problems in Staging

In order to plan adequate treatment and to compare results of treatment in large series of patients with bladder cancer it is necessary to introduce accurate methods for determining the extension of the disease at the time of diagnosis and during the clinical course. Histological examination of endoscopic biopsy specimens may be inadequate in accurately establishing the metastatic potential, the level of invasion and the spread of the bladder cancer. A good correlation exists between the degree of malignancy of urothelial papillary cancers and their level of invasion and the clinical behaviour of the disease. On the other hand, the grade of the bladder cancer found in the biopsy specimen may be lower than that of the surgical specimen, since the tumour differentiation may vary from area to area. The bladder cancer shows a great tendency to invade lymphatic and/or blood vessels, even in the apparently localised tumours. However, the permeation of the lymphatic network is not strictly related to the presence of nodal metastases.

The clinicopathological data presented here suggest that the study of the draining nodes is always necessary, given the great difficulty in predicting the biological behaviour of the tumour. In this context, the aspiration biopsy cytology may have a fundamental role in association with radiological diagnostic means such as lymphography, sonography and CT in determining the nodal extent of biopsy-proven bladder carcinoma.

Acknowledgement. The authors wish to thank Francesco Gallo, MD, for his help in the translation of this chapter.

References

1. Di Silverio F, La Pera G, Capocaccia F, Mariotti S (1985) Mortalita' per tumori urologici in Italia 1971–1980. Acta Medica, Rome
2. Friedell GH, Jacobs JB, Nagy GM, Cohen SM (1977) The pathogenesis of bladder cancer. Am J Pathol 89:431–440
3. Maltoni C (1971) Introduction to the plenary session on cancer of the urinary apparatus, and report on occupational carcinogenesis of the urinary tract. In: Proceedings of the IV Simposio Internazionale sulla caratterizzazione istologica dei tumori umani, Heidelberg, 1971
4. Mason TJ, MacKay FW, Hoover R et al. (1975) Atlas of cancer mortality for U.S. counties, 1950–1969. Dhew Publication No. (NIH) pp. 75–78
5. Wynder EL, Goldsmith R (1977) The epidemiology of bladder cancer: a second look. Cancer 40: 1246–1268
6. Rehn L (1895) Blasengeschwulste bei Anilinarbeitern. Arch Klin Chir 50:588–600
7. Case RAM (1966) Tumors of the urinary tract as an occupational disease in several industries. Ann R Coll Surg Engl 39:213–235
8. Case RAM (1969) Some environmental carcinogens. Proc R Soc Med 62:1061–1066
9. Cole P, Hoover R, Friedell GH (1972) Occupation and cancer of the lower urinary tract. Cancer 29:1250–1260
10. Guira AO (1971) Bladder carcinoma in rubber workers. J Urol 106:548–552
11. Miller AB (1977) The etiology of bladder cancer from the epidemiological viewpoint. Cancer Res 37:2939–2942
12. Clayson DB (1975) Epidemiology of bladder cancer. In: Cooper EH, Williams RE (eds) The biology and clinical management of bladder cancer. Blackwell Scientific, Oxford, pp 65–86
13. Cole P (1971) Coffee-drinking and cancer of the lower urinary tract. Lancet I:1335–1337
14. Simon D, Yen S, Cole P (1975) Coffee drinking and cancer of the lower urinary tract. J Natl Cancer Inst 54:587–591
15. Fokken W (1979) Phenacetin abuse related to bladder cancer. Environ Res 20:192–198
16. Sadeghi A, Behmard S, Vesselinovitch SD (1975) Opium: a potential urinary bladder carcinogen in man. Cancer 43:2315–2321
17. Miller CT, Neutel CI, Nair RC et al. (1978) Relative importance of risk factors in bladder carcinogenesis. J Chronic Dis 31:51–56
18. Ferguson AR (1911) Associated bilharziasis and primary malignant disease of the urinary bladder, with observations on a series of factory cases. J Pathol 16:76–94
19. Pamukcu AM, Göksoy SK, Price JM (1967) Urinary bladder neoplasm induced by feeding bracken ferm (*Pteris aquilina*) to cows. Cancer Res 27:917–924
20. Koss LG (1977) Some ultrastructural aspects of experimental and human carcinoma of the bladder. Cancer Res 37:2824–2831
21. Hicks RM, Chowaniec J (1978) Experimental induction, histology and ultrastructure of hyperplasia and neoplasia of the urinary bladder epithelium. Int Rev Exp Pathol 18:199
22. Melicow MM (1974) Tumors of the bladder: a multifaceted problem. J Urol 112:467–478
23. Cummings KB (1980) Carcinoma of the bladder: predictores. Cancer 45:1849–1855
24. Friedell GH, Bell JR et al. (1976) Histopathology and classification of urinary bladder carcinoma. Urol Clin North Am 3:53
25. Koss LG (1975) Tumors of the urinary bladder. In: Atlas of tumor pathology. Second series, Fascicle 11. Washington DC, Armed Forces Institute of Pathology
26. Piazza M, Dalla Palma P (1979) Problemi classificativi dei tumori papillari della vescica. Morgagni XI: 47–51
27. Simon W, Cordonnier JJ, Snodgrass WT (1962) The pathogenesis of the bladder carcinoma. J Urol 88:797–802
28. Weinstein RS (1979) Origin and dissemination of human urinary bladder carcinoma. Semin Oncol 6:149–156
29. Tefft M, Hays D, Raney RB, Lawrence W, Soule E, Donaldson MH, Sutow WW, Gehen E (1980) Radiation to regional nodes for rhabdomyosarcoma of the genitourinary tract in children: is it necessary? Cancer 45:3065–3068
30. Alabaster AM, Jorden WP, Solowary MS, Shippel RM, Young JM (1981) Leiomyosarcoma of the bladder and subsequent urethral recurrence. J Urol 12:655–668
31. Flanigan RC, Wittmann RP, Huhn RG, Davis CS (1980) Malignant pheochromocytoma of urinary bladder. Urology 16:386–388
32. Wang CC, Scully RE, Leadbetter WF (1969) Primary malignant lymphoma of the urinary bladder. Cancer 21:772–776
33. Wills RA (1950) Malignant mixed tumor of bladder. Tex Med 46:627
34. Anichkov NM, Nikonov AA (1982) Primary malignant melanomas of the bladder. J Urol 128:813–814
35. Coldby ThV (1980) Carcinoid tumor of the bladder. Arch Pathol Lab Med 104:199–200
36. Cooper TP, Weelis RF, Correa RJ, Gibbons RP, Tate Mason J, Cummings KB (1977) Random mucosal biopsies in the evaluation of patients with carcinoma of the bladder. J Urol 117:46–48
37. Koss LG (1979) Mapping of the urinary bladder: its impact on the bladder cancer. Hum Pathol 10:533–548
38. Soto EA, Friedell GH, Tiltman AJ (1977) Bladder cancer as seen in giant histologic sections. Cancer 39:447–455
39. Wallace DN (1975) Total urothelial neoplasia. In: Cooper EH, Williams RE (eds) The biology and clinical management of bladder cancer. Blackwell Scientific, Oxford, pp 255–272
40. Piazza M, Dalla Palma P, Gentile HM (1980) Histoprognostic evaluation of papillary bladder tumors. Urology 16:207–211
41. Mostofi FK, Sobin LH, Torloni H (1973) Histological typing of urinary bladder tumors. WHO, Geneva
42. Potts IF, Hirst E (1963) Inverted papilloma of the bladder. J Urol 90:175–179

43. Kim YH, Reiner L (1978) Brunnian adenoma (inverted papilloma) of the urinary bladder: report of a case. Hum Pathol 9:229–231
44. Aboul Nasr AL, Gazayerli ME, Fawzi RM, El-Sibai I (1962) Epidemiology and pathology of cancer of bladder in Egypt. Acta Unio Int Contra Cancrum 18:528–537
45. Butterick SD, Schinitzer B, Abell MR (1971) Ectopic prostatic tissue in urethra. A clinicopathological entity and a significant cause of hematuria. J Urol 105:97–104
46. Molland EA, Trott PA, Paris AMI, Blandy JP (1976) Nephrogenic adenoma: a form of adenomatous metaplasia of the bladder. A clinical and electron microscopical study. Br J Urol 48: 453–462
47. Tiltman AJ (1977) The identification and significance of early stromal invasion in papillary carcinomas of the urinary bladder. J Pathol 122:91–95
48. Brawn PN (1984) Interpretation of bladder biopsies. Raven, New York (Biopsy interpretation series)
49. Yoshida H, Iwata H, Ochi N, Yoshida A, Fukunishi R (1981) Primary signet-ring cell carcinoma of urinary bladder. Urology 17:481–483
50. Pegoraro V, Cosciani-Cunico S, Graziotti PP, Dalla Palma P (1982) L'adenocarcinome mesonephrique de la vessie. J Urol (Paris) 88:531–532
51. Friedell GH, Nagy GK, Cohen SM (1983) Pathology of human bladder cancer and related lesions. In: Bryan GT, Cohen SM (eds) Pathology of bladder cancer. CRC Press, Boca Raton Fla, pp 11–42
52. Koss LG (1982) Evaluation of patients with carcinoma in situ of the bladder. Pathol Annu 17(1):353–359
53. Murphy WM, Soloway MS (1982) Developing carcinoma (dysplasia) of the urinary bladder. Pathol Annu 17(1):197–217
54. Murphy WM, Soloway MS (1982) Urothelial dysplasia. J Urol 127:849–854
55. Murphy WM (1983) Current topics in the pathology of bladder cancer. Pathol Annu 18(2):1–25
56. Pugh RCB (1981) Carcinoma in situ of the urinary bladder. In: Conolly G (ed) Carcinoma of the bladder. Raven, New York, pp 101–106 (Progress in cancer research and therapy, vol 18)
57. Weinstein RS, Coon J, Alroy J, Davidsohn I (1981) Tissue-associated blood group antigens in human tumors. In: De Lellis RA (ed) Diagnostic immunohistochemistry. Masson, New York, pp 239–261
58. Weinstein RS (1982) Evolving concepts of the cancer cell. In: Bonney WW, Prout GR (eds) Bladder cancer. Williams and Wilkins, Baltimore, pp 3–11
59. Wolf H, Hojgaard H (1982) Prognostic factors in local surgical treatment of invasive bladder cancer, with special reference to presence of urothelial displasia. Cancer 51:1710–1715
60. Dalla Palma P, Parenti A, De Caro R, Piazza M (1983) Urinary Intraepithelial Neoplasia (U.I.N.). In: Proceedings of the IX European Congress of Pathology, September 1983. Pathol Res Pract 178:121 (abstract)
61. Dalla Palma P, Parenti A, Piazza M (1983) Neoplasia intraepiteliale delle vie urinarie (U.I.N.): aspetti morfologici. Atti V Congresso Nazionale S.I.A.P., Perugia, 8–10 June, 1983
62. Shevchun M, Fenoglio CM, Richart RM (1981) Carcinoembryonic antigen localization in benign and malignant transitional epithelium. Cancer 47:899–905
63. Dalla Palma P, Parenti A, Poletti A (1984) CEA and ABO(H) in upper urinary tract transitional tumors. Appl Pathol 2:146–152
64. Wiley EL, Mendelsohn G, Droller MJ, Eggleston JC (1982) Immunoperoxidase detection of carcinoembryonic antigen and blood group substance in papillary transitional cell carcinoma of bladder. J Urol 128:276–280
65. Jacobs JB, Cohen SM, Farrow GM, Friedell GH (1981) Scanning electron microscopic features of human urinary bladder cancer. Cancer 48: 1399–1409
66. Pauli BU, Weinstein RS et al. (1977) Ultrastructure of cell junctions in fanft-induced urothelial tumours in urinary bladder of Fisher rats. Lab Invest 37:609–621
67. Pauli BU, Friedell GH, Weinstein RS (1978) Topography and numerical densities of intramembrane particles in chemical carcinogen-induced urinary bladder carcinoma in Fisher rats. Lab Invest 39:565–573
68. Piazza M, Dalla Palma P, Poletti A (1983) Carcinoembryonic antigen and transitional carcinomas of the upper urinary tract. Pathol Res Pract 178:156 (abstract)

Chapter 24

Clinical and Pathological Staging of Bladder Carcinoma

C. Selli, C. Bartolozzi, N. Villari and G. Masini

Accurate staging of bladder carcinoma is of paramount importance in assessing the prognosis and planning the therapy. Staging procedures must provide information regarding the extent of infiltration of the bladder wall (T status), the involvement of the pelvic and abdominal lymph nodes (N status) and the presence of distant metastases (M). The two staging systems most frequently used are TNM [1] (Table 24.1) and the system proposed by Marshall and Jewett [2] (Table 24.2).

Table 24.1. TNM classification of bladder carcinoma[a]

Tis	Carcinoma in situ
Ta	Papillary, non-invasive carcinoma
T1	Lamina propria infiltration
T2	Extension to the superficial muscle layer
T3a	Extension to deep muscle layer
T3b	Infiltration through bladder wall
T4a	Invasion of prostate, uterus or vagina
T4b	Invasion of pelvic or abdominal wall
N0	No regional lymph node metastases
N1	One homolateral solitary regional node
N2	Contralateral or bilateral or multiple regional nodes
N3	Fixed regional lymph node metastases
N4	Juxtaregional lymph node metastases

[a] See UICC 1978 [1].

Many different techniques are currently available for the preoperative evaluation of T status; they can be divided into clinical procedures (endoscopy with biopsy and bimanual examination) and imaging techniques, such as intravenous pyelography (IVP), dynamic cystographical studies, sonography, computed tomography (CT) and magnetic resonance imaging (MRI). Preoperative N status is evaluated by lymphangiography and/or CT scan, while chest X-ray, bone scan and liver sonography are usually performed to rule out distant metastases.

Table 24.2. Jewett–Marshall classification of bladder carcinoma[a]

0	Confined to mucosa
A	Infiltration of submucosa
B_1	Infiltration of superficial muscle
B_2	Infiltration of deep muscle
C	Perivesical infiltration
D_1	Adjacent organs and pelvic lymph nodes
D_2	Distant metastases or nodes above aortic bifurcation

[a] See Marshall [2].

Preoperative assessment of the extent of local diffusion of tumour using IVP, cystoscopical evaluation with biopsy and bimanual palpation with the patient under general anaesthesia is misleading in over 40% of the cases, with a marked tendency to understage the tumour. This inaccuracy has basically remained unchanged over the past 30 years. Cystoscopy provides information on the size, number and location of bladder tumours, but cold cup biopsy or transurethral resection are required to assess the depth of infiltration. Bimanual palpation with the patient under general anaesthesia is greatly influenced by the experience of the examiner, and the results are inadequate in obese patients and in those who have undergone previous surgery or radiation therapy. Conventional IVP still represents the basic step in the diagnostic pathway, but has a limited role in evaluating the depth of infiltration of bladder tumours, since the intracavitary portion of the neoplasm is well demonstrated in some locations, such as the lateral walls and the dome, and less satisfactorily in others, like the trigone and the posterior wall. This can be partially obviated by obtaining oblique and lateral views. IVP can also demonstrate hydronephrosis, indicating neoplastic infiltration of the intramural portion of the ureter up to complete functional exclusion of the kidney.

Dynamic radiological studies of the bladder walls (fractionated cystogram and cineradiography) have been employed to document intramural extension of bladder tumours. Both can demonstrate a circumscribed parietal rigidity, indicating neoplastic infiltration. Therefore their reliability decreases greatly in the presence of wall fibrosis caused by previous open or endoscopic surgery and radiotherapy. Fractionated cystography consists of retrograde filling of the bladder through a urethral catheter, documenting the passive distension of the organ with multiple exposures on the same film [3]. Voiding cineradiography or spot-camera study is an active evaluation of bladder wall contraction and can be performed at the end of an IVP. Under normal circumstances a symmetrical and concentric diminution of bladder volume is documented [4]. Prostatic and urethral obstruction has a negative influence on this technique, preventing complete emptying of the bladder.

Sonographic studies of the vesical area can be performed either suprapubically or transurethrally. In the first case access is provided by the distended anterior wall of the bladder which comes in close contact with the abdominal wall. In transurethral sonography, probes (rotating or linear) are introduced through the sheath of a resectoscope after endoscopic evaluation. Their power has so far been limited, and therefore the reliability of endocavitary sonography has been confined mainly to superficial tumours, but rapid technological improvements are underway [5].

Suprapubic sonography is based on the principle that the bladder wall presents uniform echoes and that infiltrating tumours interrupt this aspect, allowing an evaluation of the depth of infiltration and also of superficial extension along the mucosa. Landmarks are represented by the prostate and the uterus, but it is difficult to separate the anterior wall from the subcutaneous tissues and the dome is often impossible to distinguish from intraperitoneal viscera. Tumours must be greater than 0.5 cm to be recognisable. A tendency towards overstaging has been reported for superficial tumours because of some difficulty in distinguishing the mucosa from the muscularis, while diagnostic accuracy is greater for high-stage tumours [6]. In about 20% of the studies the quality of the image is unsatisfactory because of obesity, low-capacity bladders and the technical skill of the examiner, which represents a major variable.

In the evaluation of vesical tumours CT scan is used, whereby the bladder is filled with urine, contrast medium, oil or gas in order to obtain a clearer definition of its walls [7–11]. Neoplasms may appear as sessile or pedicled masses projecting inside the lumen or as a thickening and rigidity of the bladder wall. The contribution of CT scan is not substantial for superficial tumours, but its accuracy increases in higher stages. Increased density of the perivesical fat caused by P3B tumours infiltrating the entire wall is easily demonstrated. CT scan also clearly defines tumour extension to the abdominal wall, to the seminal vesicles, with obliteration of the prostatoseminal angle, to the rectum and to the bony pelvis. Previous surgery and radiotherapy, causing fibrosis and oedema, result in reduced specificity and overstaging. CT finally provides a spatial balance of the entire abdomen, evaluating at the same time the status of pelvic nodes and the presence of visceral metastases. Diagnostic accuracy of CT scan in defining the local extent of tumour in patients with surgically confirmed disease varies between 64% and 81% [11, 12]. The diagnostic accuracy rate in the detection of lymph node metastases varies between 25% and 40% [13, 14]. This is due to the fact that only enlarged nodes are clearly demonstrated, taking a diameter of 1.5 cm as an arbitrary limit for positivity. In some cases nodal enlargement may be due to hyperplastic lymphadenitis (false positives), while nodes of normal size may already harbour micrometastases (false negatives).

Dynamic Vesical Studies, Sonography and CT Scan

We compared the accuracy of different imaging techniques in a series of 30 patients who subsequently underwent surgical treatment for bladder carcinoma without prior radiotherapy. They were examined by CT scan, suprapubic sonography (26/30), fractionated cystography and cineradiographic voiding studies (Figs. 24.1, 24.2). CT

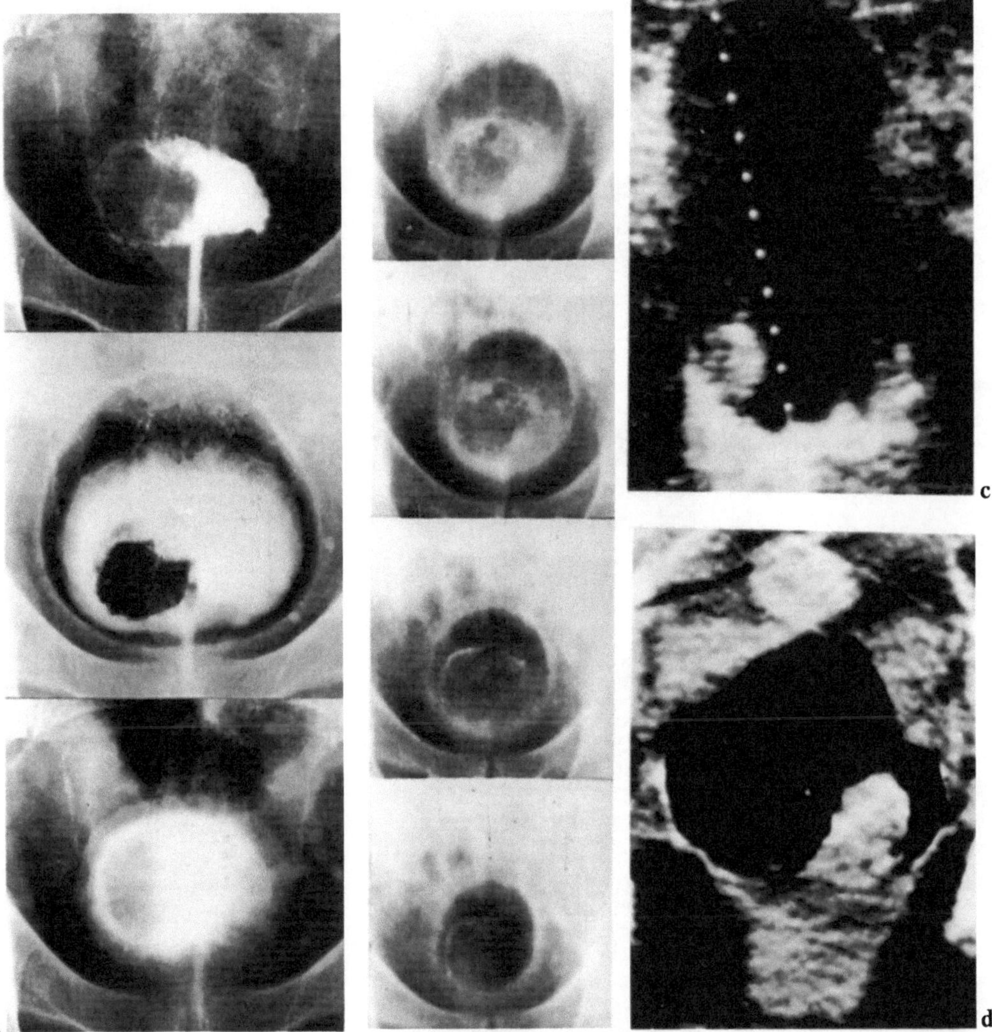

Fig. 24.1a–d. Superficial tumour arising from the right wall of the bladder (T1). **a** Cystography and fractionated cystography documenting lack of infiltration of the bladder wall. **b** Voiding cineradiography showing symmetrical and concentric diminution of the bladder volume. **c** Suprapubic sonography: marked bladder distension is evident. **d** CT scan, supine position, with gas-filled bladder.

scan of the pelvis was performed using a second generation device with 18 s scanning time. The bladder was emptied through a catheter and subsequently inflated with approximately 100 cm^3 air, since gas improves the definition of the bladder walls, which in axial section are delineated internally by air and externally by low-enhancement pelvic fat. With the patient in the supine position 13-mm scans were obtained from the pubic bone to the bladder dome. The same scans were repeated in the prone position in order to evaluate more accurately the posterior bladder wall and its relationship to the adjacent organs. Sonography was performed suprapubically in 26 cases using a manual digital device without the use of endocavitary probes. Fractionated cystography was done by filling the bladder with contrast medium through a urethral catheter and documenting its distension with four exposures on the same film. Double contrast studies were performed in 18 cases. Cineradiographic evaluation was obtained at the end of a conventional IVP: the patient was asked to void in the erect position under fluoroscopic control and images were recorded every 1.5 s with a spot-camera system on a 7-cm film.

In 27 cases a radical cystectomy was performed and the bladder was inflated with 100 ml 10% buffered formalin solution, placed in a bath of the same

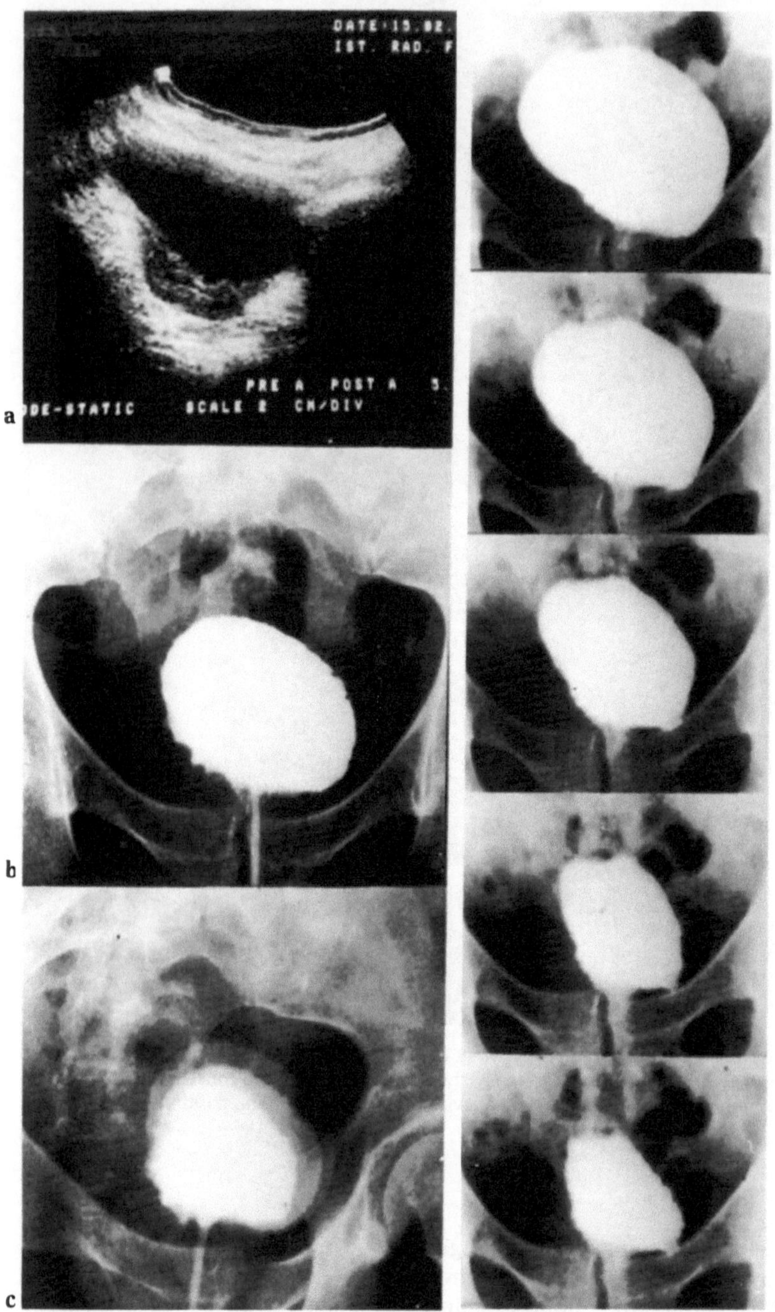

Fig. 24.2a–d. Infiltrating tumour of the right bladder wall and base (T3B). **a** Suprapubic sonography showing full-thickness wall infiltration by the neoplasm. **b** Cystography. **c** Fractionated cystography revealing rigidity of the right bladder wall. **d** Voiding cineradiography showing asymmetrical diminution of the bladder volume.

fluid and sectioned 48 h later along the axial and sagittal plane. Multiple full-thickness sections of the neoplasm were obtained and processed for histological examination in order to achieve a precise definition of wall infiltration. In the remaining three patients, who had superficial tumours, staging was performed on a specimen obtained by transurethral resection.

We tried to differentiate between non-infiltrating (P0, Pis, Pa, P1) and infiltrating tumours (P2, P3a, P3b, P4). In 20 cases there was an agreement between clinical and pathological staging of tumour extension, while in 10 cases conflicting results were observed. In detail, CT scan led to overstaging in two patients and understaging in three, sonography gave understaging in three patients and cineradiography in four. Fractionated cystography was uninterpretable in 25% of the cases, but provided correct information in the remaining patients (Tables 24.3, 24.4).

Table 24.3. Pathological staging of 30 bladder cancers

Stage	No. of cases	
Pa–P1	5	Not over lamina propria
P2	8	Superficial muscle
P3	15	Deep muscle—perivesical fat
P4	3	Adjacent organs
Total	30	

Table 24.4. Percentage of error of different imaging techniques

Cineradiography	13%
Sonography	11%
CT scan	16%
Fractionated cystography	25% doubtful (limits of the method)

Although no single modality appears completely reliable for the definition of T status, the combined use of different imaging techniques can greatly improve the diagnostic accuracy, but is to some extent time consuming and unpractical for routine clinical use. Therefore, whenever IVP, followed by dynamic studies, suggests the presence of a superficial bladder tumour, the patient is examined further by suprapubic sonography. When an infiltrating tumour is suspected, a CT scan is made as a guideline for further treatment.

Magnetic Resonance Imaging

The physical principles of MRI are beyond the purpose of the present review. This imaging technique presents some drawbacks in the evaluation of abdominal organs because of long data acquisition times, since the quality of the images may be negatively affected by breathing movements and vascular pulsations. The introduction of respiratory and cardiac "gating" has resulted in an improvement in image quality, but requires a longer scanning time. All these technical problems are relevant in the upper and middle abdomen, but are of less concern in the pelvis, since its anatomical structures are virtually still during breathing. The use of MRI in the pelvis is facilitated by the natural contrast between its various structures in relation to the different kinds of signal with different sequences [15, 16]. For instance, with the "spin echo" technique pelvic fat gives a high-intensity signal in comparison with muscle (low intensity) and circulating blood, which gives virtually no signal. Parenchymatous organs such as the prostate yield a signal intermediate between those of fat and muscle. The appearance of bowel loops varies according to their content.

A distinct advantage of MRI over CT scan is the possibility of displaying a given area on various planes (axial, sagittal, coronal, oblique), and this is particularly useful when dealing with neoplasms since a more precise spatial balance is achieved.

We have studied with MRI 23 patients with bladder cancer using a superconductive device (0.5 tesla), covering a region of interest of 50 cm diameter, with a 256 × 256 matrix, a viewing field of 350–500 mm and slice thickness of 10 mm. As for data acquisition techniques, we generally used spin echo (multi-echo, TE 50 ms) with short, medium and long TR intervals (700, 1000, 1500 ms) in function of the region of interest and the different relaxation times of the examined structures (Figs. 24.3–24.5). All patients had pathological confirmation of the extent of tumour infiltration following surgery (either transurethral resection or radical cystectomy), in analogy with the previously described protocol for the other imaging techniques.

The discriminating ability of MRI in bladder cancer is good, since we have been able to identify small papillary tumours of 1 cm diameter in the anterior bladder wall, which is notoriously difficult to evaluate with sonography. Greater difficulties are encountered with tumours of the bladder base, particularly with coexistent benign prostatic hypertrophy. However, the main diagnostic advantage of this technique is in the definition of intra- or extraparietal extension of the neoplasm. For this purpose we found it useful to use the spin echo technique with a medium or long TR interval and multiple echoes. With this technique the first echo visualises the tumour protruding inside the bladder lumen, since urine in this phase gives no signal. With medium to late echoes urine has a signal inversion and therefore appears "white". During this phase the contrast between the tumour, the bladder wall and the perivesical fat is enhanced, thus allowing a precise evaluation of the depth of infiltration [17] (Fig. 24.5).

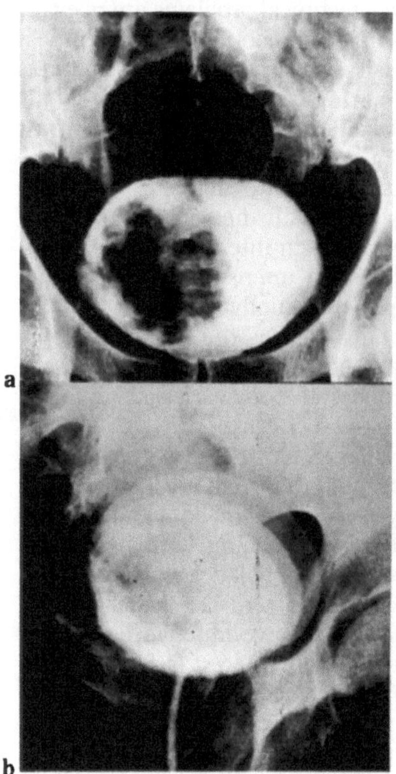
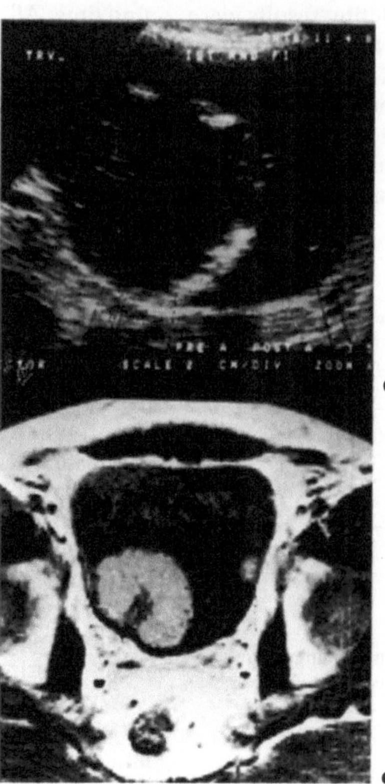

Fig. 24.3a–d. Superficial tumours of the bladder (T1). **a** Cystography. **b** Fractionated cystography. **c** Suprapubic sonography. **d** MRI, spin echo technique, axial view, clearly demonstrating the small left tumour and the fibrovascular stalk of the right one.

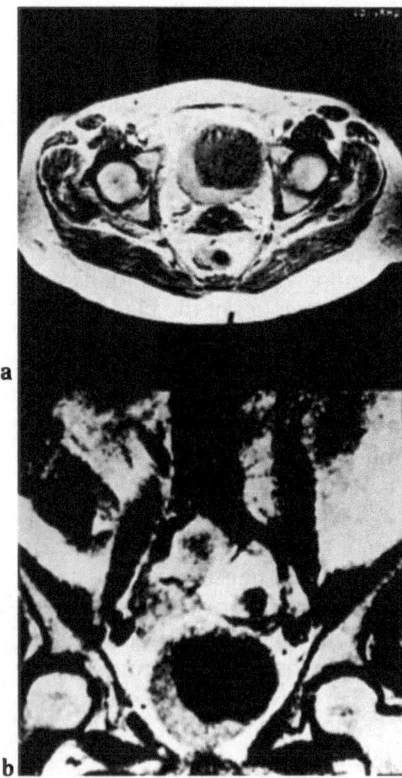
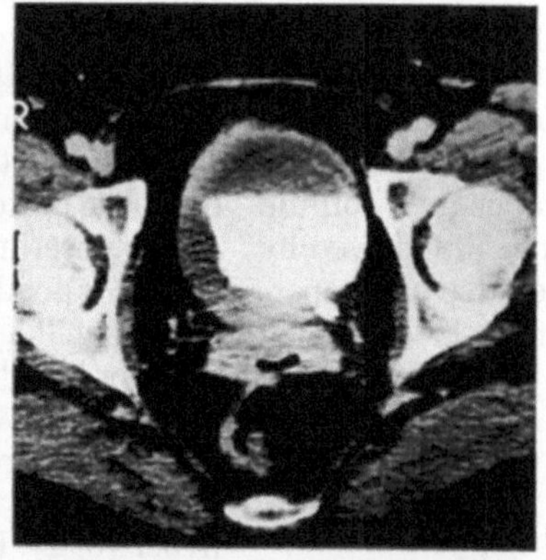
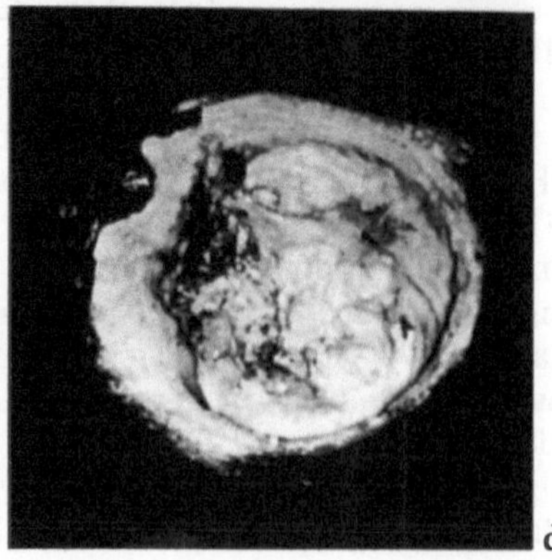

Fig. 24.4a–d. Infiltrating tumour (same as in Fig. 24.2). **a** MRI, axial view. **b** CT scan with endovesical contrast medium. **c** MRI, coronal view. **d** Pathological specimen of the formalin-distended bladder.

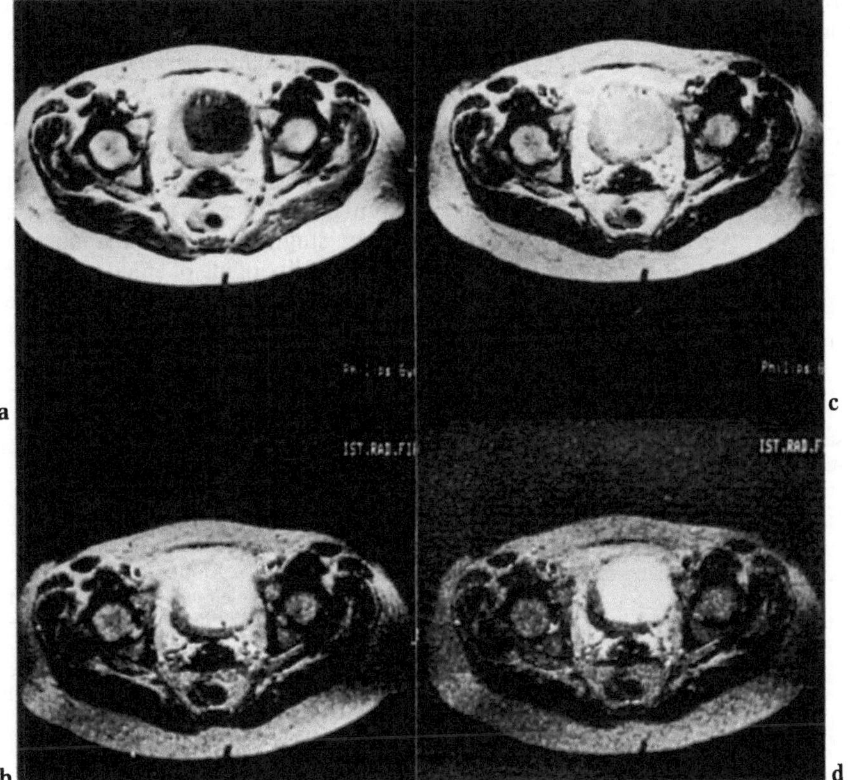

Fig. 24.5a–d. Infiltrating tumour (same as in Fig. 24.2). MRI, spin echo technique with multiple echoes (see text).

The ability to define a normal bladder wall makes it possible to recognise T1 tumours and to separate them from T2–T3, where parietal thickening and tumour extension into the perivesical fat are clearly visible. MRI is particularly useful in defining extravesical tumour extension, since the relationship of the bladder to the surrounding structures such as the prostate, the seminal vesicles, the rectum and the abdominal wall are clearly demonstrated.

In the present series there was a correct classification of P_1 tumours (5/5), but a slight tendency towards overstaging infiltrating forms. In fact, by MRI staging nine cases were T2–T3 and nine were T4, while pathological staging following surgery gave 11 P2–P3 and seven P4 (Table 24.5). Greater difficulties were met with tumours of the bladder base, since there is no significant signal variation between bladder cancer and normal prostatic tissue.

Correct staging also requires evaluation of the nodal status. For this purpose the multi-slice technique is useful, but its long TR reduces contrast between pelvic lymph nodes and fat. There are also interpretation difficulties, since the nodal signal is very similar to that of small bowel loops.

Table 24.5. MRI staging of 23 bladder cancers

	Ta–T1	T2–T3	T4
Pathological stage	5	5 6 11	7
MRI stage	5	9	9

Discussion

An accurate preoperative knowledge of bladder cancer diffusion is of great importance in defining the prognosis (Table 24.6), since as soon as the muscularis is infiltrated there is a marked decrease in survival [18]. Using conventional clinical procedures, considerable inaccuracy is inevitable, even in the more qualified centres, with a tendency to understage bladder cancer even in "virgin" cases (Table 24.7) [19]. Preoperative radiation renders even more useless all attempts at precise definition of tumour invasion. The advent of CT scanning has to some extent improved our preoperative efforts to assess T status, and accuracy increases as wall infiltration becomes more pronounced.

Table 24.6. Relationship of clinical stage and survival rate[a]

Stage	5-year survival		
0–A	78.6%		
B_1	39.9%		$P<0.01$
B_2	40.4%	All B 40%	
C	19.7%		
D	6.2%		

[a] Richie et al. [18]; reproduced by kind permission of the Editor of *Journal of Urology*.

Table 24.7. Pathological vs clinical stage in 137 patients treated by radical cystectomy without radiation[a]

Clinical stage (no. of patients)		Pathological stage						
		is	A	B_1	B_2	C	D_1	D_2
		Understaging						
is	(7)	7	—	—	—	—	—	—
A	(17)	—	15	2	—	—	—	—
B_1	(42)	—	—	28	9	1	3	1
B_2	(39)	—	—	—	16	8	6	9
C	(26)	—	—	—	6	11	4	5
D_1	(6)	—	—	—	—	—	3	3
		Overstaging						

[a] Whitmore [19]; reproduced by kind permission of the Editor of *Current Problems in Cancer (Chicago)*.

Assessment of nodal involvement is also of great prognostic significance: numerous studies show that 18%–25% of all patients with bladder cancer undergoing surgical treatment have positive nodes [20–23] and this profoundly affects survival, since patients with nodal metastases have a 5%–35% chance of being alive after 5 years [20–22, 24]. It is also known that with increasing stage of bladder tumours, there is a higher incidence of positive nodes, ranging from 5% for Pis to 50% for P4 [24, 25].

All these considerations compel the clinician to make extensive preoperative use of different imaging techniques in order to obtain the greatest possible amount of information, but surgical confirmation has so far been considered necessary in the vast majority of cases when radical cystectomy appeared suitable.

At the present stage of technological progress, CT scanning is the single technique most widely used, since it provides information both about local tumour extension and nodal involvement. MRI is currently proving very promising, probably superior to CT scanning in the definition of T status, but its role in the evaluation of regional nodes is still to be determined. On the other hand, aspiration cytology of the pelvic lymph nodes under sonographic or fluoroscopic guidance is rapidly expanding, and therefore we can foresee that in the coming years a long expected advance in accuracy of evaluation of bladder cancer will become possible. Whether or not this will result in increased patient survival is not known, but there is no doubt that a greater accuracy in staging will bring about an improvement in the quality of life of bladder cancer patients, who are affected by one of the more aggressive neoplasms in human pathology.

References

1. UICC (1978) Union Internationale Contre le Cancer. TNM classification of malignant tumors. Bladder, 3rd edn. International Union against Cancer, Geneva, pp 113–117
2. Marshall VF (1952) The relation of the preoperative estimate to the pathologic demonstration of the extent of vesical neoplasms. J Urol 68:714–723
3. Temeliescu I (1958) Cystopolygraphy in the diagnosis of vesical tumors. Urol Int 7:285–290
4. Masini GC, Bozza A, Carini M (1982) Contributo della cistografia minzionale alla diagnostica delle neoplasie vescicali infiltranti e non infiltranti la parete (documentazione alla spot-camera). Diagn per Immag 2:216–220
5. Nakamura S, Nijima T (1980) Staging of bladder cancer by ultrasonography: a new technique by transurethral intravesical scanning. J Urol 124:341–344
6. Itzchak Y, Singer D, Fischelovitch Y (1981) Ultrasonographic assessment of bladder tumors. I. Tumor detection. J Urol 126:31–33
7. Ahlberg RG, Calissendorff B, Wijkstrom H (1982) Computed tomography in staging bladder carcinoma. Acta Radiol [Diagn] (Stockh) 23:47–53
8. Colleen S, Ekelund L, Henrikson H, Karp W, Mansson W (1981) Staging of bladder carcinoma with computed tomography. Scand J Urol Nephrol 15:109–113
9. Hildell JG, Nyman UR, Noordlindh ST, Hellsten SFJ, Stenberg PBA (1981) New intravesical contrast medium for CT: preliminary studies with arachis (peanut) oil. AJR 137:777–780
10. Sager EM, Talle K, Fôssa SD, Ous S, Stenwig AE (1983) The role of CT in demonstrating perivesical tumor growth in the preoperative staging of carcinoma of the urinary bladder. Radiology 146:443–446
11. Vock P, Haertel M, Fuchs WA, Karrer P, Bishop MC, Zingg EJ (1982) Computed tomography in staging carcinoma of the urinary bladder. Br J Urol 54:158–163
12. Koss JC, Arger PH, Coleman BG, Mulhern CB Jr, Pollack HM, Wein AJ (1981) CT staging of bladder carcinoma. AJR 137:359–362
13. Morgan CL, Calkins RF, Cavalcanti EJ (1982) Computed tomography in the evaluation, staging and therapy of carcinoma of the bladder and prostate. Radiology 137:759–761
14. Weinerman PM, Argez PH, Coleman BG, Pollack HM, Banner MP, Wein AJ (1984) Pelvic adenopathy from bladder

and prostate carcinoma: detection by rapid-sequence computed tomography. AJR 140:95–99
15. Hricak H, Williams RD, Spring DB, Moonk L Jr, Hedgcock MW, Watson RA, Crooks LE (1983) Anatomy and pathology of the male pelvis by magnetic resonance imaging. AJR 141:1101–1110
16. Hricak H, Williams RD (1984) Magnetic resonance imaging and its applications in urology. Urology 23:442–454
17. LiPuma J, Bryan P (1985) Magnetic resonance imaging of the urinary tract. In: Kressel HY (ed) Magnetic resonance annual 1985. Raven Press, New York, p 149
18. Richie JP, Skinner DG, Kaufman JJ (1975) Radical cystectomy for carcinoma of the bladder: 16 years experience. J Urol 113:186–189
19. Whitmore WF Jr (1979) Management of bladder cancer. Curr Probl Cancer 4:1–48
20. Laplante M, Brice M (1973) The upper limits of hopeful application of radical cystectomy for vesical carcinoma: does nodal metastasis always indicate incurability? J Urol 109:261–264
21. Reid EC, Oliver JA, Fishman IJ (1976) Preoperative irradiation and cystectomy in 135 cases of bladder cancer. Urology 8:247–250
22. Skinner DG (1982) Management of invasive bladder cancer: a meticulous pelvic node dissection can make a difference. J Urol 128:34–36
23. Whitmore WF Jr, Batata MA, Hilaris BS, Reddy GN, Unal A, Ghoneim MA, Grabstald H, Chu F (1977) A comparative study of two preoperative radiation regimens with cystectomy for bladder cancer. Cancer 40:1077–1086
24. Smith JA, Whitmore WF Jr (1981) Regional lymph node metastasis from bladder cancer. J Urol 124:591–593
25. Skinner DG, Tift JP, Kaufman JJ (1982) High-dose, short course preoperative radiation therapy and immediate single stage radical cystectomy with pelvic node dissection in the management of bladder cancer. J Urol 127:671–674

Chapter 25

TNM Staging System and Bladder Carcinoma: the Problems of the N Categorisation

C. C. Schulman and E. Wespes

The accurate clinical staging of bladder carcinoma is of the foremost importance in determining the prognosis and modality of treatment of the patient. The prognosis is directly related to the extent of spread of the disease. Therefore, clinical determination of spread provides the best information regarding the potential curability of the patient [1], and the choice of treatment modality is regarded as less important than the stage of the tumour in determining the prognosis [2].

Clinical staging requires a physical examination, intravenous urography, cystoscopy with transurethral biopsies and bimanual examination with the patient under anaesthesia. Complementary diagnostic regional evaluation includes lymphangiography, liver scintigraphy, double or triple contrast cystography, ultrasonography and selective pelvic arteriography. Evaluation of distal metastases requires chest and skeletal radiography and isotope liver and bone scintigraphy.

Currently available clinical methods of staging bladder tumours have limited value and their pitfalls have been frequently underlined [2]. Clinicopathological correlation reveals that the staging of bladder tumours is inaccurate despite the numerous diagnostic procedures available. An overall inaccuracy of 56% is reported, 23% being clinically overstaged and 33% understaged [3]. In particular, tumours of the urinary bladder have been poorly staged, and recent experience has demonstrated errors ranging from 20% to 50%, the majority being errors of understaging [4]. Though computed tomography (CT; Fig. 25.1) and ultrasonography will assess extravesical involvement, the stage of the disease may often be evident on bimanual examination. Similarly, these techniques have not been useful in assessing tumour involvement of the bladder wall, and urologists continue to rely on biopsy to make this determination [5, 6]. Recent advances in transurethral ultrasonography indicate that this modality is the most promising available for determining tumour involvement of the bladder wall [7].

The presence of positive pelvic wall lymph nodes upgrades bladder carcinoma from a local to a systemic disease that is no longer amenable to any form of local treatment. Nodal assessment is therefore critical to treatment planning in order to spare N+ patients unnecessary radical surgery. In the past, pelvic lymph nodes have been evaluated radiologically by bipedal lymphography or CT scanning. The high false positive and negative rates of these procedures [8] preclude their routine use [9].

Assessment of the pelvic lymph nodes for metastatic disease has been unsatisfactory in evaluating patients with bladder tumours. The false negative and false positive rates remain substantial, primarily because of the inability of CT and ultrasonography to distinguish reliably hyperplastic benign nodes from those containing metastatic tumour and, secondly, because of the inability of these techniques to reveal lymph nodes less than

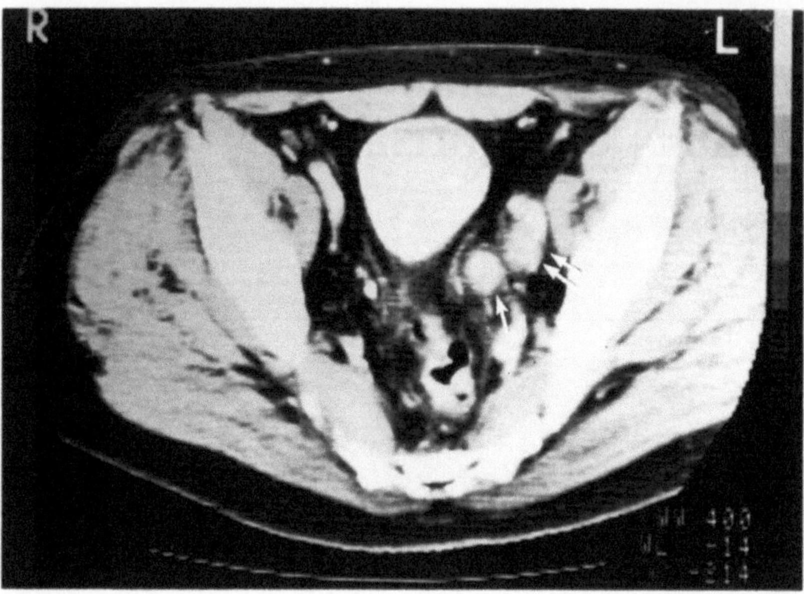

Fig. 25.1. CT scan: presence of metastatic nodes (*double arrows*) close to the dilated ureter (*single arrow*) in a patient with a left invasive bladder tumour.

1.5–2.0 cm in size. Studies have indicated that sensitivity of CT examination approaches 65%, while specificity is 90% and accuracy is only 80% [9,10].

Percutaneous fine needle biopsy of lymph nodes guided by fluoroscopy after lymphangiography was emphasised as an accurate and reliable method in ascertaining possible metastatic involvement [11, 12]. This technique appears to be an improvement over lymphography alone, but it remains limited by the accuracy of lymphangiography itself, which is reported to be around 60%, and is also influenced by the experience of the investigator [13]. Inflammatory changes, micrometastases or total replacement of lymphoid tissue by the metastatic process are responsible for the frequency of false positive and negative images. Moreover, negative cytological findings do not exclude the presence of metastases. This technique might be improved by the use of ultrasonography or CT scanner to guide the needle, especially when the nodes are not visible at lymphography but sufficiently large to be detected (more than 1.5–2 cm in size) [9, 14].

As there are no false positive results, a positive biopsy finding clearly indicates the presence of metastatic pelvic lymph nodes which turn the bladder tumour into a systemic disease, no longer amenable to any form of local radical treatment. Thus these unfortunate patients can be spared this unnecessary radical surgery. In this respect this

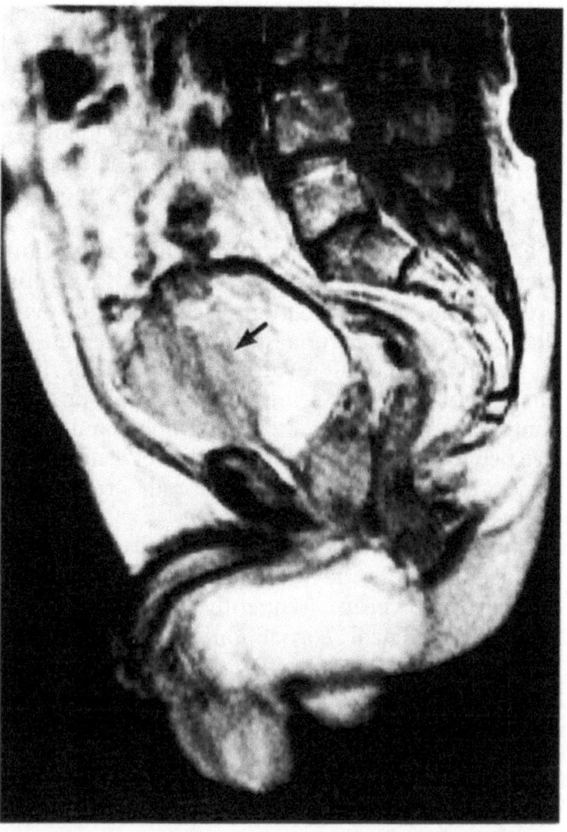

Fig. 25.2. MRI in a patient with a bladder carcinoma (*arrow*).

technique should prove cost-effective in the future. When the biopsy finding is negative, radical lymphadenectomy should be performed whenever radical cystectomy is the treatment option, and, although frozen section of nodes has a 10% false negative rate [15], routine histological examination of the lymphadenectomy specimens will disclose 5%–10% of micrometastases [12]. Bearing these limitations in mind, lymphography with percutaneous fine needle aspiration biopsy definitely has a role in the staging of bladder carcinoma.

Pelvioscopy with lymph node biopsy has been recently added to the armamentarium of techniques for N staging of pelvic malignancies [16]. This procedure is based on the same principle as mediastinoscopy: it is performed with the patient under general anaesthesia and allows biopsy of pelvic lymph nodes guided by direct vision.

Magnetic resonance imaging (MRI) is a promising new technique which is likely to improve the accurate staging of cancer (Fig. 25.2). Unfortunately, however, preliminary reports relating to bladder tumour do not appear to provide better information than conventional CT scanning [17].

References

1. Jewett JH (1973) Cancer of the bladder. Diagnosis and staging. Cancer 32:1072–1074
2. Schmidt JD, Weinstein SH (1976) Pitfalls in clinical staging of bladder tumours. Urol Clin North Am 3: 107–127
3. Vock P, Haertel M, Fuchs WA, Karrer P, Bishop MC, Zingg EF (1982) Computed tomography in staging of carcinoma of the urinary bladder. Br J Urol 54:158–163
4. Nelson RP (1983) New concepts in staging and follow-up of bladder carcinoma. Urology 21:105–112
5. Hodson NJ, Husband JE, MacDonald JS (1979) The role of computed tomography in the staging of bladder cancer. Clin Radiol 30:389–395
6. Seidelman FE, Cohen WN, Bryan PJ (1977) Computed tomographic staging of bladder neoplasms. Radiol Clin North Am 15:419–440
7. Nakamura S, Niijima T (1980) Staging of bladder cancer by ultrasonography: a new technique by transurethral intravesical scanning. J Urol 124:341–344
8. Benson KH, Watson RA, Spring DB, Agee RE (1981) The value of computed tomography in evaluation of pelvic lymph nodes. J Urol 126:63–70
9. Walsh JW, Amendola MA, Konerding KF, Tisrado J, Hazra T (1980) Computed tomographic detection of pelvic and inguinal lymph-node metastases from primary and recurrent pelvic malignant disease. Radiology 137:157–166
10. Sager EM, Talle K, Fosa S, Ous S, Stenwig AE (1983) The role of CT in demonstrating perivesical tumour growth in the preoperative staging of carcinoma of the urinary bladder. Radiology 146:443–446
11. Wallace S, Jing BS, Zornoza J (1977) Lymphangiography in the determination of the extent of metastatic carcinoma. The potential value of percutaneous lymph node biopsy. Cancer 39:706–718
12. Wajsman Z, Gamarra M, Park J, Beckley S, Pontes J (1982) Transabdominal fine needle aspiration of retroperitoneal lymph node in staging of genitourinary tract cancer (correlation with lymphography and lymph node dissection findings). J Urol 128:1238–1240
13. Farah RN, Cerny JC (1978) Lymphangiography in staging patients with carcinoma of the bladder. J Urol 119:40–41
14. Weinerman PM, Arger PH, Coleman BG, Pollack HM, Banner MP, Wein AJ (1983) Pelvic adenopathy from bladder and prostate carcinoma: detection by rapid-sequence computed tomography. Am J Radiol 140:95–99
15. Correa RJ, Kidd R, Burnett L, Brannen GE, Gibbons RP, Cummings KB (1981) Percutaneous pelvic lymph node aspiration in carcinoma of the prostate. J Urol 126:190–191
16. Hald T (1985) Pelvioscopy with lymph node biopsy. In: Controversies in urological surgery. University of Leiden Press, Leiden, pp 59–62
17. Resnick MI, Kursh ED, Bryan PD (1984) Nuclear magnetic resonance imaging and bladder cancer. In: Kuss R, Khowy S, Demis LJ, Murphy GP, Karr JP (eds) Bladder cancer, Part A. Pathology, diagnosis and surgery. Alan R Liss, New York, pp 255–265

Chapter 26

Bladder Carcinoma: the Impact of Pretreatment Knowledge of Nodal Involvement

L. Boccon-Gibod and A. Steg

The involvement of pelvic lymph nodes by metastases from bladder carcinoma upgrades the tumour from a local to a systemic disease. To what extent pretreatment knowledge of nodal involvement should alter therapeutic policies remains to be determined. Three questions at least should be answered:

1. What is the impact of positive nodes on survival?
2. Are positive nodes amenable to any form of local treatment?
3. Should the patient with positive nodes be offered combined local and systemic therapy?

The occurrence of positive nodes increases with the tumour stage from 5% in P1 to 50% in P4 tumours [1–4]. The impact of these nodal metastases on survival is dramatic: 5-year survival figures for N+ patients vary from 0% to 18% [4–6]. Although minimal disease, i.e. micrometastases, to less than three nodes can be associated with a 35% 5-year life expectancy [1], the presence of metastatic nodes on the pelvic wall remains the main deciding feature for survival, as shown by Studer and Zingg [7]. It can therefore be assumed that the presence of positive nodes carries an ominous prognosis regardless of tumour grade and stage.

In the past, treatment of metastatic nodes has resorted first to radical lymphadenectomy and then to various doses and schedules of radiotherapy followed by radical lymphadenectomy. Radical lymphadenectomy alone has yielded poor results and 5-year survival figures vary from 0%, when nodes above the pelvic brim are involved, to 18%–20% in cases of "minimal" nodal disease [4–6]. The 35% 3-year survival rate reported by Skinner remains a unique figure and it should be emphasised that these patients were mainly recruited from among the minimal disease group. As pointed out by Whitmore [6], these figures could reflect the pathologist's dedication to the detection of micrometastatic disease as well as the therapeutic efficacy of a meticulously performed extended lymphadenectomy.

Preoperative radiotherapy using various doses and schedules from 1600 rad in 1 week to 4000 rad in 4 weeks followed by radical cystectomy in conjunction with pelvic lymphadenectomy was considered in the late 1970s to represent a major advance in the treatment of invasive bladder carcinoma: Bloom [8], considering the low incidence (8%) of positive nodes in patients down-staged by radiotherapy, assumed that the therapeutic regimen was able to sterilise metastatic nodes. However, the curative effect of radiotherapy on nodes was questioned by Whitmore himself [6] as his comparatively historic series showed a constant incidence of positive nodes regardless of the pre-cystectomy radiotherapy regimen (from 0 to 4000 rad in 4 weeks), a finding confirmed by Skinner [2]. Moreover, preoperative radiotherapy had no apparent effect on survival or recurrence patterns, local or distant, of patients with metastatic nodes [6]. It is therefore apparent that the presence of positive nodes upgrades the tumour from a local to a systemic disease no longer amenable to local treatment, be it surgery, radiotherapy or both.

These considerations can lead one to question imposing the burden of mutilating, extensive, extirpatory surgery on patients with a minimal life expectancy. Considering the dismal prognosis of invasive bladder tumours and positive nodes, a reasonable option would be to offer these unfortunate patients combined local and systemic treatment.

Postcystectomy doxorubicin-based protocols (DDP) of chemotherapy regimens have not so far led to any definitive advance. On the other hand, impressive preliminary results have been obtained by radiotherapy/chemotherapy (DDP) in stage D patients [9, 10]. However, it should be noted that the high response and survival figures are balanced by a far from negligible morbidity, an issue worth considering in these often debilitated and elderly patients.

In conclusion, the importance of pretreatment knowledge of the status of pelvic wall nodes cannot be over emphasised: lymphangiography and percutaneous fine needle lymph node biopsy [11] can help to select with 75% accuracy the negative node patients, who can be submitted to local treatment, from the positive node patients, who should be treated by combined radiochemotherapy regimens. Repeat node biopsy is used to test the efficacy of the treatment of nodal metastasis. However, it should be kept in mind that nodes are not the only issue to be considered: 45%–55% of negative node patients ultimately die of metastatic disease, suggesting the possibility of haematogenous spread of the tumour and emphasising once again the need for a systemically effective chemotherapy.

References

1. Skinner DG (1982) Management of invasive bladder cancer: a meticulous pelvic node dissection can make a difference. J Urol 128:34–36
2. Skinner DG, Lieskowsky G (1984) Contemporary cystectomy with pelvic node dissection compared to pre-operative radiation therapy plus cystectomy in management of invasive bladder cancer. J Urol 131:1069–1072
3. Skinner DG, Daniels JR, Lieskowsky G (1984) Current status of adjuvant chemotherapy after radical cystectomy for deeply invasive bladder cancer. Urology 24:46–52
4. Smith JA, Whitmore WF (1981) Regional lymph node metastasis from bladder cancer. J Urol 126:591–593
5. Dretler SP, Ragsdale RD, Leadbetter WF (1973) The value of pelvic lymphadenectomy in the surgical treatment of bladder cancer. J Urol 109:414–416
6. Whitmore WF (1981) Integrated irradiation and cystectomy for bladder cancer. In: Connolly JG (ed) Carcinoma of the bladder. Raven Press, New York, pp 235–249
7. Studer UE, Ruchti E, Zingg EJ (1982) The regional lymph node metastasis in invasive bladder cancer: the most important prognostic factor. In: Proceedings of the 19th congress of the Société Internationale d'Urologie, San Francisco, 6–9 September 1982, p 133
8. Bloom HJG, Hendry WF, Wallace DM, Skeet RG (1982) Treatment of T 3 bladder cancer: controlled trial of preoperative radiotherapy and radical cystectomy versus radical radiotherapy. Br J Urol 54:136–151
9. Schaeffer A, Grayhack JT, Merrill JM, Kies MS, Bulkley GJ, Shetty RM, Chmiel JS (1984) Adjuvant doxorubicin hydrochloride and radiation in stage D bladder cancer: a preliminary report. J Urol 131:1073–1076
10. Shipley WU, Coombs LJ, Einstein AB, Soloway MS, Wajsman Z, Prout GR Jr, National Bladder Cancer Collaborative Group A (1984) Cisplatin and full dose irradiation for patients with invasive bladder carcinoma: a preliminary report of tolerance and wall response. J Urol 132:899–903
11. Boccon-Gibod L, Katz M, Le Portz B, Cochand-Priolet B, Steg A (1984) Lymphangiography and percutaneous fine needle biopsy in the staging of infiltrating bladder carcinoma. J Urol 132:24–26

Chapter 27

Bladder Cancer Staging Procedures: the Role of Lymph Node Aspiration Biopsy

Z. Wajsman

The role of proper clinical staging in the case of bladder cancer is even more important than in prostatic cancer. In contrast to prostatic cancer, for which radiation therapy is an alternative choice of treatment, total cystectomy with or without preoperative radiation therapy is at present the most effective treatment of invasive disease. However, it is still a formidable procedure with significant perioperative and postoperative morbidity. The effort to increase the accuracy of staging in bladder tumours is mainly directed, therefore, to ruling out extravesical disease and to avoiding unnecessary mutilating surgery for many patients with invasive bladder neoplasm.

Intravenous pyelography, cystoscopy and bimanual evaluation with the patient under anaesthesia remain the most important and time-honoured ways of establishing the operability of bladder tumours. Although a high incidence of understaging is known, it is not as significant as in prostatic cancer, since all stage B and C tumours still require radical surgery. Computed tomography (CT) scanning has been used mainly to define the extravesical extension of the disease, and it is of value in defining tumour extension towards the pelvic wall and the extension or infiltration into perivesical fat, the prostate and seminal vesicles. False positive diagnosis of extravesical extension is known to many surgeons and it has been recently reported [1]. The value of CT in the detection of regional lymph node involvement is quite disappointing. The reported accuracy of CT in the diagnosis of lymph node metastasis ranges from 70% to 90% with false negative rates of 25%–40% [2, 3].

Pelvic lymph nodes, as discussed in Chapter 3, are involved in a significant number of patients treated with radical cystectomy. The incidence varies between 18% and 25% in the series of Skinner [4] and Smith [5]. In 39 consecutive radical cystectomies performed at the University of Florida from 1983 to 1985 (18 months), seven patients, or 18%, were found to have regional nodal invasion. All seven patients had muscle-invasive tumours (five pT2, one pT3a, one pT4b; Tables 27.1, 27.2). However, all seven patients had grade III tumours and one patient was clinically staged to be T1. The CT scan was suspicious for two of these seven patients (Table 27.3 describes the incidence of pelvic lymph node involvement in relation to pathological staging of bladder cancer as reported by Skinner [4]; in patients with pathological stage P2, one-third of the patients were found to have lymph node metastasis).

Lymphography was also extensively performed in bladder cancers and, not surprisingly, the results are similar to those observed in prostate cancer. In addition to the previously mentioned disadvantages of this test, it does significantly reduce pulmonary function, however temporarily—a significant event since many of these patients are, indeed, very heavy smokers.

Table 27.1. Classification of clinical-diagnostic staging of the primary tumour (T)

cTis	Carcinoma in situ
cTa	Papillary non-invasive carcinoma
cT1	Carcinoma with microscopic evidence of invasion of the lamina propria, but not beyond. On bimanual examination prior to transurethral resection, a freely mobile mass may be felt. This mass should not be felt after complete transurethral resection of the lesion
cT2	Microscopic evidence of invasion of superficial muscle of bladder. On bimanual examination prior to transurethral resection there may be induration of the bladder wall, which is mobile. There is no residual induration after complete transurethral resection of the lesion
cT3	On bimanual examination prior to transurethral resection, there is induration of the bladder wall, or a nodular mobile mass is palpable in the bladder wall, which persists after transurethral resection
cT3a	Microscopic evidence of invasion of muscle, but not beyond
cT3b	Microscopic evidence of invasion of perivesical fat
cT4	Tumour fixed or invading neighbouring structures and there is microscopic evidence of muscle invasion
cT4a	Tumour invading substance of prostate (histologically proved), uterus or vagina
cT4b	Tumour fixed to the pelvic wall and/or infiltrating the abdominal wall

Table 27.2. Pathological classification of primary tumours

pTO	No microscopic evidence of tumour
pTis	Carcinoma in situ
pTa	Papillary non-invasive carcinoma
pT1	Papillary invasive carcinoma that extends into, but not beyond, the lamina propria
pT2	Microscopic evidence of tumour invasion of superficial muscle
pT3a	Microscopic evidence of tumour invasion of deep muscle
pT3b	Microscopic evidence of tumour invasion of perivesical fat
pT4a	Microscopic evidence of tumour invading the substance of the prostate, uterus or vagina
pT4b	Microscopic evidence of tumour invading the abdominal or pelvic wall

Table 27.3. Relation of depth of bladder wall invasion at time of cystectomy (pathological stage) to lymph node involvement [a]

Pathological stage	*No. of patients*	*% + Nodes*
P1 and Pis	41	5
P2	20	30
P3A	13	31
P3B	28	64
P4	8	50

[a] See Skinner [6].

In my experience, gross nodal involvement is frequently detected by CT scanning. Thus, the role of lymphography with fine needle aspiration is limited to patients at high risk of metastatic spread who have negative CT scan findings. These "high-risk" patients are those with palpable masses on bimanual examination and patients with high-grade tumours. Lymph node involvement does not imply incurability—36% 5-year survival was recently reported in patients with lymphatic invasion [6]. How important is it then, in view of these data, to detect microscopic disease in the pelvic nodes? Will it or should it change the treatment decision? One possible advantage of preoperative radiation therapy is the chance of sterilisation of the occult pelvic lymph node metastases. Detection of microscopic lymph node spread by fine needle aspiration and lymphography should not change our treatment decision if preoperative radiation therapy is applied. The chance of detection of these occult spreads is only 15% in patients with negative lymphography and therefore the need for this procedure is limited. Fine needle aspiration, however, has an important role in verification of positive CT scan findings in patients who are candidates for cystectomy.

In our practice the indication for lymphography with fine needle aspiration is limited to patients who are borderline surgical candidates, when one tries to find a good reason not to operate. An additional use for fine needle aspiration in conjunction with lymphography may be in very thin patients whose CT scans are extremely difficult to interpret, particularly in patients with high-grade and large tumour burdens, when there is a high chance of gross nodal involvement undetectable by CT scanning.

References

1. Koss JC, Arger PH, Coleman BG, Mulhern CB Jr, Pollack HM, Wein AJ (1981) CT staging of bladder carcinoma. AJR 137:359–362
2. Lee JKT, Sageb SS, Stanley RJ (eds) (1983) Computerized body tomography. Raven Press, New York
3. Moss (eds) (1983) Computerized tomography of the body. Saunders, Philadelphia
4. Skinner DG, Tift JP, Kaufman JJ (1982) High dose, short course preoperative radiation therapy and immediate single stage radical cystectomy with pelvic node dissection in the management of bladder cancer. J Urol 127:671–674
5. Smith JA, and Whitmore WF Jr (1981) Regional lymph node metastasis from bladder cancer. J Urol 126:591–593
6. Skinner DG (1982) Management of invasive bladder cancer: a meticulous pelvic node dissection can make a difference. J Urol 128:34–36

Chapter 28

Lymph Node Aspiration Biopsy Cytology and Nodal Status in Bladder Cancer

F. Piscioli, P. Scappini, S. Bosetti and L. Luciani

Introduction

Clinical staging of bladder carcinoma is reported to be less reliable, the more the neoplasm infiltrates the bladder wall [1–6]. When a deeply invasive bladder carcinoma is present the possibility of nodal metastases becomes higher, but the status of the regional draining lymph nodes is not always easy to establish. Given the evident therapeutic and prognostic implications related to accurate staging of bladder tumours, numerous staging procedures have been proposed in order to define the nodal extension of the disease. Among these, aspiration biopsy cytology (ABC) of intra-abdominal nodes guided by lymphography and fluoroscopy has proved to be a safe, rapid, reliable means of determining the presence of nodal metastases. The aim of this chapter is to evaluate the role and accuracy of ABC of the pelvic nodes in the management of patients with bladder carcinoma.

Materials and Methods

A total of 31 patients with biopsy-proven bladder carcinoma underwent post-lymphographic fluoroscopy-guided ABC of the pelvic nodes. The needle used was a long-bevelled, side-holed, modified Chiba needle[1] of 22–24 gauge in diameter and 18, 21, 24 cm in length, applied to a plastic syringe handle[2], which provides optimal aspirating capabilities. The technique of aspiration has been described elsewhere [7–10].

The nodes were aspirated regardless of their lymphographic appearance. The cytological and lymphographic findings were classified only as positive or negative without any intermediate category (Figs. 28.1, 28.2). The patients were evaluated before surgery in order to exclude the presence of metastases other than those in the pelvic nodes.

All patients underwent pelvic node dissection during surgical procedures (exploratory laparatomy, urinary diversion, partial or radical cystectomy). The pathological stage was determined according to the TNM system [12].

Results

A total of 102 lymph nodes were aspirated (Table 28.1). Cytological findings were consistent with metastases in 6 nodes from 5 patients, while 13

[1] Cyto-Aspir (patent pending). Cook Urological Inc., Spencer, Ind., USA.

[2] Cameco, Sweden.

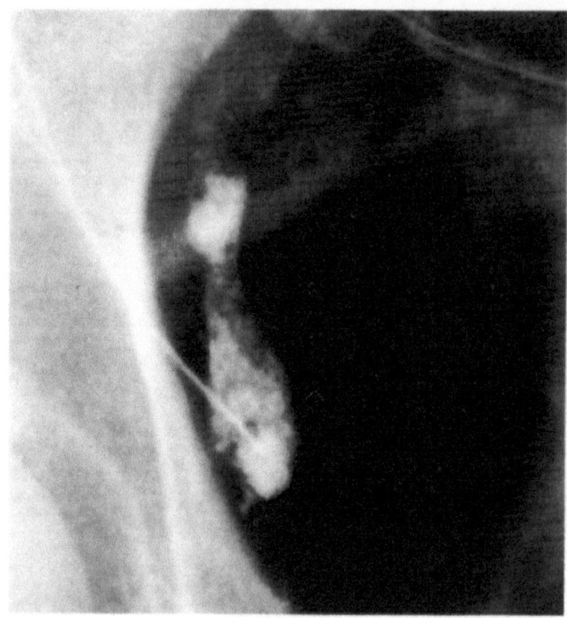

Fig. 28.1. Transcutaneous ABC of a pelvic node. The needle is positioned through the entire node. (Piscioli et al. [11]; reproduced by kind permission of the Editor of *Acta Cytologica*)

nodes from 11 patients gave positive findings on lymphography (Table 28.2).

The cytological specimens were histologically controlled in 66 nodal chains. Histological and cytological findings were negative in 60 nodes and positive in 5. There was one false negative result and no false positive findings. ABC showed an accuracy in determining the status of the single nodal chains of 98%, with a specificity of 100% and a sensitivity of 83%. By contrast, lymphography showed an accuracy of 86%, a sensitivity of 67% and a specificity of 82% (Table 28.3). In evaluating only the presence or absence of nodal involvement, ABC and lymphography showed an accuracy of 98% and 71%, a specificity of 100% and 72%, and a sensitivity of 83% and 67%, respectively (Table 28.4). No significant complications were encountered.

Cytological Findings

The cellular composition of each aspirate reflected the corresponding nodal histology. Inadequate aspi-

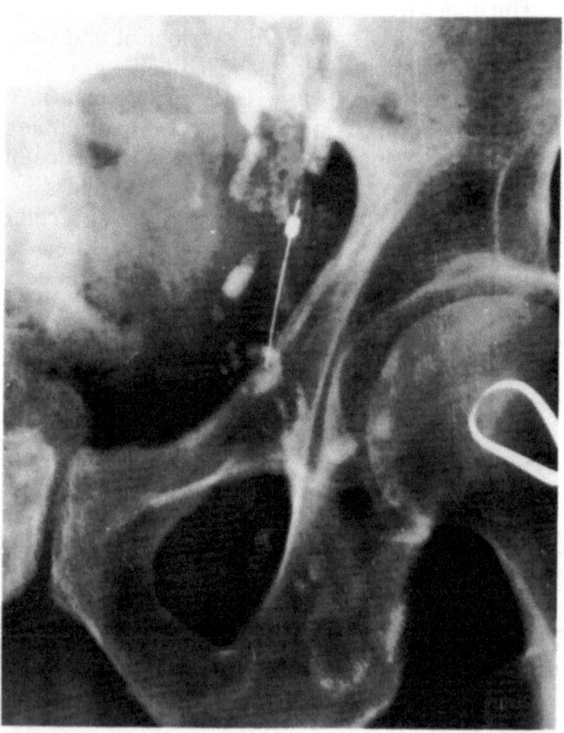

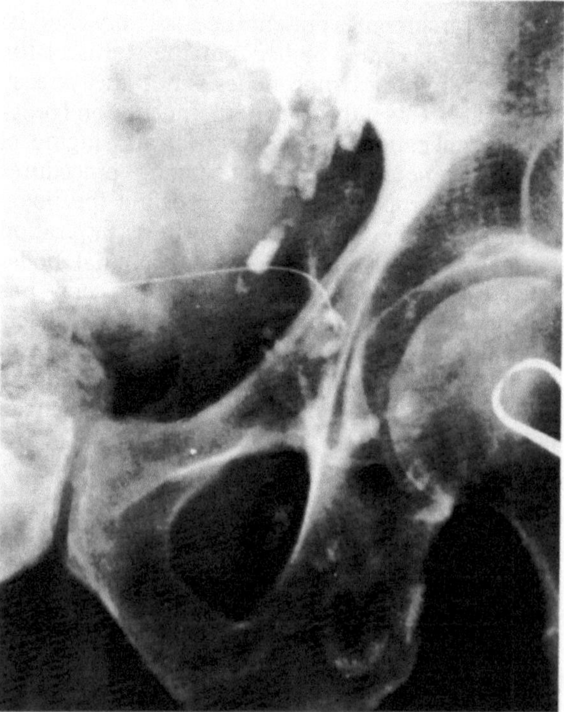

Fig. 28.2. **a** Transcutaneous ABC of a pelvic node. **b** Proper placement of the needle is demonstrated by large dislocation of the node provoked by the instrument.

Table 28.1. Results of ABC in determining the pelvic nodal status in 31 patients with bladder carcinoma[a]

Nodal chains	Cytological finding		No. of nodal chains with histology	True negative	True positive	False negative	False positive
	Negative	Positive					
External iliac							
Right	25	1	22	21	1	—	—
Left	24	1	20	18	1	1	—
Common iliac							
Right	5	—	1	1	—	—	—
Left	5	1	1	—	1	—	—
Internal iliac							
Right	5	1	—	—	—	—	—
Left	5	—	2	2	—	—	—
Obturator							
Right	16	1	12	11	1	—	—
Left	11	1	8	7	1	—	—
Total	96	6	66	60	5	1	—

[a] Modified from Piscioli et al. [13]; reproduced by kind permission of the Editor of *Cancer* and the publisher, J. B. Lippincott Co., Philadelphia.

Table 28.2. Results of lymphography in diagnosing pelvic nodal metastases in 31 patients with bladder carcinoma[a]

Nodal chains	No. of nodal chains examined	True negative	True positive	False negative	False positive
External iliac					
Right	24	19	1	—	4
Left	22	16	—	1	4
Common iliac					
Right	2	2	—	—	—
Left	1	—	1	—	—
Internal iliac					
Right	2	2	—	—	—
Left	4	4	—	—	—
Obturator					
Right	14	12	1	1	1
Left	11	10	1	—	—
Total	80	65	4	2	9

[a] Modified from Piscioli et al. [13]; reproduced by kind permission of the Editor of *Cancer* and the publisher, J. B. Lippincott Co., Philadelphia.

Table 28.3. Reliability of ABC and lymphography in determining the status of single nodal chains in bladder carcinoma[a]

	No. of patients	Sensitivity	No. of false negative results	No. of false positive results	Specificity	Accuracy
ABC	31	83%	1	—	100%	98%
Lymphography		67%	2	9	88%	86%

[a] Modified from Piscioli et al. [13]; reproduced by kind permission of the Editor of *Cancer* and the publisher, J. B. Lippincott Co., Philadelphia.

Table 28.4. Reliability of ABC and lymphography in determining the true stage of bladder carcinoma[a]

	No. of patients	Sensitivity	No. of false negative results	No. of false positive results	Specificity	Accuracy
ABC	31	83%	1	—	100%	98%
Lymphography		67%	2	7	72%	71%

[a] Modified from Piscioli et al. [13]; reproduced by kind permission of the Editor of *Cancer* and the publisher, J. B. Lippincott Co., Philadelphia.

rates usually contained only blood and fibrin. A negative diagnosis indicated an adequately cellular aspirate that was characterised by the presence of the cellular components of benign lymphadenopathy following lymphography: multinucleated giant cells, vacuolated macrophages (with fat globules emanating from the oily contrast solution), eosinophils and neutrophils (Fig. 28.3).

The aspirates from metastatic nodes showed a large number of malignant cells, singly and in clusters. The cancer cells showed variability in shape and size, with scanty cytoplasm and prominent abnormal hyperchromatic nuclei (Fig. 28.4). Multiple and large nucleoli were occasionally found.

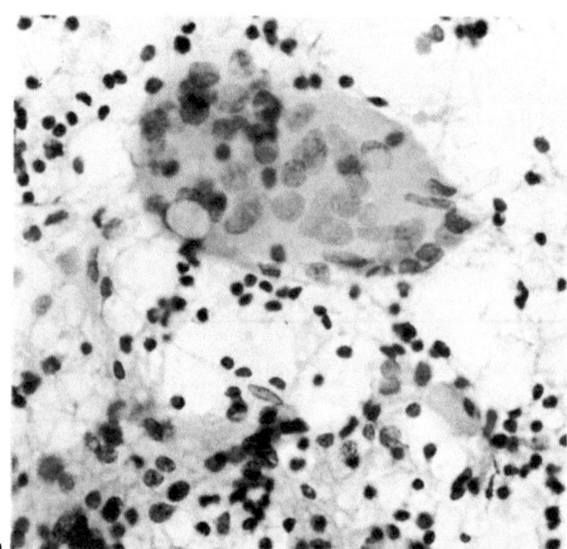

a

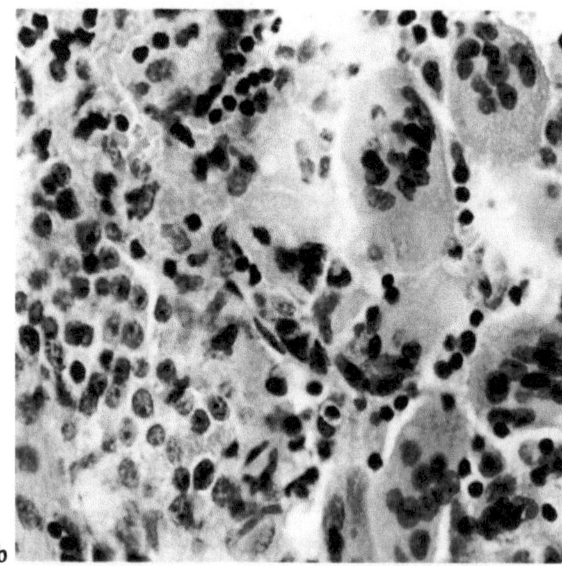

b

Fig. 28.3. **a** Cytological and **b** histological patterns of benign lymphadenopathy following lymphography showing multinucleated giant cells. (**a** Papanicolaou stain, ×275; **b** H&E, ×275.) (Piscioli et al. [11]; reproduced by kind permission of the Editor of *Acta Cytologica*)

Discussion

As in other human malignancies, the adequate treatment of bladder cancer depends strictly on accurate knowledge of the extension of the disease at diagnosis. Bipedal lymphography is certainly one of the most widely used staging methods in urological oncology. Nevertheless, its reliability is limited by its false positive and false negative results [14–18]. The main source of errors in the interpretation of the lymphograms is the lack of specificity of the lymphographic appearance of the metastatic nodes. Fibrosis, fat, infections or errors of interpretation may all lead to an erroneous diagnosis of metastatic deposits [19]. Lymphadenectomy remains the only procedure to provide accurate information on the nodal status, but routine surgical staging is impracticable, given the associated morbidity and mortality.

As in other urological cancers, ABC can be considered a safe, minimally invasive procedure, an alternative to surgery in the staging of patients with bladder cancer. Some preliminary observations should be made regarding the therapeutic implications of clinical staging.

Lymphographic studies demonstrated the presence of an extremely well-developed lymphatic network in the bladder wall [20, 21]. It is arguable that tumour cells of the bladder cancer may easily invade the lymphatic system and may metastasise to perivesical and regional lymph nodes early in the evolution of this tumour. These anatomical data suggest that the knowledge of the status of the draining nodes is crucial for adequate management of the disease, even in apparently localised bladder cancer.

The perivesical nodes are not visualised by conventional radiological means (lymphography, CT, echography); however, the preoperative investigation of these nodal chains is of limited importance, since the perivesical nodes are included in the field of primary radiotherapy or radical surgical treatment. Therefore, the main determinant in the preoperative management of bladder cancer is the study of the nodes beyond those in the immediate vicinity of the bladder wall.

Diffuse extension of advanced bladder cancer to common iliac, external iliac or para-aortic nodes

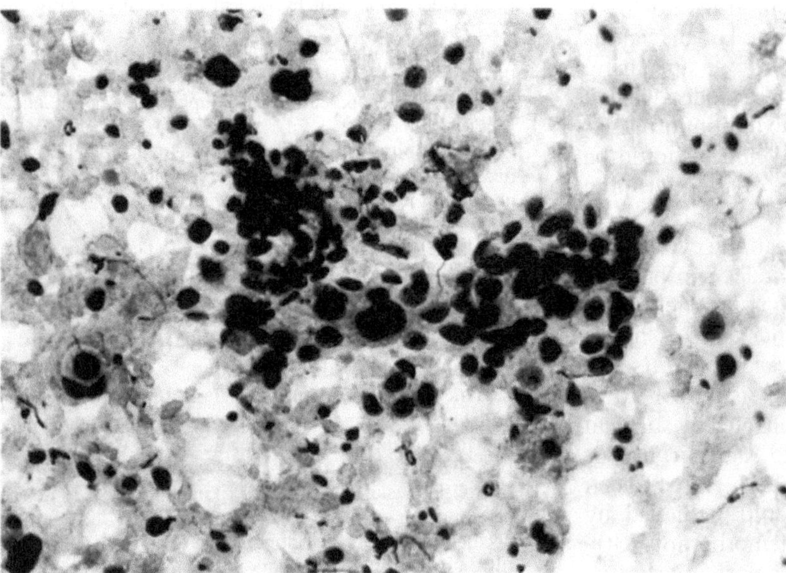

Fig. 28.4. Aspirate from nodal metastasis shows clusters of malignant cells amidst necrotic material and leucocytes. Nuclear enlargement and hyperchromasia and irregular nuclear outline are clearly seen. (Papanicolaou stain, × 272.) (Piscioli et al. [13]; reproduced by kind permission of the Editor of *Cancer* and the publisher, J.B. Lippincott Co., Philadelphia)

is generally a contraindication to major surgery because the disease is regarded as systemic [22]. On the other hand, lymphadenectomy could be regarded as curative when the metastatic involvement is limited to few pelvic nodes (minimal metastatic disease) [23]. In this connection ABC has proved to be particularly useful in the study of these nodal chains since they are all opacified by pedal lymphography and easily punctured, with minimal morbidity for the patients.

Experience accumulated over the past few years has shown that significant improvements in staging

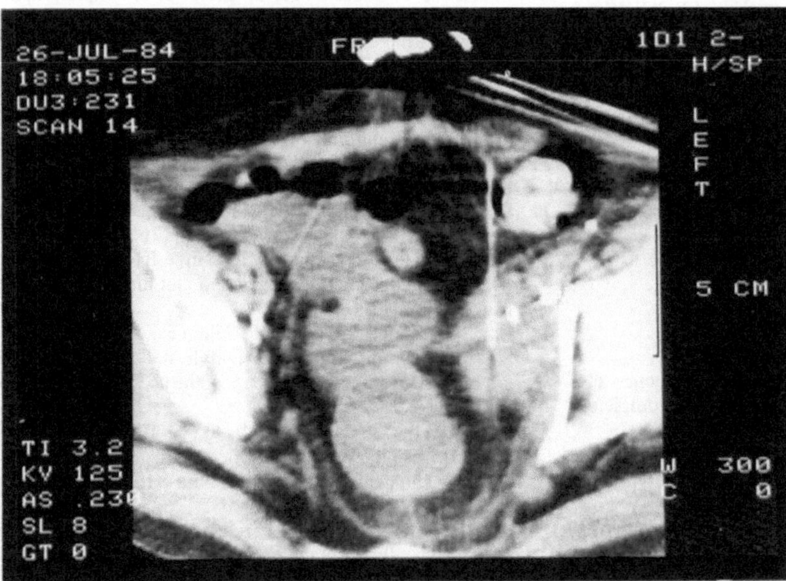

Fig. 28.5. Restaging after radical cystectomy for invasive bladder cancer. CT scan showing a mass in the left side of the true pelvis. ABC revealed recurrence of bladder carcinoma.

accuracy can be achieved by aspirating not only lymphographically positive, abnormal or suspicious lymph nodes [24, 25] but even those appearing "normal", since 10%–25% of these nodes yielded positive results on cytological examination [26–29]. For this reason, all opacified nodes should be aspirated, regardless of their lymphographic appearance.

The presence of malignant cells in the cytological specimens from more than two nodes can be generally accepted as evidence of massive metastatic nodal involvement and the bladder cancer should be considered systemic. In such cases, patients can be spared elective radical surgery since the 3- to 5-year survival is not significantly affected by surgical or radiation therapy or combined radiosurgical or chemicosurgical treatments. There have been recent reports of interesting results achieved by radiotherapy combined with chemotherapy [30–32]; this form of treatment seems particularly justified by the fact that 45%–55% of the patients undergoing radical cystectomy die from distant metastasis [33]. By contrast, negative cytological findings have no diagnostic value, and pelvic lymphadenectomy should always be associated with radical cystectomy because of the possibility of micrometastases (5%–10%) [29, 34, 35].

In conclusion, ABC can be of great help in the pretherapeutic evaluation of patients with bladder carcinoma. This effective, minimally invasive procedure can provide useful information on the status of the pelvic lymph nodes, permitting an accurate staging and prognostic evaluation of the disease. Since the radiopaque contrast medium continues to opacify the nodes for 6–9 months, ABC is the procedure of choice for a careful follow-up of patients who have undergone combined radio-chemotherapeutic treatment and for restaging of surgically treated patients (Fig. 28.5).

References

1. Jewett HJ (1952) Carcinoma of the bladder: influence of depth of infiltration on the 5-year results following complete extirpation of the primary growth. J Urol 67:672–676
2. Marshall VF (1952) The relation of the preoperative estimate to the pathologic demonstration of the extent of vesical neoplasms. J Urol 68:714–723
3. Kenny GM, Hardner GJ, Murphy GP (1970) Clinical staging of bladder tumors. J Urol 104:720–723
4. Varkarakis MJ, Gaeta J, Moore RH, Murphy GP (1974) Superficial bladder tumor: aspects of clinical progression. Urology 4:414–417
5. Richie JP, Skinner DG, Kaufman JJ (1975) Carcinoma of the bladder: treatment by radical cystectomy. J Surg Res 18:271–275
6. Wallace S, Jing BS, Zornoza J (1977) Lymphangiography in the determination of the extent of metastatic carcinoma. The potential value of percutaneous lymph node biopsy. Cancer 39:706–708
7. Zornoza J, Johnson K, Wallace S, Lukeman JM (1977) Fine needle aspiration biopsy of retroperitoneal lymph nodes and abdominal masses: an updated report. Radiology 125:87–88
8. Göthlin JH (1976) Post lymphographic percutaneous fine needle biopsy of lymph nodes guided by fluoroscopy. Radiology 120:205–207
9. Göthlin JH (1978) Percutaneous transperitoneal fluoroscopy-guided fine-needle biopsy of lymph nodes. Acta Radiol [Diagn] (Stockh) 20:660–664
10. Luciani L, Piscioli F, Menichelli E, Pusiol T (1983) The value and role of percutaneous pelvic lymph node aspiration biopsy in definitive staging of prostatic and bladder carcinoma. Eur Urol 9:216–226
11. Piscioli F, Pusiol T, Leonardi E, Luciani L (1983) Role of percutaneous pelvic node aspiration cytology in the management of bladder carcinoma. Acta Cytol 29:37–43
12. UICC (1978) Union Internationale Contre le Cancer. TNM classification of malignant tumor, 3rd edn. International Union Against Cancer, Geneva
13. Piscioli F, Scappini P, Luciani L (1985) Aspiration cytology in the staging of urologic cancer. Cancer 56:1173–1180
14. Collard M, Timmermans L (1973) Intérêt et limite de la lymphadènographie dans l'exploration des tumeurs des vies urinaries. J Urol Nephrol 79:368–369
15. Grasset D, Navratic H (1973) Limites de la lymphographie dans le diagnostic de l'extension ganglionnaire des cancer vèsicaux. J Urol Nephrol 79:369–370
16. Turner AG, Hendry WF, MacDonald JS, Wallace DM (1976) The value of lymphography in the management of bladder cancer. Br J Urol 48:579–586
17. Essed E, Mulder JD (1979) Reliability of lymphography as a diagnostic aid in the planning of treatment for bladder carcinoma. Urol Res 7:93–96
18. Juimo AG, Masselot J, Markovitz P, Amar MH, Rebibo Lorimy G, Sassoon C, Budet C (1982) Intérêt de la lymphographie dans le bilan pre-therapeutique des cancers vèsicaux de l'adulte. J Radiol (Paris) 63:415–421
19. Schmidt JD, Weinstein SH (1976) Pitfalls in clinical staging of bladder tumors. Urol Clin North Am 3:107–127
20. Wirtnanen GW, Miller RC (1973) Bladder lymphatics and tumor dissemination. J Urol 109:58–59
21. Fiorelli C, Luciani L (1977) La linfografia indiretta. Nota di tecnica originale. Riv Med Trentina 15:1–8
22. Smith JA, Jr, Whitmore WF Jr (1981) Regional lymph node metastasis from bladder cancer. J Urol 126:591–593
23. Skinner DG (1982) Management of invasive bladder cancer. A meticulous pelvic node dissection can make a difference. J Urol 128:34–36
24. Dan SJ, Efremidis SC, Trains JS, Cohen BA, Mitty HA (1982) Equivocal lymphogram and lymph node aspiration: its importance in staging carcinoma of the prostate. Urol Radiol 4:215–219
25. Dan SJ, Wulsohn MA, Efremidis SC, Mitty HA, Brendler H (1982) Lymphography and percutaneous lymph node biopsy in clinically localized carcinoma of the prostate. J Urol 127:695–698
26. Göthlin JH, Hiem L (1981) Percutaneous fine needle biopsy of radiologically normal lymph nodes in the staging of prostatic carcinoma. Radiology 141:351–354
27. Göthlin JH, Rupp N, Rothenberger KH, MacIntosh PK (1981) Percutaneous biopsy of retroperitoneal lymph nodes: a multicentric study. Eur J Radiol 1:46–50

28. Rothenberger KH, Hofstetter A, Rupp N, Pfeifer KJ (1981) Transabdominal fine needle biopsy of lymph nodes for urologic cancer staging. In: Schulman CC (ed) Advances in diagnostic urology. Springer, Berlin Heidelberg New York, pp 113–117
29. Wajsman Z, Gamara M, Park JJ, Berkley SA, Pontes JE (1982) Transabdominal fine needle aspiration of retroperitoneal lymph nodes in staging of genitourinary tract cancer (correlation with lymphography and lymph node dissection findings). J Urol 128:1238–1240
30. Studer UE, Ruchti E, Zingg EJ (1982) The regional lymph node metastasis in invasive bladder cancer: the most important prognostic factor. In: Proceedings of the XIX Congress of the International Society of Urology, San Francisco, 6–9 September 1982, p 133
31. Schaeffer A, Grayhack JT, Merrin JM, Kies MS, Buckley GJ, Shetty RM, Schniel JS (1984) Adjuvant doxorubicin hydrochloride and radiation in stage D bladder cancer. A preliminary report. J Urol 131:1073–1077
32. Shipley WU, Coombs LJ, Einstein AB, Soloway MS, Wajsman Z, Prout G (1984) Cis-platin and full dose irradiation for patients with invasive bladder carcinoma: a preliminary report of tolerance and wall response. J Urol 132:899–904
33. Boccon-Gibod L, Katz M, Cochand D, La Porta B, Steg A (1984) Lymphography and percutaneous fine needle node aspiration biopsy in the staging of bladder carcinoma. J Urol 132:24–26
34. Boccon-Gibod L, Steg A (1985) Cancer de vessie. Importance de la détermination dans l'envahissement ganglionnaire, dans la choix de la thérapeutique. Ann Urol 19 (4):223–224
35. Wajsman Z, Gamara M, Park JJ, Beckley SA, Pontes JE, Murphy JP (1982) Fine needle aspiration of metastatic lesions and regional lymph nodes in genitourinary cancer. Urology 4:356–360

Section V

Penile Carcinoma

Chapter 29

Cabanas Approach: Is It a Reliable Sentinel Node for Staging Penile Carcinoma?

E. Wespes and C. C. Schulman

Introduction

The evaluation and management of the regional lymph nodes of patients with carcinoma of the penis remains the subject of controversy [1–3]. Prophylactic inguinal node dissection has been advocated by some authors, while others advise a "wait and see" policy for patients with clinically negative nodes [4, 5]. It is well documented that clinical evaluation of the inguinal lymph nodes is inaccurate and various studies have reported clinically false positive and false negative examinations [6–8].

In 1977 Cabanas [9] suggested, after performing lymphography of the dorsal lymphatic vessels of the penis, that a node located on the anteromedial aspect of a superficial epigastric vein just above and medial to the junction of the epigastric and saphenous veins is the first metastatic lymph node relay in carcinoma of the penis [9]. He concluded that if this node is not infiltrated by carcinoma no further immediate surgical therapy is indicated. We report the case of two patients with penile carcinoma treated by partial penectomy who within 1 year developed palpable, positive, unresectable inguinal and iliac lymph nodes, although the sentinel lymph node biopsies had been negative at initial treatment.

Clinical Report

Two patients aged 54 and 68 years presented with an indurated ulcer in the glans penis. Biopsies of these lesions confirmed invasive squamous cell carcinoma. At that time there were no palpable groin lymph nodes. They both underwent partial penectomy 2 cm proximal to the lesion and bilateral sentinel lymph node biopsies were performed as advocated by Cabanas. These node biopsies were reported as negative on histological examination. Ten months later one patient developed palpable nodes in the right groin. Surgical exploration revealed infiltration of deep inguinal lymph nodes by squamous cell carcinoma, considered unresectable at this level. The patient received chemotherapy with bleomycin but died from rupture of the femoral artery which was infiltrated by the carcinomatous process. The other patient presented with palpable nodes in the left groin 6 months later. Percutaneous cytological examination revealed metastasis of the penile carcinoma. The patient underwent bilateral ileoinguinal lymphadenectomy. Histological examination showed metastasis in the nodes at the bifurcation of the aorta. He received a full course of chemotherapy with bleomycin and, at the time of writing, was still alive 2 months after treatment.

Discussion

Prophylactic management of the inguinal nodes in carcinoma of the penis still remains controversial because 20% of the patients with clinically negative nodes have micrometastases, and 50% of the patients with clinically positive nodes have inflammatory lesions but no tumour on histological study [1–8, 10, 11]. Although inguinal lymphadenectomy is the most accurate diagnostic procedure, its attendant morbidity and mortality are too great to be inflicted on the patients with truly negative nodes [6, 9, 12–14].

Cabanas [9] performed lymphography via the dorsal lymphatics of the penis and postulated that the initial lymphatic node relay is located adjacent to the junction of the saphenous and femoral veins [15]. He performed biopsy of this sentinel node to document the regional lymphatics in patients with carcinoma of the penis. Only 3 of the 15 patients with positive findings from sentinel node biopsies had additional ipsilateral inguinal node metastases, while metastases were not found in other lymph nodes when the sentinel node biopsy findings were negative [9].

In a recent study, Fowler supports the conclusion that the sentinel lymph node is often the first site of regional lymphatic metastasis in penile cancer [16]. However, such an approach may not be a reliable indicator on which to base the management of the regional nodes in patients with penile carcinoma, because in the series of Cabanas, while 28 of 31 patients with negative sentinel node biopsy results were alive, apparently free of tumour, after 5 years, the cause of death in the other 3 patients (10%) was not specified [9]. This possible false negative percentage is dangerously close to the incidence of occult metastases in patients with clinically negative nodes. On the other hand, direct drainage into the iliac nodes, bypassing the inguinal nodes, is also possible [17]. Bypassing the sentinel node with drainage into the inguinal or pelvic nodes may also be considered.

In the series of Fowler one patient developed a superficial inguinal recurrence infiltrated with tumour 4 months after a negative sentinel node biopsy result [16]. Perinetti et al. reported a patient in whom unresectable pelvic lymph node metastases developed 6 months after negative bilateral sentinel node biopsy findings [18]. One of our two patients with negative sentinel node biopsy results developed unresectable inguinoiliac lymph node metastases and the other presented metastases in the iliac nodes considered to have a very poor prognosis (Fig. 29.1). Although we recognise that theoretically we could have missed the sentinel lymph nodes in these two patients like Perinetti et al. [18]

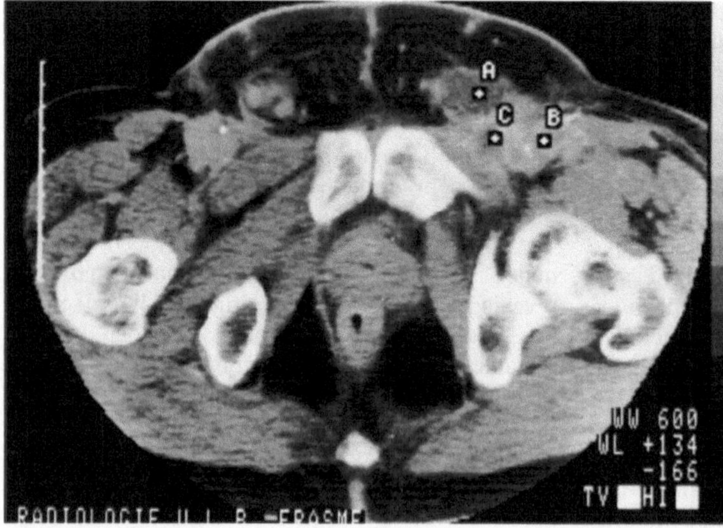

Fig. 29.1. CT scan in a patient with a penile carcinoma and previously negative Cabanas ganglia: *A*, hydrocele; *B*, metastatic nodes; *C*, inguinal vessels.

we consider that this approach is not sufficiently reliable to give better management of patients with carcinoma of the penis.

Recently Luciani et al. [19] proposed regional node aspiration cytology after penile and pedal lymphography in the management of penile carcinoma [19]. This technique seems to be effective and it spares some patients the risk of an unnecessary ilioinguinal lymph node dissection. Nevertheless, we believe that only positive sentinel node biopsy or cytology findings may be meaningful, while a negative biopsy result must not be considered as an accurate guide in the assessment of patients with penile carcinoma.

References

1. Hoppman HJ, Fraley EE (1978) Squamous cell carcinoma of the penis. J Urol 120:373–398
2. Nelson RP, Derrick FC, Allen UR (1982) Epidermoid carcinoma of the penis. Br J Urol 54:172–175
3. Krieg RM, Luck KH (1981) Carcinoma of the penis. Urology 18:149–154
4. Edwards R, Sawyers J (1968) The management of carcinoma of the penis. South Med J 61:843
5. Khezrei AA, Dunn M, Smith PJB, Mitchell JP (1978) Carcinoma of the penis. Br J Urol 50:275–279
6. Beggs JH, Spratt JS (1964) Epidermoid carcinoma of the penis. J Urol 91:166–172
7. deKernion JB, Tynberg P, Persky L, Fegen JP (1973) Carcinoma of the penis. Cancer 32:1256–1262
8. Hardner GI, Bhanalaph T, Murphy GP, Albery DJ, Moore RH (1972) Carcinoma of the penis: analysis of therapy in 100 consecutive cases. J Urol 108:428–435
9. Cabanas RM (1977) An approach for the treatment of penile carcinoma. Cancer 39:456–466
10. Catalona WJ (1980) Role of lymphadenectomy in carcinoma of the penis. Urol Clin North Am 7:785
11. Grabstald H (1980) Controversies concerning lymph node dissection for cancer of the penis. Urol Clin North Am 7:785
12. Fraley EE, Hutchens HC (1972) Radical ileo-inguinal node dissection-skin bridge technique. J Urol 108:279–281
13. Ekstrom T, Edsmyr F (1958) Cancer of the penis at the Norwegian Radium Hospital. Acta Chir Scand 115:25–45
14. Uehling DT (1973) Staging laparotomy for carcinoma of penis. Urology 110:213–215
15. Riveros M, Garcia R, Cabanas R (1967) Lymphadenography of the dorsal lymphatics of the penis. Cancer 20:2026–2031
16. Fowler JE (1984) Sentinel lymph node biopsy for staging penile cancer. Urology 23 (4): 352–354
17. Schellhammer PF, Grabstald H (1979) Tumors of the penis and urethra. In: Harrison JH, Gittes RD, Perlmutter AD, Stamey TA, Walsh PC (eds) Campbell's urology, vol 2. Saunders, Philadelphia, pp 1177–1190
18. Perinetti E, Crane DB, Catalona WJ (1980) Unreliability of sentinel lymph node biopsy for staging penile carcinoma. J Urol 124:734–735
19. Luciani L, Piscioli F, Scappini P, Pusiol T (1984) Value and role of percutaneous regional node aspiration cytology in the management of penile carcinoma. Eur Urol 10:294–302

Chapter 30

Lymph Node Aspiration Biopsy and Nodal Status in Penile Cancer

K. H. Rothenberger

Introduction

In patients with penile cancer, as in those with other urological malignancies, treatment and prognosis are related to tumour staging. The main factor limiting the 5-year survival rate is the presence of nodal metastasis [1, 2]; therefore knowledge of the nodal status is decisive. As previously mentioned (see Chap. 29), lymphadenectomy cannot be proposed as a routine procedure because of the morbidity and mortality accompanying this surgery [3, 4]. Pedal lymphography is not completely satisfactory, and even penile lymphography in conjunction with surgical biopsy of the sentinel node, as proposed by Cabanas [4]. has shown some limitations [5]. For these reasons, we use aspiration cytology to complete preoperative staging of the disease.

Material and Methods

The present study includes 22 patients with penile cancer who have undergone irradiation of the primary lesion with neodymium, yttrium aluminium garnet (YAG) and laser [6]. We carried out only pedal lymphography, not penile, so the sentinel lymph node could not be visualised. Aspiration biopsy was performed on this node only in the case of palpable lesions. The particular technique of aspiration biopsy has been described previously (see Chap. 7). In our series of 22 patients aspiration biopsy was performed in 21 patients and lymph node dissection in 14. No significant complications were seen.

Results

The main clinical, lymphographic, cytological and histological features are summarised in Table 30.1. The cytological and histological findings were identical. In one case, the lymphographic diagnosis of nodal metastases was not confirmed on histological examination.

In a follow-up study we lost two patients, classified as N0 patients, because of lymph node metastases. In one patient lymphographic findings were positive and cytological findings were negative and in the other lymphographic and cytological findings were negative. We did not perform staging operation in these two patients.

Table 30.1. Penile carcinoma

TNM	*n*	Lymphography	Aspiration biopsy	Staging operation	Follow-up (months)
T1 N0 M0	11	9 negative	11 negative	7 negative	31–59, 7 no relapse
		2 positive		4 not done	36–65, 3 no relapse
					53, lymph node metastases
T2 N0 M0	6	5 negative	6 negative	4 negative	
		1 positive		2 not done	30–62, 6 no relapse
T3 N0 M0	1	Positive	Negative	Not done	34, lymph node metastases
T1 N1 M0	1	Positive	Positive	Positive	59, no relapse
T2 N1 M0	2	2 positive	Not done	2 positive	50, no relapse
			1 positive		24, lymph node metastases
T2 N3 M1	1	Positive	Positive	Not done	30, general metastases

Conclusion

Based on our experience we think that positive cytological findings are conclusive of metastatic nodal involvement. In contrast, in cases of negative aspiration cytological findings surgical staging must be performed. Aspiration biopsy of lymph nodes along the possible pathway of metastasis in cancer of the penis is a simple method which provides sufficient information in tumour staging. Only positive findings are considered conclusive. If no tumour cells are found, further diagnostic steps (lymph node dissection) are recommended.

References

1. deKernion JB, Tymberg P, Perky L, Fegen JP (1973) Carcinoma of the penis. Cancer 32:1256–1262
2. Kuruvilla JI, Garlick FH, Mammen KE (1971) Results of surgical treatment of carcinoma of the penis. Aust NZ J Surg 41:157–159
3. Beggs JH, Spratt JS (1964) Epidermoid carcinoma of the penis. J Urol 91:166–172
4. Cabanas RM (1977) An approach for the treatment of penile carcinoma. Cancer 39:456–466
5. Perinetti EP, Crane DB, Catalona WJ (1980) Unreliability of sentinel lymph node biopsy for staging penile carcinoma. J Urol 124:734–735
6. Rothenberger K, Hofstetter A, Pensel J, Keiditsch E (1982) Neodym-YAG-Laser-Behandlung maligner Tumoren des Penis. Fortschr Med 100:1806–1808

Chapter 31

Aspiration Cytology in the Staging of Penile Cancer

P. Scappini, F. Piscioli, E. Menichelli and L. Luciani

Introduction

Carcinoma of the penis is relatively uncommon in developed countries, accounting for less than 1% of the cancer-bearing male population in the USA. However, penile cancer has a higher incidence in underdeveloped countries or in areas in which early circumcision is not routinely practised [1]. This neoplasm is therefore a significant world health problem. Because of the frequent presence of inflammatory changes in the primary tumour, it is extremely difficult to establish the extent of local nodal disease. Nevertheless, 10%–30% of patients without clinical evidence of adenopathy have occult metastases (Table 31.1) [2–17]. From a review of the series reported in the literature, Catalona [19] found that these percentages were strictly parallel with the proportion of patients with clinically negative nodes who ultimately died of cancer.

The clinical assessment of lymph node involvement is largely inaccurate (Table 31.2) [3, 6, 9, 13, 20, 25] so that prophylactic radical superficial and deep groin node dissection would be the procedure of choice in determining the true status of the regional nodes. Controversy exists, however, because criteria for appropriate timing and extent of surgery have not been clearly defined and the procedure itself has been associated with a high incidence of complication (Table 31.3) [6, 10, 13, 22, 26, 27] and a 1%–3% mortality [3, 4].

Aspiration biopsy cytology (ABC) has been proposed as an accurate, safe, reliable diagnostic procedure which is an alternative to surgical staging for determining the regional node status in patients with bladder and prostatic cancer. The aim of this chapter is to delineate the role of ABC in the staging of penile cancer.

Materials and Methods

A total of 8 patients, ranging in age from 51 to 79 years (mean 64 years) with histologically proved squamous cell carcinoma of the penis, underwent ABC at the Division of Urology, S. Chiara Hospital from 1977 to 1985. None of these patients were circumcised at birth.

Pedal lymphography was performed in every case, while combined pedal and penile lymphography were accomplished in four cases only. Fluoroscopy was the guidance mechanism in seven cases and computed tomography (CT) in the remaining one.

The aspiration set consisted of a long-bevelled, side-holed, modified Chiba needle of 22–23 gauge

Table 31.1. Mortality of patients with occult regional nodal metastases from penile carcinoma[a]

Reference	No. of patients with clinically negative nodes	Patients who died from cancer	%
Engelstad (1948) [2]	36	3	8%
Ekström and Edsmyr (1958) [3]	112	11	9%
Beggs and Spratt (1964) [4]	12	2	17%
Murrel and Williams (1965) [5]	63	28	44%
Edward and Sawyers (1968) [6]	25	8	32%
Fegen and Persky (1969) [7]	22	3	13%
Hanash et al. (1970) [8]	106	24	22%
Kuruvilla et al. (1971) [9]	59	18	31%
Skinner et al. (1972) [10]	24	3	12%
deKernion et al. (1973) [11]	22	5	23%
Derrick et al. (1973) [12]	60	14	23%
Gursel et al. (1973) [13]	10	2	20%
Johnson et al. (1973) [14]	125	52	42%
Baker et al. (1976) [15]	102	27	27%
Khezri et al. (1978) [16]	43	2	4%
Salaverria et al. (1979) [17]	73	24	32%
Total	894	226	25%

[a] Scappini et al. [18]; reproduced by kind permission of the Editor of *Cancer* and the publisher, J. B. Lippincott Co., Philadelphia.

in diameter and 18–22 cm in length, attached to a plastic syringe handle. After bland sedation and local anaesthesia had been administered, the needle was inserted from a transabdominal approach. When a particular node was thought to be punctured, up and down movements and rotations were made with the needle before accomplishing aspirations (Fig. 31.1). Generally, six to eight smears could be prepared from a single aspiration.

The cytological material was immediately fixed in 95% ethyl alcohol, and one slide was processed according to the rapid Papanicolaou method while the patient remained in the fluoroscopy suite. If the sample proved inadequate the procedure was repeated. The remaining slides were stained according to the standard Papanicolaou technique (Fig. 31.2). The nodes were punctured regardless of their lymphographic appearance. The cytological as well as the lymphographic findings were classified as either positive or negative; no equivocal diagnostic category was used.

In an attempt to investigate the inguinal and pelvic lymphatic pathways better, additional pedal and penile lymphoscintigraphy was performed in three of the cases reported and in the small number of cases of prostatic [5] and bladder [4] carcinoma (Fig. 31.3). In addition, 100 pedal lymphograms previously performed for staging purposes in cases of other urological malignancies were also reviewed.

Local excision was performed in two cases, partial amputation in three cases and total penectomy in the remaining three cases; one patient in the last group also underwent radiotherapy. Histological examinations of the nodes removed were performed in six cases (Fig. 31.4). No significant complications were encountered.

Results

In Table 31.4 the main cytological, lymphographic and histological findings of our series are summarised. Positive cytological findings were observed in four patients. There were no false positive or false negative results in the staging of the single patients. On the other hand, pedal lymphography did not show any metastasis in one patient whose nodes contained malignant cells on histological examination.

As regards the detection of malignant cells in any single node, ABC never failed to identify metastatic disease, while lymphography showed signs of nodal involvement in two nodes which were negative on histological examination. No significant details of the lymphatic pathways were disclosed on radionuclide studies. In 15 of the 100 pedal lymphograms reviewed, the same inguinal nodes radiologically located over the junction of the femoral head and the ascending ramus of the pubis, which are reported to be opacified only via penile lymphatics, were visualised (Fig. 31.5). The follow-up period ranged from 4 months to 64 months (mean 27 months). Three patients with metastatic nodal involvement died within 4, 5 and 36 months respectively. At the time of writing, the patient on whom radiotherapy was attempted is still living and well after 22 months. The other patients are also living and well.

Table 31.2. Discrepancy between clinical and surgical stage in penile carcinoma[a]

Reference	Total no. of patients examined	No. of patients with clinically and surgically positive nodes	No. of patients with clinically positive and surgically negative nodes	Predictive value of clinical examination (%)	False positive rate of clinical examination (%)
Ekström and Edsmyr (1958) [3]	229	63	42	60	40
Marcial et al. (1962) [20]	258	61	45	57	43
Edward and Sawyers (1968) [6]	77	14	13	52	48
Kuruvilla et al. (1971) [9]	177	37	85	30	70
Hardner et al. (1972) [21]	100	28	14	67	33
Skinner et al. (1972) [10]	34	6	4	60	40
deKernion et al. (1973) [11]	48	16	10	62	38
Derrick et al. (1973) [12]	87	12	13	48	52
Gursel et al. (1973) [13]	64	15	19	39	61
Kossow et al. (1973) [22]	100	18	17	51	49
Uehling (1973) [23]	4	2	2	50	50
Williams (1975) [24]	83	17	27	39	61
Cabanas (1977) [25]	100	25	52	32	68

[a] Scappini et al. [18]; reproduced by kind permission of the Editor of *Cancer* and the publisher, J. B. Lippincott Co., Philadelphia.

Table 31.3. Complications occurring after groin dissection[a]

Complication	Incidence (%)
Skin flap necrosis	14–65
Leg oedema	2–50
Seroma/lymphoedema	19–45
Necessity for skin grafting	20–41
Wound infection	10–20
Haemorrhage	2–16
Thrombophlebitis	3–6

[a] Data from references [6, 10, 13, 22, 26, 27]

Discussion

Clinicopathological staging

Staging systems are commonly used in planning treatment and determining prognosis in patients with urologic cancer. Unfortunately, all systems for cancer of the penis suffer from the small number of cases. Nevertheless, most American urologists use Jackson's classification [29]:

Stage I tumours are those that are confined to either glans or penis or both.

Stage II tumours include those extending on to the shaft of the penis or the corpora, but for which there is no clinical evidence of lymphatic or distant metastases.

Stage III tumours are those presenting malignant, but operable, lymph nodes.

Stage IV tumours are those extending off the shaft of the penis, or which are accompanied by inoperable nodal or distant metastases.

There is a significant correlation between clinicopathological spread and survival rates (Table 31.5) [6, 9, 10, 12, 14, 16, 24, 30]. In the series reported by Edwards and Sawyers [6] there was a 5-year survival rate for patients with stage I disease of 68%, whereas for patients with nodal involvement the rate was only 25%. The survival of patients with stage II and III or IV penile carcinoma reported by Kuruvilla et al. [9] was 65% and 27%, respectively. Skinner et al. [10] found that patients with pathologically negative ileoinguinal nodes had an absolute 5-year survival rate of 75%. By contrast, the patients with histologically positive nodes had a 5-year survival rate of only 20%. In the series reported by Johnson et al. [14] the 5- and 9-year survival rates for patients without clinical evidence of metastatic disease were 64.4% and 50.1%, whereas the treatment for the patients with clinically suspicious or evident regional nodal metastases was varied, with 5- and 9-year survival rates of only 21.8%. Gursel et al. [13] observed a 5-year survival rate of 58% in stage I cancer and of 48% in combined stage II and III, whereas none of the patients with stage IV penile carcinoma survived more than 3 years.

Derrick et al. [12], who considered patients with penile carcinoma who died within 5 years of other causes and those still living as cured, obtained a cure rate of 76% for stage I and II, 55% for stage III and none for stage IV disease. In Williams' series [24] the 5-year survival rate was 76% in stage I, 50% in stage II, 40% in stage III and none in stage IV disease. Khezri et al. [16] reported a 5-year survival rate for stage I, II, III and IV penile carcinoma of 59%, 58%, 44% and 0%, respectively.

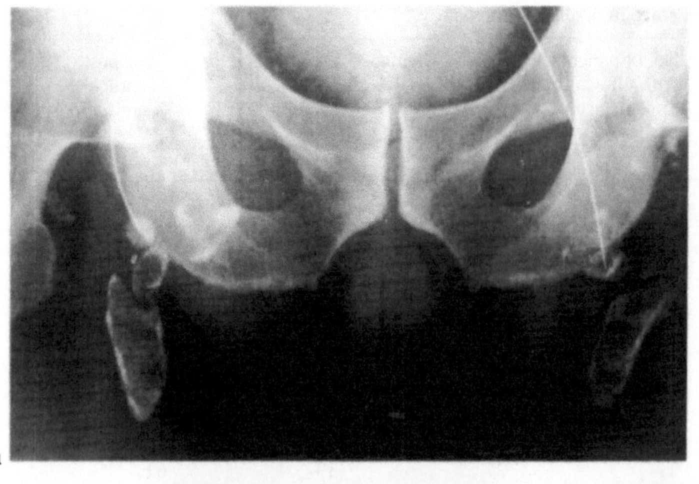

a

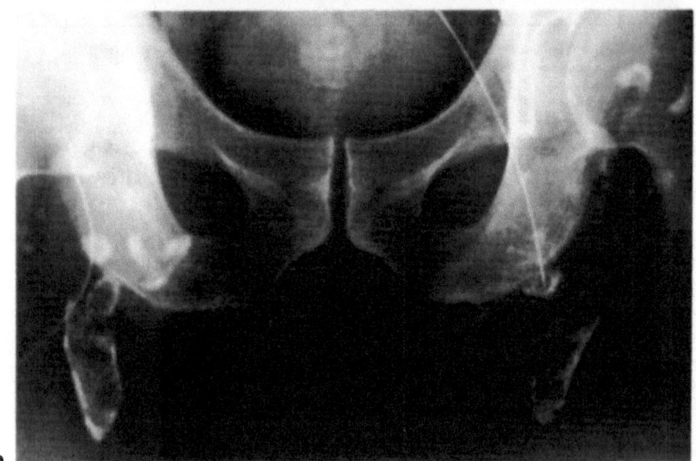

b

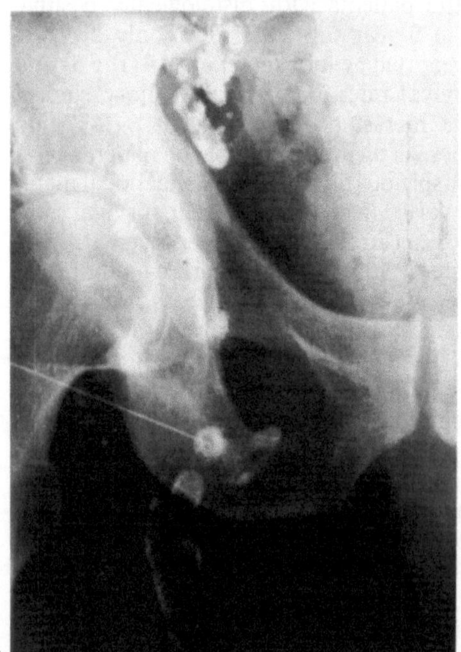

c

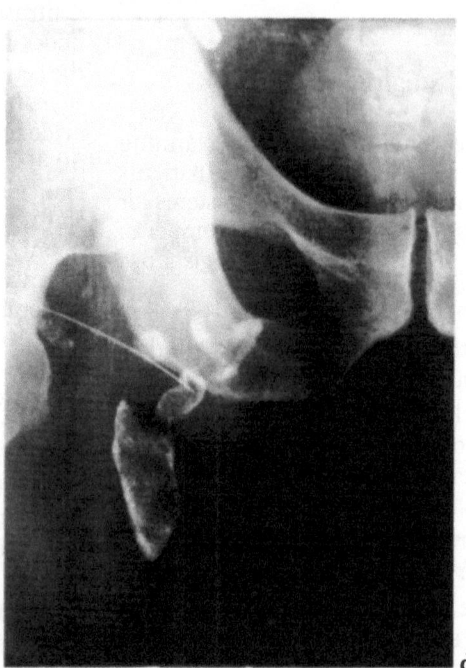

d

Fig. 31.1

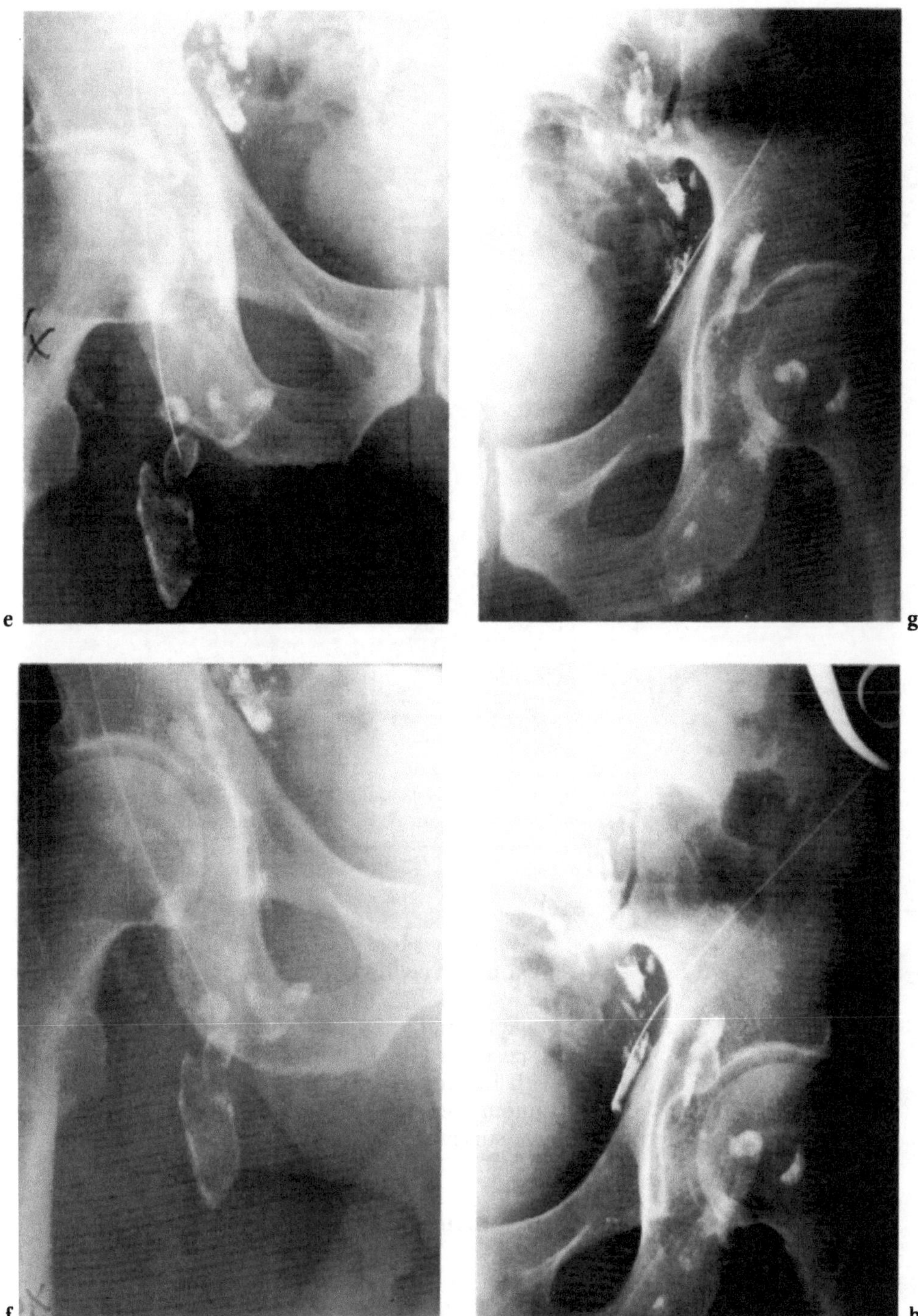

Fig. 31.1. a, b Bilateral synchronous ABC of pelvic and inguinal nodes. **c–h** The correct placement of the needle provokes dislocations of the nodes synchronous with the aspiration manoeuvre. (Luciani et al. [28]; reproduced by kind permission of the Editor of *European Urology* and the publisher, S Karger AG, Basel)

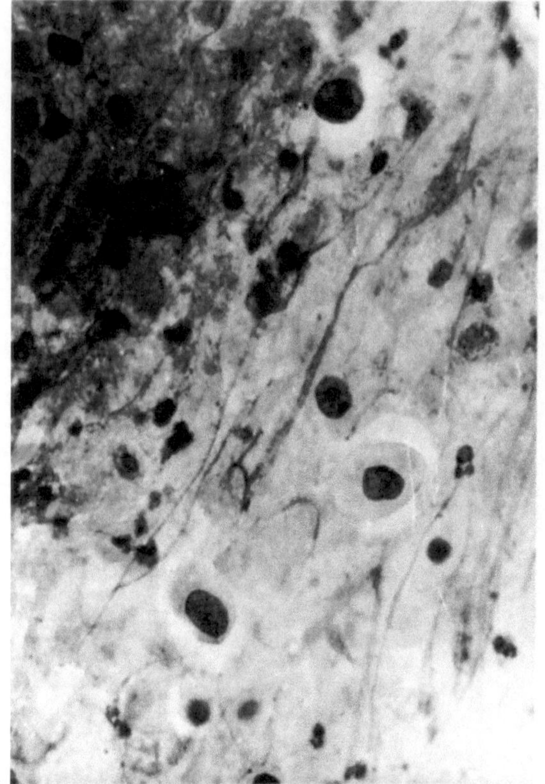

a

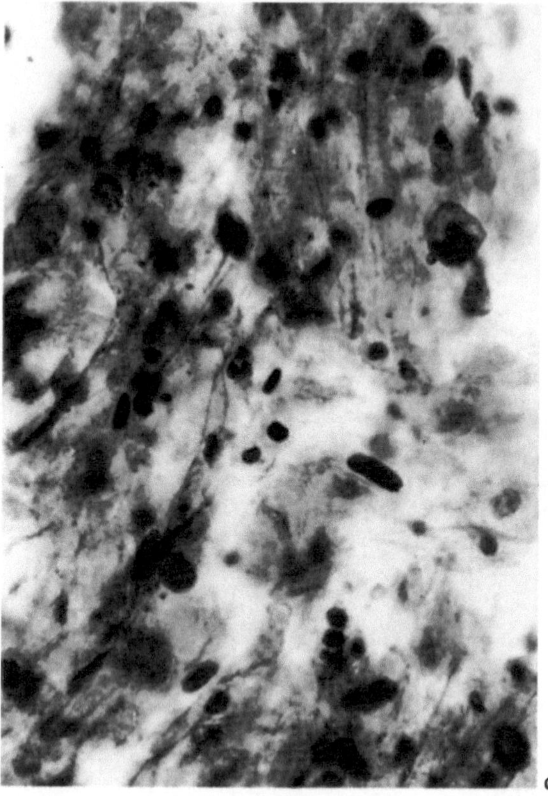

c

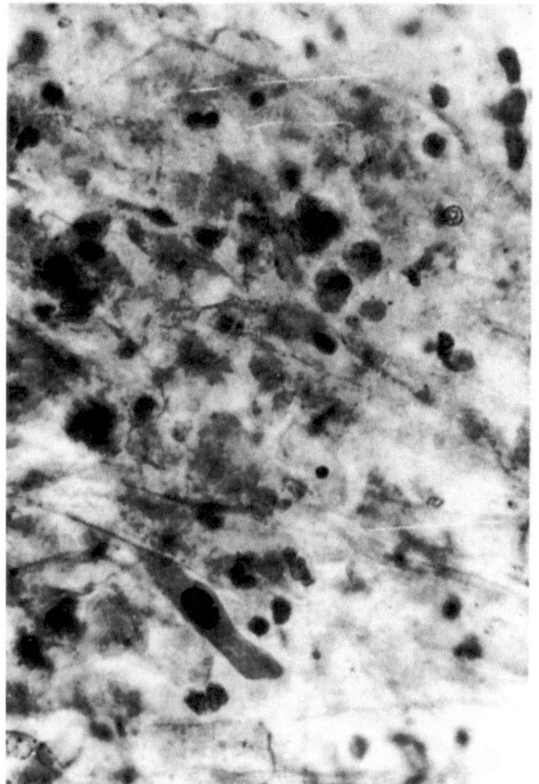

b

Fig. 31.2 a–c. Cytological findings from metastatic nodes showing malignant keratinised cells (Papanicolaou stain, × 332)

Matveyev et al. [30], without considering the treatment, reported 3- and 5-year survival rates of 98.6% and 97.5% in stage I, 84.5% and 83.3% in stage II, and 26.2% and 24.9% in stage III disease, respectively. In Leisering's series [31] no patients with T1 tumours died, whereas no patients with T4 tumour were alive after 5 years, with an overall 5-year survival rate of 68%.

Treatment

Treatment of patients with penile carcinoma can be divided into that for the primary lesion and that directed toward control of the regional lymph nodes. As regards the primary tumour, the selection of the proper therapy depends upon the assessment by the clinician of the extension of the local disease. Naturally, other factors such as general medical condition of the patients, morbidity of the procedure and overall cost-effectiveness also have to be considered.

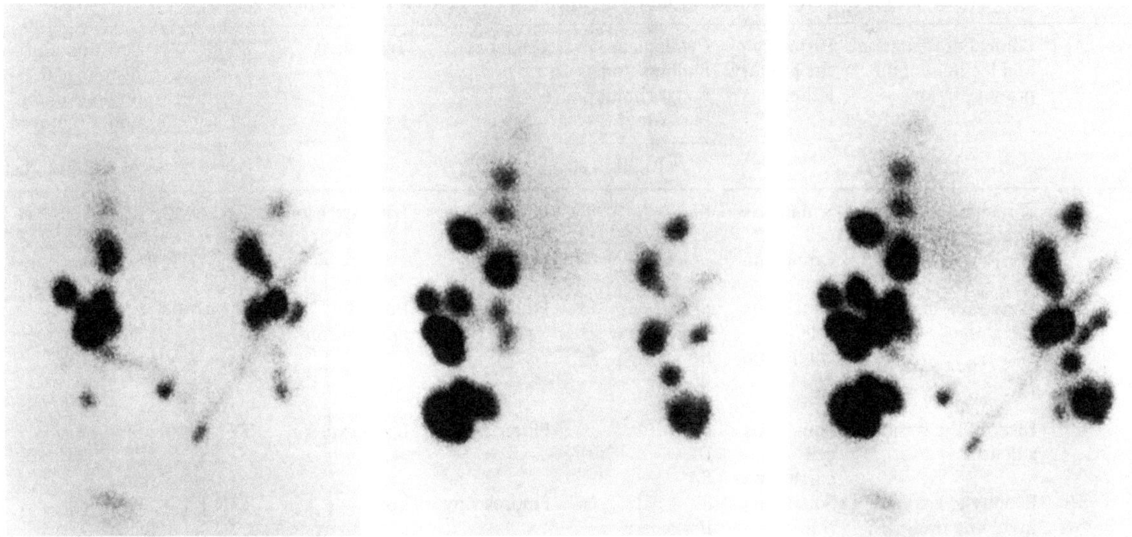

Fig. 31.3a–c. Sequential penile (**a**) and pedal (**b**) lymphoscintigraphies, performed using ^{99m}Tc sulphur microcolloid to visualise different lymphatic drainages. A prepubic node (rare variant of the superficial inguinal lymphatic system) is detected in the area of the pubis (**a**, **c**). In **c**, penile and pedal lymphoscintigrams are superimposed, giving a complete picture of both inguinal and pelvic nodes.

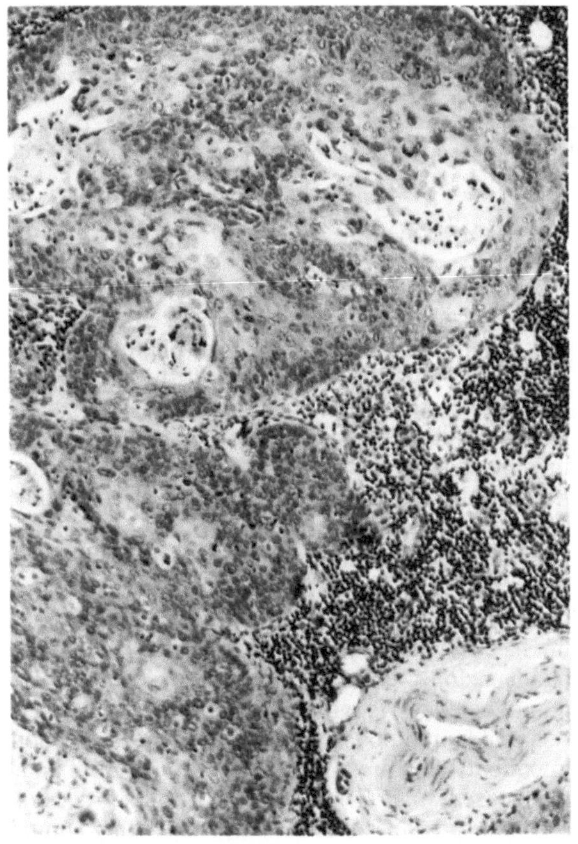

Topical chemotherapy, using 5-fluorouracil, should only be considered in young patients with erythroplasia of Queyrat or Bowen's disease of the penis, confirmed after extensive biopsy. Systemic bleomycin has been successfully used by Ichikawa et al. [32] to obtain the regression of most tumours and the cure of several patients. Kyalwazi et al. [33], using the same drug, reported complete healing of the tumour in about 7 out of 15 patients, with minor side-effects in more than 50% of the individuals, but without any case of pulmonary fibrosis. Unfortunately, although these studies are encouraging, one must not forget the high morbidity and even mortality of pulmonary fibrosis occurring in about 1% of patients given this drug. Therefore, systemic bleomycin therapy is too dangerous to be used in the uncomplicated cases of penile carcinoma.

Several reports have confirmed the value of radiation in treating the early stages of penile carcinoma, especially in young men [2, 5, 17, 29, 34–39]. Controversy exists, however, regarding the optimal technique; radium or iridium mould tech-

Fig. 31.4. Histological appearance of surgical specimen of lymph nodes consistent with metastases from squamous penile carcinoma.

Table 31.4. Comparison of clinical features, histology and ABC in eight patients with penile cancer

Case no.	Age	Clinical presentation and location of the primary lesion	Histology of the primary lesion	Cytological findings and nodal chains punctured			Guidance	Treatment	Stage	Location and histological findings of the nodes removed		
					Right	Left					Right	Left
1	51	Spreading ulcer of foreskin and glans	Squamous cell carcinoma	SI	–	⊕	Fluoroscopy	Local excision	T2N+M0	SI	–	+
										DI	–	–[a]
										SA	–	+
										EI	–	+
2	58	Fungating mass of the glans	Squamous cell carcinoma	SI	–	–	Fluoroscopy	Partial amputation	T1 N0 M0	SI	–	–
				DI	–	–				DI	–	–
				SA	–	–				SA	–	–
				EI	–	–						
				II	–	–						
3	69	Ulcer of the coronal sulcus	Squamous cell carcinoma	SI	–	–	Fluoroscopy	Local excision	T2 N0 M0	SI	–	–
				DI	–					DI	–	–
				SA	–	–				SA	–	–
4	56	Exophytic growth involving urethra and corpus	Squamous cell carcinoma	EI	–	⊕	Fluoroscopy	Total penectomy, radiotherapy	T3N+M0	EI	–[a]	+[a]
				II	–	–						
				SI	–	⊕						
5	60	Ulcerated growth of glans	Squamous cell carcinoma	SI	–	–	Fluoroscopy	Total penectomy	T2 N0 M0	SI	–	–
				DI	–	–				DI	–	–
6	79	Exophytic growth involving glans and urethra	Squamous cell carcinoma	SI	–	–	Fluoroscopy	Partial amputation	T2 N0 M0			
				EI	–							
				PA	–							
				Obt	–							
				IE	–							
				CI	–							
7	78	Ulcer of the glans	Squamous cell carcinoma	SI		⊕	CT	Total penectomy	T3N+M0			
8	62	Growth involving glans urethra and corpus	Squamous cell carcinoma	SI		⊕	Fluoroscopy	Partial amputation	T4N+M0	SI		+

SI, superficial inguinal; DI, deep inguinal; SA, sentinel area; EI, external iliac; II, internal iliac; PA, para-aortic; Obt, obturator; CI, common iliac.
[a] Positive lymphogram.

nique, orthovoltage and megavoltage external beam therapy have all been employed with favourable results. The current use of external beam megavoltage therapy is producing far better results.

The major complications with megavoltage X-rays are meatal stenosis, urethral stricture and radionecrosis as late sequelae. Phimosis, paraphimosis, posthitis and adhesive balanitis may develop early instead. Long time intervals for regression of the irradiated lesion may be required, and persistence of disease may not be ascertained until up to 12 months after radiotherapy. When suspicious inguinal nodes are present such a delay would be intolerable. Thus, although radiation can be used to treat some young patients with early superficial lesions, the advantage over excision has not been demonstrated, even if radiation can be used successfully for treating recurrence after surgical excision.

Surgical treatment includes (1) circumcision for non-invasive lesions limited to the foreskin; (2) partial amputation for tumours involving the glans and distal shaft, when 2 cm of grossly tumour-free margin can be assured with a sufficient penile length to allow the patient upright directing of the urinary stream; and (3) total penectomy with perineal urethrostomy for the most extensive lesions. As carcinoma of the penis spreads via the lymphatic pathways, involvement of the inguinal lymph nodes may first be expected, although direct metastasis can occur in iliac glands or in internal organs. Since no clear evidence exists that prophylactic lymphadenectomy is superior to a more conservative management, the guiding principle for treating regional lymph nodes is to perform groin dissection in patients who present initially with regional node metastases (stage III) or in those who later develop regional spread. Therefore, surgical staging should

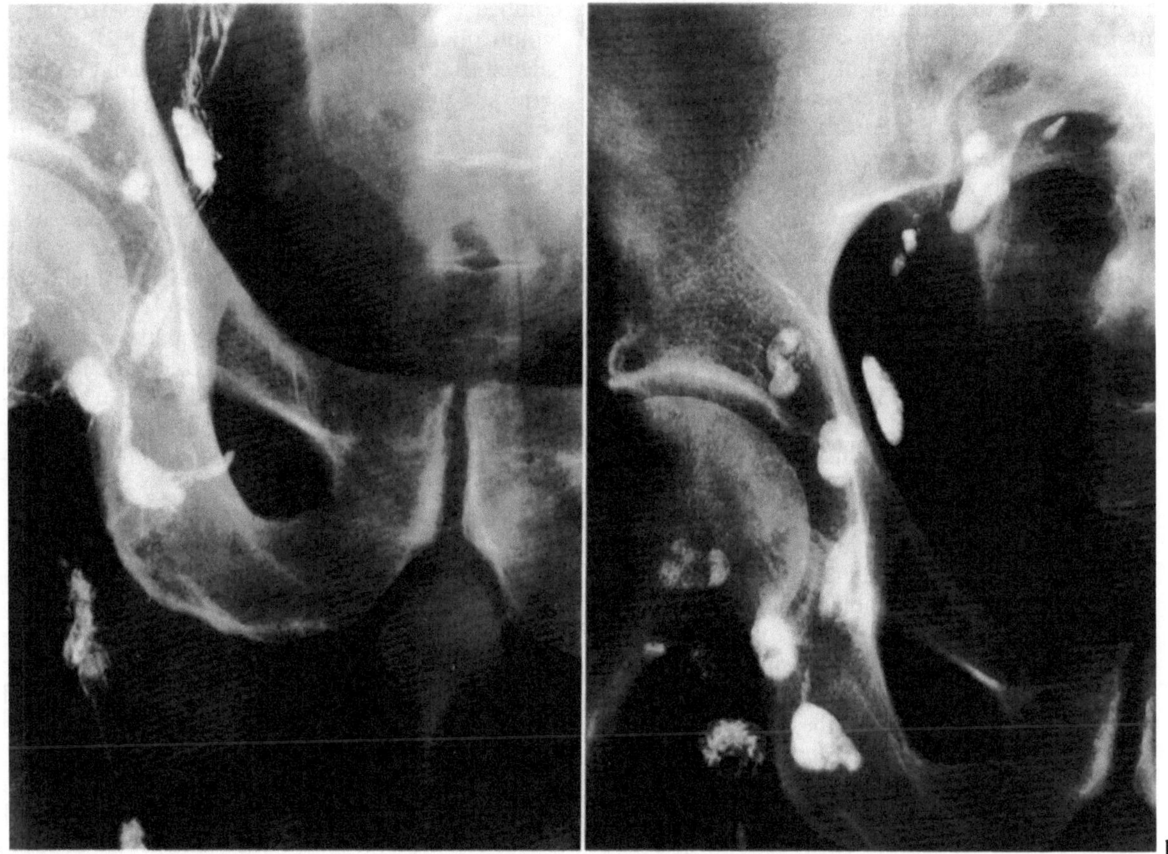

Fig. 31.5 a, b. Lymphograms from pedal lymphography showing opacifications of the nodes located between the ascending ramus of the pubis and the femoral head.

be considered useless and dangerous in the 80% of patients who have true negative nodes and the 50% of patients who have false positive nodes. Lymphadenectomy is to be performed after a waiting period of several weeks, during which the primary lesion has to be excised and the associated regional infection properly treated with appropriate antibiotics. Bilateral lymphadenectomy is generally advisable, except in those cases in which a single groin appears to be involved many months or even years after treatment of the primary tumour.

The appropriate extension of regional node dissection is still controversial. A poorer prognosis is to be expected when pelvic nodes are involved, so that some surgeons believe that palliative therapy is sufficient [13, 15, 17], while others routinely dissect up to the aortic bifurcation. It seems arguable, however, that a nodal dissection carried out up to the common iliac bifurcation is therapeutic [7, 10, 21, 22, 40].

Table 31.5. Survival rate (5-year) of patients with penile carcinoma[a]

Reference	Patients with negative nodes (%)	Patients with positive nodes (%)
Edward and Sawyers (1968) [6]	68	25
Kuruvilla et al. (1971) [9]	69	33
Skinner et al. (1972) [10]	75	20
Johnson et al. (1973) [14]	64	21.8
Derrick et al. (1973) [12]	53	22
Williams et al. (1975) [24]	71.7	40
Khezri et al. [16]	58	43.7
Matveyev et al. [27]	90	24.9

[a] Scappini et al. [18]; reproduced by kind permission of the Editor of *Cancer* and the publisher, J. B. Lippincott Co., Philadelphia.

Diagnostic Procedures

Proper recognition of the extent of the disease is essential in assessing the appropriate management and determining the prognosis. Besides clinical examination, numerous imaging diagnostic procedures have been proposed to determine the pelvic nodal status in patients with urological malignancies. CT, ultrasonography and lymphography are routinely used for staging purposes in cases of prostatic and bladder carcinoma. The accuracy of these procedures in detecting nodal involvement in patients with penile cancer has never been assessed. Therefore, their value in the staging of penile cancer is hard to state, and the role of lymphography in this disease is not yet well defined.

The lymphatics of the prepuce and skin of the penis drain into the superficial and deep inguinal lymph nodes on either side. The lymphatics of the glans and those from the penile urethra and corpora cavernosa drain directly into the deep inguinal and external iliac nodes [10, 13, 20].

Direct drainage into the iliac nodes has also been reported by some authors [13, 39, 41]. Embolisation rather than permeation of the lymph channels seems to be the mechanism of lymphatic metastasis for this tumour [22]. Penile and pedal lymphography may be unreliable when performed separately, since in most cases they opacify different lymphatic sites and some nodes may not be visualised by one or the other of the methods. Both procedures should therefore be performed in the same patient for a complete demonstration of the lymphatic pathways that may be involved by penile carcinoma. However, lymphography suffers from defined limitations: the inability of detecting microscopic involvement; the lack of opacification of the nodes that are extensively involved with disease; and the necessity of excluding all those benign or inflammatory diseases that can affect the lymphatic system, to avoid false positive diagnosis.

Since routine pelvic lymphadenectomy cannot be performed because of the high rate of treatment-related morbidity and even mortality, Cabanas [25] suggested performing bilateral sentinel lymph node (SLN) surgical biopsy. In the radiographic antero-posterior view this lymph node is seen near the saphofemoral junction projecting over the junction of the femoral head and the ascending ramus of the pubis. According to Cabanas [25] the SLN is the first site of metastasis and may be the only lymph node involved, so that positive or negative biopsy findings in this node should be a reliable indicator of the presence or absence of nodal disease. Further anatomical studies [42] show that one to seven nodes may be found between superficial epigastric and superficial external pudendal veins, where these anastomose with the great saphena vein. Therefore we prefer the term "sentinel node area" to that of "sentinel lymph node".

It should be mentioned, however, that recently Perinetti et al. [43] described the case of a patient with bilateral negative SLN biopsy results who succumbed 6 months later from subsequent regional lymph node metastases. It should be further considered that 3 out of 31 patients (10%) of Cabanas' series did not survive 5 years. Although the causes of death were not reported, we can argue that the biopsy results in these three patients were probably false negative because this percentage closely resembles that of occult metastases in patients with clinically negative nodes (15%–20%) [43].

The results reported by Fowler [42] parallel those of Cabanas [25]. In this series of 10 patients, 5 had bilateral inguinal lymphadenopathy and 5 had unilateral inguinal adenopathy. One patient out of five biopsy-negative cases (20%) developed a superficial inguinal recurrence in 4 months, and a later ipsilateral ilioinguinal lymphadenectomy was performed.

The Role of ABC

In our review of a series of 100 lymphograms, we observed that, in some cases, pedal lymphography surprisingly visualised the same nodes described by Cabanas [25] as SLN, which, according to this author, is opacified only by penile lymphography. In this connection pedal lymphography should be performed first and penile lymphography should follow if the pelvic lymph nodes appear free from disease and the nodes of the sentinel area are not visualised. Since lymphography opacifies deep inguinal and pelvic nodes but is unreliable in establishing whether they are involved by the tumour or not, ABC may be suggested as the diagnostic procedure of choice, as an alternative to surgery, for determining the nodal extension of penile carcinoma.

Although numerous reports have documented the use of ABC in detecting the lymphatic spread of urologic malignancies [45–66], surprisingly few cases of penile carcinoma have been reported in the literature as having been investigated with ABC of the lymph nodes. Zadala Ramos [67] first advocated the use of this procedure in some selected cases with suspicious lymph nodes. Only Rothenberger et al. [59] and Wajsman et al. [62, 63] included cases of penile carcinoma in their series of transabdominal fine needle aspiration biopsy in the staging of genitourinary cancers, but they placed no particular emphasis on the staging of penile cancer.

We believe that pedal lymphography should be performed first in every case. If the same nodes located in the so-called sentinel area are opacified, ABC can be performed without resorting to penile lymphography, which remains mandatory in the remaining cases (Fig. 31.6). Since a careful dis-

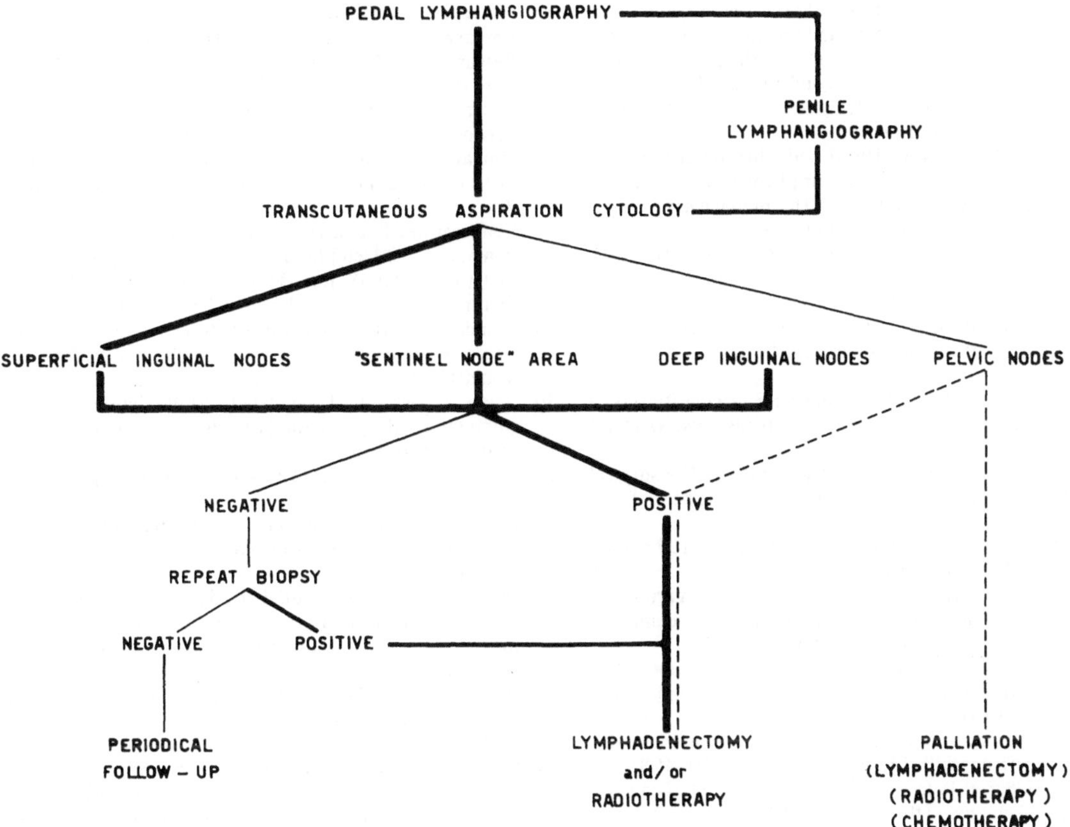

Fig. 31.6. Protocol for the management of penile cancer.

section of the sentinel node area seems to be more reliable than single node removal, preliminary ABC on most opacified nodes in this area is much less invasive and could limit the excisional false negative rate. Furthermore, these nodes are easy to puncture.

Most superficial, deep inguinal and pelvic nodes should be biopsied. No problem exists in puncturing the superficial inguinal nodes, whereas deep inguinal and pelvic nodes must be aspirated under fluoroscopic or CT guidance. Since lymphography is inaccurate in determining the true extension of metastatic disease, the visualised nodes should be punctured regardless of the lymphographic findings.

The presence of cancer cells in the specimens is conclusive of stage III disease, and further surgery may be performed. On the other hand, a negative cytological finding is of no value. This limitation is due to the difficulty in determining if the needle is within a particular node [68] and the possibility of micrometastasis. In our own series no false negative results were obtained, and we believe that to repeat the aspirations from different parts of the node may enhance the significance of negative cytological findings.

In cases with negative aspiration results a "watch and wait" policy may be adopted since radiopaque contrast medium remains in the nodes for 6–9 months and sequential aspirations may be performed for detecting delayed nodal metastases with minimal discomfort for the patients. However, local bilateral surgical biopsy of the nodes of the sentinel area could also be performed, given the minimal morbidity of the procedure.

In conclusion, we believe that preliminary ABC has a fundamental role in clinical staging of penile cancer in order to limit the morbidity and mortality associated with staging lymphadenectomy to those patients at risk of disease progression.

References

1. Wynder EL (1964) Symposium de cancer del pene. Some epidemiological observations on cancer of the penis. Rev Inst Nacl Cancerol (Mex) 15:273–285

2. Engelstad RB (1948) Treatment of cancer of the penis at the Norwegian Radium Hospital. Am J Roentgenol 60:801–806
3. Ekström T, Edsmyr F (1958) Cancer of the penis: a clinical study of 229 cases. Acta Chir Scand 115:25–45
4. Beggs JH, Spratt JS Jr (1964) Epidermoid carcinoma of the penis. J Urol 91:166–172
5. Murrel DS, Williams JL (1965) Radiotherapy in the treatment of carcinoma of the penis. Br J Urol 37:211–222
6. Edward RH, Sawyers JL (1968) The management of carcinoma of the penis. South Med J 61:843–845
7. Fegen P, Persky L (1969) Squamous cell carcinoma of the penis: its treatment with special reference to radical node dissection. Arch Surg 99:117–120
8. Hanash KA, Furlow WL, Utz DC, Harrison EG Jr (1970) Carcinoma of the penis: a clinicopathologic study. J Urol 104:291–297
9. Kuruvilla JT, Gorlick FH, Mammen KE (1971) Results of surgical treatment of carcinoma of the penis. Aust NZ J Surg 41:157–159
10. Skinner DG, Leadbetter WF, Kelley SB (1972) The surgical management of squamous cell carcinoma of the penis. J Urol 107:273–277
11. deKernion JB, Tynberg P, Persky LM, Fegen P (1973) Carcinoma of the penis. Cancer 32:1256–1262
12. Derrick FC Jr, Lynch KM Jr, Kretkowski RC, Yarbrough WJ (1973) Epidermoid carcinoma of the penis: computer analysis of 87 cases. J Urol 110:303–305
13. Gursel EO, Georgountzos C, Uson AC, Melicow MM, Veenema RJ (1973) Penile cancer: clinicopathologic study of 64 cases. Urology 1:569–578
14. Johnson DE, Fuerst DE, Ayala AG (1973) Carcinoma of the penis: experience with 153 cases. Urology 1:404–408
15. Baker VH, Spratt JS Jr, Perez-Mesa C, Watson FR, Leduc RJ (1976) Carcinoma of the penis. J Urol 116:458–462
16. Khezri AA, Dunn M, Smith PJB, Mitchell JP (1978) Carcinoma of the penis. Br J Urol 50:275–279
17. Salaverria JC, Hope-Stone HF, Paris AMI, Molland EA, Blandy JP (1979) Conservative treatment of carcinoma of the penis. Br J Urol 51:32–37
18. Scappini P, Piscioli F, Pusiol T, Hofstetter A, Rothenberger KH, Luciani L (1986) Penile cancer aspiration biopsy cytology for staging. Cancer 58:1526–1533
19. Catalona WJ (1980) Role of lymphadenectomy in carcinoma of the penis. Urol Clin North Am 7:785–792
20. Marcial VA, Figueroa-Colon J, Marcial-Rojas RA, Colon JE (1962) Carcinoma of the penis. Radiology 79:209–220
21. Hardner GJ, Bhanalaph T, Murphy GP, Albert DJ, Moore RH (1972) Carcinoma of the penis: analysis of therapy in 100 consecutive cases. J Urol 108:428–430
22. Kossow JH, Hotchkiss RS, Morales PA (1973) Carcinoma of the penis treated surgically: analysis of 100 cases. Urology 2:169–172
23. Uehling DT (1973) Staging laparatomy for carcinoma of penis. J Urol 110:213–215
24. Williams JL (1975) Surgical treatment of carcinoma of the penis. Proc R Soc Med 68:781–785
25. Cabanas RM (1977) An approach for the treatment of penile carcinoma. Cancer 39:456–466
26. Johnson DE, Lo RK (1984) Complications of groin dissection in penile cancer. Experience with 101 lymphadenectomies. Urology 24:312–314
27. Crawford ED (1984) Radical ilio-inguinal lymphadenectomy. Urol Clin North Am 11:543–552
28. Luciani L, Piscioli F, Scappini P, Pusiol T (1984) Value and role of percutaneous regional node aspiration cytology in the management of penile carcinoma. Eur Urol 10:294–302
29. Jackson SM (1966) The treatment of carcinoma of the penis. Br J Surg 53:33–35
30. Matveyev BP, Zak BI, Gotsadze DT (1984) Treatment of primary cancer of the penis. The Centre of Oncology Research of the USSR Academy of Medical Sciences, Moscow. Vopr Onkol 1:71–76
31. Leisering W (1984) Therapie und Prognose des Peniskarzinom. Z Urol Nephrol 77:93–100
32. Ichikawa T, Nakano J, Hirokawa J (1969) Bleomycin treatment of the tumor of penis and scrotum. J Urol 102:699–707
33. Kyalwazi SK, Bhana D (1973) Bleomycin in penile carcinoma. East Afr Med J 60:331–336
34. Kelley CD, Arthur K, Ragoff E, Grabstald H (1974) Radiation therapy of penile cancer. Urology 4:571–573
35. Kmudsen OS, Brennhoval JO (1967) Radiotherapy in the treatment of the primary tumor in penile cancer. Acta Chir Scand 133:69–71
36. Vaeth JM, Green JP, Lowy RO (1970) Radiation therapy of carcinoma of the penis. Am J Roentgenol Radium Ther Nucl Med 108:130–135
37. Grabstald H, Kelley CD (1980) Radiation therapy of penile cancer. Urology 15:575–576
38. Duncan W, Jackson SM (1972) The treatment of early cancer of the penis with megavoltage X-rays. Clin Radiol 23:246–248
39. Delclos L (1982) Interstitial irradiation of the penis. In: Johnson DE, Boileau MA (eds) Genitourinary tumors: fundamental principles and surgical techniques. Grune and Stratton, New York 219–225
40. Hoppman HJ, Fraley EE (1978) Squamous cell carcinoma of the penis. J Urol 120:393–398
41. Pack GT, Rekers P (1942) The management of malignant tumors in the groin. Am J Surg 56:545–565
42. Daseler EH, Anson BJ, Reimann AF (1948) Radical excision of the inguinal and iliac lymph glands. A study based upon 450 anatomical dissections and upon supportive clinical observations. Surg Gynecol Obstet 87:679–694
43. Perinetti EP, Grane DB, Catalona WJ (1980) Unreliablility of sentinel lymph node biopsy for staging penile carcinoma. J Urol 124:734–735
44. Fowler JE (1984) Sentinel lymph node biopsy for staging penile cancer. Urology 23:352–354
45. Correa RJ, Kidd RC, Burnett L, Brannen GE, Gibbons RP, Cummings KB (1981) Percutaneous pelvic lymph node aspiration in carcinoma of the prostate. J Urol 126:190–191
46. Dan SJ, Wulsohn MA, Efrimidis SC, Mitty HA, Brendler H (1982) Lymphography and percutaneous lymph node biopsy in clinically localized carcinoma of the prostate. J Urol 127:695–698
47. Efremidis SC, Dan JS, Nieburgs HI, Mitty HA (1981) Carcinoma of the prostate: lymph node aspiration for staging. AJR 136:489–492
48. Göthlin JH (1976) Post-lymphographic percutaneous fine needle biopsy of lymph node guided by fluoroscopy. Radiology 120:205–207
49. Göthlin JH (1978) Percutaneous transperitoneal fluoroscopic guided fine needle biopsy of lymph nodes. Acta Radiol [Diagn] (Stockh) 20:660–664
50. Göthlin JH, Hoeim L (1980) Percutaneous transperitoneal fine needle biopsy of normal looking lymph nodes and small lesion at lymphography: a preliminary report. Urol Radiol 1:237–239
51. Göthlin JH, Hoeim L (1981) Percutaneous fine needle biopsy of radiographically normal lymph nodes in the staging of prostate carcinoma. Radiology 141:351–354
52. Göthlin JH, Rupp N, Rothenberger KH, MacIntosh PK (1981) Percutaneous biopsy of retroperitoneal lymph nodes: a multicentric study. Eur J Radiol 1:46–50
53. Khan O, Pearse E, Bowley N, Williams G, Krausz T (1983)

Combined bipedal lymphangiography, CT scanning and transabdominal lymph node aspiration cytology for node staging in carcinoma of the prostate. Br J Urol 55:538–541
54. Luciani L, Piscioli F (1983) Accuracy of transcutaneous aspiration biopsy in definitive assessment of nodal involvement in the prostatic carcinoma. Report of 24 cases and review of the literature. Br J Urol 55: 321–325
55. Piscioli F, Scappini P, Luciani L (1985) Aspiration cytology in the staging of urologic cancer. Cancer 56:1173–1180
56. Luciani L, Scappini P, Pusiol T, Piscioli F (1985) Comparative study of lymphography and aspiration cytology in the staging of prostatic carcinoma. Urol Int 40:181–189
57. Luciani L, Piscioli F, Pusiol T, Scappini P (1986) The value of aspiration cytology in the definitive staging of bladder carcinoma. Br J Urol 58:26–30
58. Rothenberger KH, Hofstetter A, Pfeifer KJ, Rupp N (1979) Transperitoneale Feinnadel Biopsie retroperitonealer. Lymphknoten in der Karzinomadiagnostik. Prog Med Virol 97:2218–2222
59. Rothenberger KH, Hofstetter A, Rupp N, Pfeifer KJ (1981) Transabdominal fine needle biopsy of lymph nodes for urologic cancer staging. In: Schulman CC (ed) Advances in diagnostic urology. Springer, Berlin Heidelberg New York, pp 113–117
60. Rupp N, Rothenberger KH, Bajer-Pietsch E, Feuerbach ST, Esch U (1979) Die perkutane Feinnadelbiopsie von Lymphknoten. Fortschr Geb Rontgenstr Nuklearmed Erganzungsband 130:328–331
61. Von Eschenbach A, Zornoza J (1982) Fine-needle percutaneous biopsy: a useful evaluation of lymph node metastases from prostate cancer. Urology 6:589–590
62. Wajsman Z, Gamarra M, Park JJ, Beckley SA, Pontes JE, Murphy GP (1982) Fine-needle aspiration of metastatic lesions and regional lymph nodes in genitourinary cancer. Urology 4:356–360
63. Wajsman Z, Gamarra M, Park JJ, Beckley SA, Pontes JE (1982) Transabdominal fine-needle aspiration of retroperitoneal lymph nodes in staging of genitourinary tract cancer (correlation with lymphography and lymph node dissection findings). J Urol 128:1238–1240
64. Wallace S, Jing BS, Zornoza J (1977) Lymphangiography in the determination of the extent of metastatic carcinoma. The potential value of percutaneous lymph node biopsy. Cancer 39:706–708
65. Zornoza J, Johnson K, Wallace S, Lukeman JM (1977) Fine needle aspiration biopsy of retroperitoneal lymph nodes and abdominal masses: an updated report. Radiology 125:87–88
66. Zornoza J, Wallace S, Goldstein H, Lukeman JM, Jing BS (1977) Transperitoneal percutaneous retroperitoneal lymph node aspiration biopsy. Radiology 122:111–115
67. Zadala-Ramos J (1964) Symposium de cancer del pene. Tratamento del cancer del pene. Rev Inst Nacl Cancerol (Mex) 15:329–333
68. Frable WJ (1983) Fine-needle aspiration biopsy: a review. Hum Pathol 14:9–28

Section VI

Renal and Adrenal Masses

Chapter 32

Staging of Renal Cell Carcinoma

E. Mukamel and J. B. deKernion

The Robson's modification of the system of Flocks and Kadesky [1] is by far the most frequently used staging system for renal cell carcinoma (Fig. 32.1). According to this classification, patients with involvement of the renal vein, vena cava and lymph nodes are included in the same stage—stage III—although their prognosis is different. The TNM system [2] proposed by the American Committee for Cancer Staging and End Results separates venous involvement from lymph node metastases (Table 32.1); thus the T3 category in this system indicates tumour with perinephric fat extension or extension into the renal vein with or without infradiaphragmatic vena cava involvement. These stages are designated by the following subgroups: T3a, T3b and T3c, respectively.

Accurate staging of renal tumours with determination of the extent of local invasion and distant metastases is essential for the treatment of renal tumours. Various imaging techniques are being used for staging, including computed tomography (CT) scanning, renal angiography, venocavography, ultrasonography and magnetic resonance imaging (MRI). Currently, the CT scan is the preferred technique to determine the extent of local invasion [4–6]. On the CT scan, tumour extension beyond the capsule (Fig. 32.2), renal vein and vena cava extension (Fig. 32.3) and lymph node enlargement (Fig. 32.4) can be seen.

Numerous studies have reported the accuracy of CT scanning for staging as compared with other imaging techniques. In their comprehensive study, Weyman et al. [4] showed that CT scanning predicted the presence and absence of tumour extension beyond the capsule in 45 of 54 (83%) of patients. In two patients the CT scan failed to detect tumour extension and in three it suggested tumour extension beyond the capsule, which was not confirmed by histological examination. These false positive findings in the three patients were caused by perinephric haemorrhage and abnormal blood vessels. In the same study, CT scanning was accurate in determining lymph node involvement in 44 of 55 patients (80%). In three patients it showed enlarged lymph nodes which, on histological examination, were free of metastases. In another three patients CT scanning failed to detect microscopic involvement of lymph nodes. The presence or absence of renal vein invasion and inferior vena cava extension was accurate in 82% and 93%, respectively.

When the staging accuracy of CT scanning was compared to that of renal angiography it was found that CT scanning was more accurate than angiography in predicting capsular invasion (83% vs 68%, respectively), more sensitive than angiography in detecting lymph node metastases (73% vs 33%, respectively) and slightly less accurate in predicting renal vein and vena cava invasion (93% vs 100%, respectively).

Richie et al. [5] also showed that CT scanning is superior to angiography for detection of perinephric invasion and lymph node involvement and equal to angiography for renal vein involvement. In this

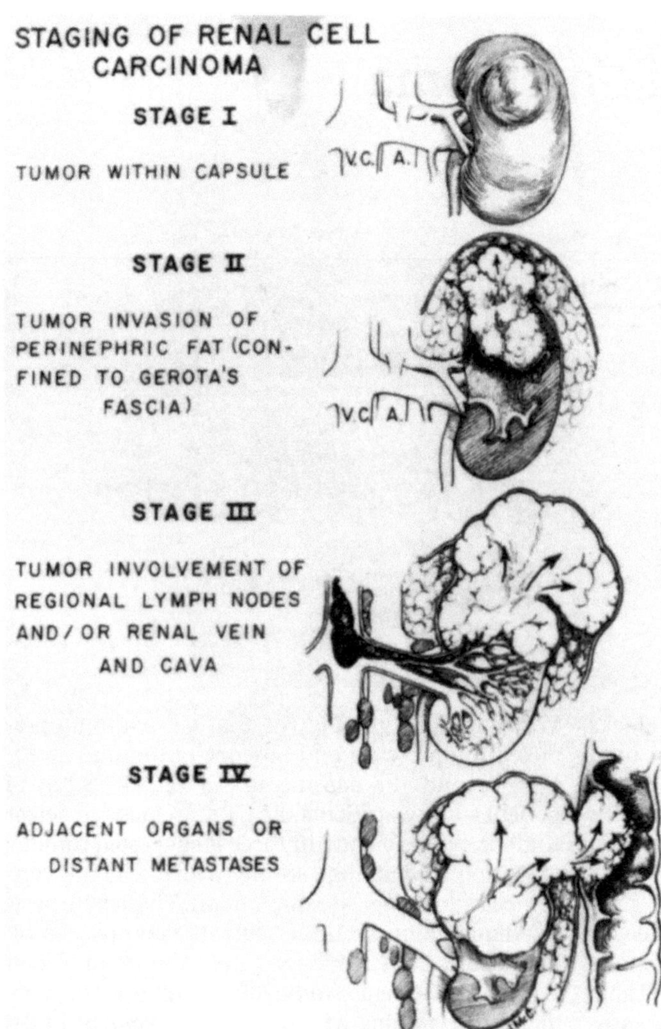

Fig. 32.1. Staging of renal cell carcinoma according to the schemes of Robson, Murphy, and Flocks and Kadesky. (Holland [3]; reproduced by kind permission of the Editor of *Cancer* and the publisher, J.B. Lippincott Co., Philadelphia)

Table 32.1 Stage classification of renal cell carcinoma

Tumour stage	Robson [1]	TNM [2]
No primary	–	T0
Small primary, minimal distortion	I	T1
Large tumors, renal distortion	I	T2
Involving perinephric fat	II	T3a
Involving renal vein	III	T3b
Involving renal vein and infradiaphragmatic vena cava	III	T3c
Invading adjacent structures	IV	T4
Involving superior vena cava	III	–
No nodes involved	I, II	N0
Single, ipsilateral node involved	III	N1
Involvement of multiple regional nodes	III	N2
Fixed regional nodes	III	N3
Involved juxtaregional nodes	III	–
Distant metastases	IV	M1

study only three patients had liver involvement, two were detected by CT scanning and one with angiography. These authors concluded that CT scanning is the preferred procedure for staging renal cell carcinoma. Cronan et al. [6] showed that a correct staging was achieved in 91% of the lesions with CT scanning, in 70% with ultrasound, and in 61% with angiography. Thus in this study the ultrasound was superior to renal angiography for staging.

In recent years, MRI was introduced for the diagnosis and staging of renal mass lesions [7, 8]. The clarity of visualisation of anatomical structures and the lack of irradiation exposure have made this technique attractive for the diagnosis and staging of renal tumours. Although recent studies have shown improved imaging of renal masses with MRI, the value of the technique for staging is not yet

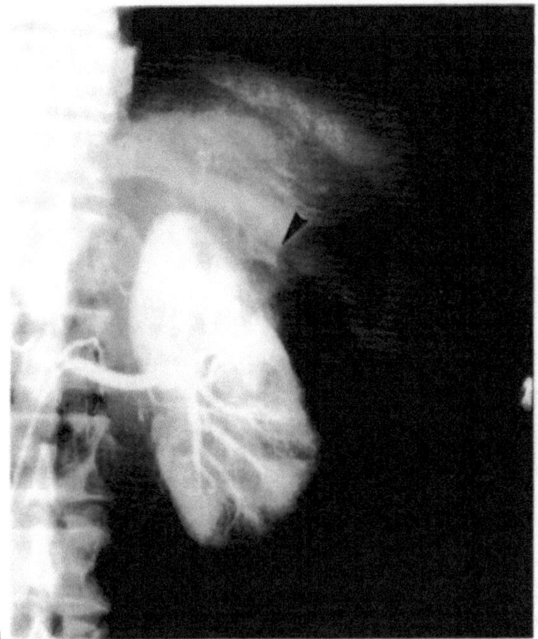

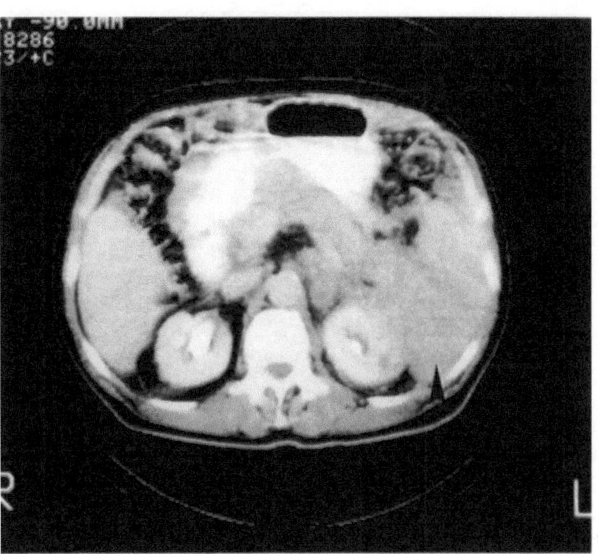

Fig. 32.2. a Renal angiogram of 70-year-old patient showing mild hypervascularity (*arrowhead*). **b** CT scan showing diffuse extension of the tumour into the retroperitoneum (*arrowhead*). Fine needle aspiration of the mass disclosed lymphoma.

established. According to our initial experience, MRI may improve the preoperative assessment of renal vein and vena cava in patients with renal tumours (Fig. 32.5). We have been able to visualise vena cava tumour thrombus and to determine preoperatively its proximal and distal extension (Fig. 32.6). Further studies will determine the role of MRI in the work-up of patients with a renal mass.

Although regional staging with CT scanning is valuable and should be performed on every patient with a renal mass, patient selection for surgery should not be based only on the findings of the CT scan. In our experience, many patients in whom the CT scan has suggested liver or abdominal wall invasion have been found during surgery to have well-confined tumours with no invasion of adjacent

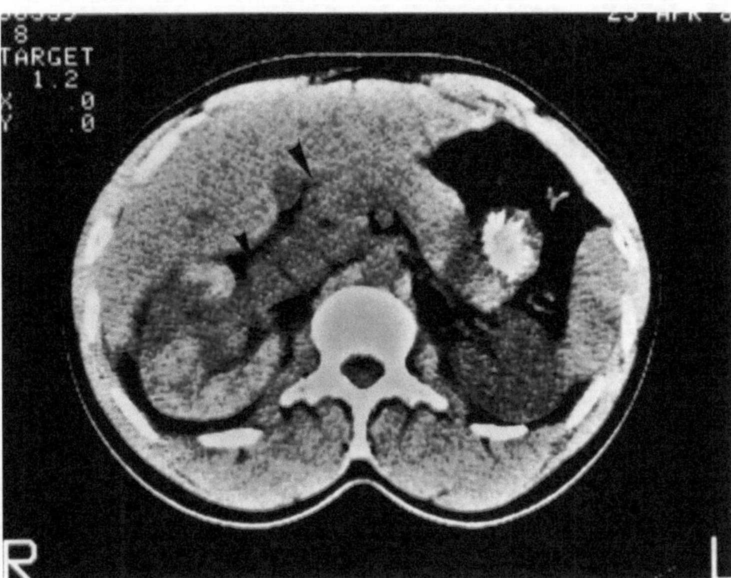

Fig. 32.3. CT scan of right renal cell carcinoma showing the tumour and enlarged lymph nodes (*arrowheads*).

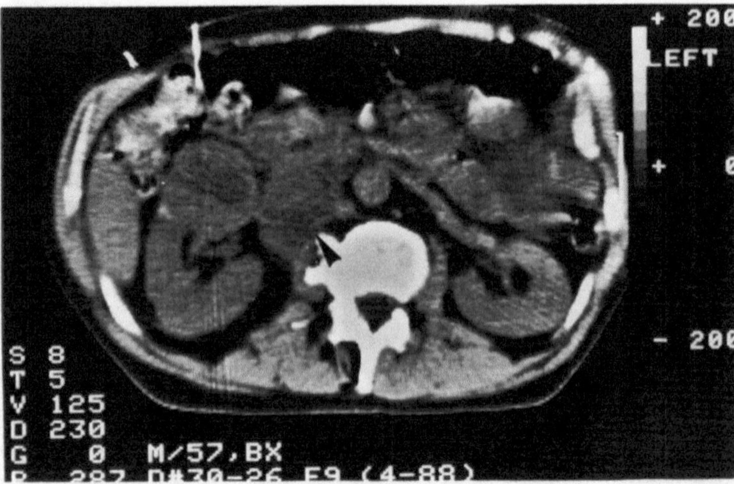

Fig. 32.4. CT scan of right renal cell carcinoma showing the tumour and enlarged lymph nodes (*arrowheads*).

organs or tissues (Fig. 32.7). Furthermore, enlargement of lymph nodes on the CT scan does not necessarily mean lymphatic spread. In these patients, transcutaneous fine needle aspiration of the lymph nodes should be performed before denying them for surgery. Every patient with indeterminate CT scan results should be referred for renal angiography and/or venocavography.

The commonest sites of metastases for renal cell carcinoma are the lungs, liver, subcutaneous tissues and central nervous system. Preoperative staging should include careful search for distant metastases. Chest X-ray films should be obtained routinely for every patient with renal tumour. As to liver scan, it was shown that in all patients with non-palpable liver with normal liver function test, the liver/spleen scan was normal [9]. Similarly, all patients with no skeletal pain and with normal alkaline phosphatase and serum calcium values had normal radionuclide scan results. Our current approach for determining distant metastases is based on history and physical examination, routine chest X-ray films, liver func-

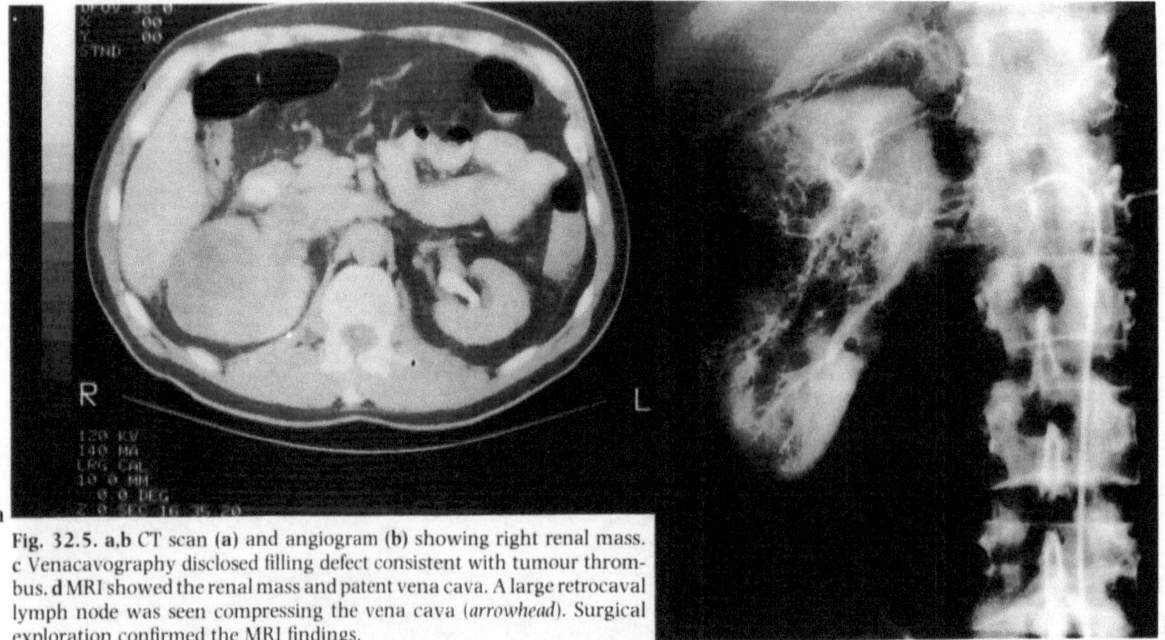

Fig. 32.5. a,b CT scan (a) and angiogram (b) showing right renal mass. c Venacavography disclosed filling defect consistent with tumour thrombus. d MRI showed the renal mass and patent vena cava. A large retrocaval lymph node was seen compressing the vena cava (*arrowhead*). Surgical exploration confirmed the MRI findings.

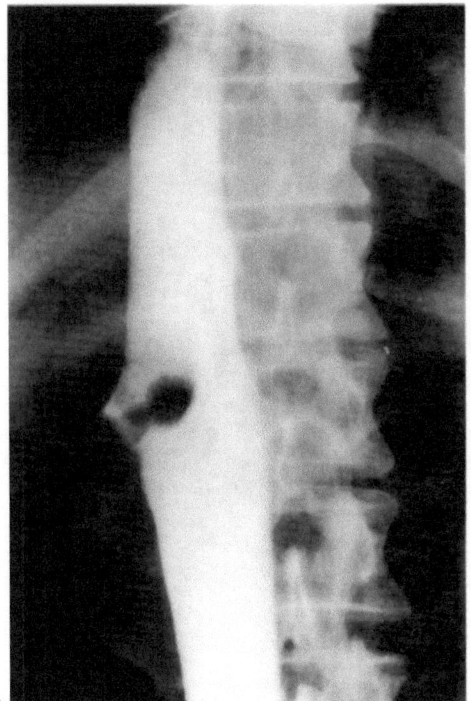

c

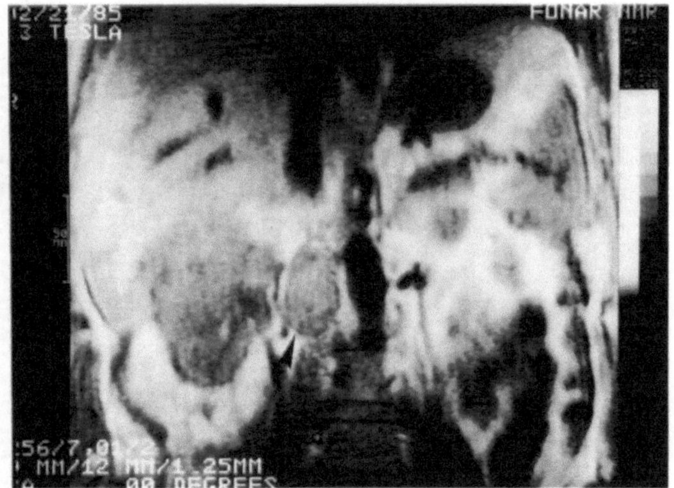

d

Fig. 32.5 (*continued*)

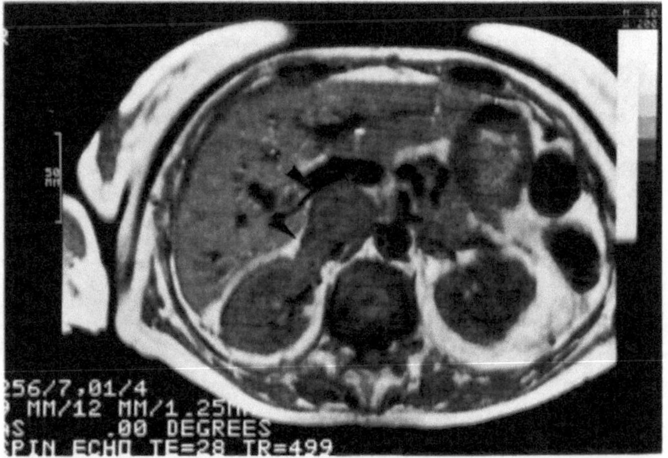

a

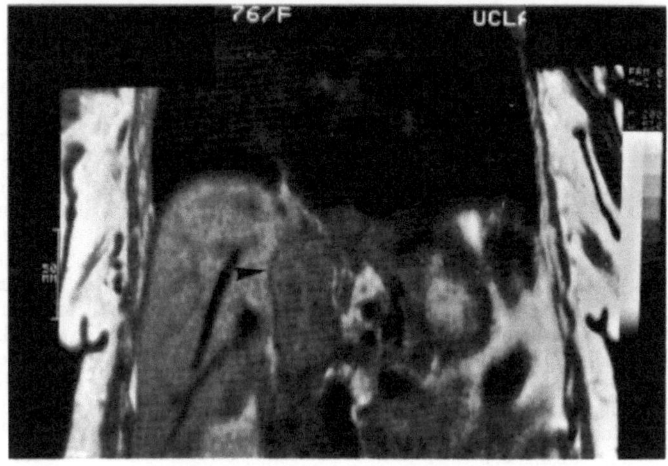

b

Fig. 32.6. a,b MRI of a patient with right renal cell carcinoma. **a** Axial cut showing the tumour in the renal vein and vena cava (*arrowheads*). **b** Coronal cut showing the vena cava thrombus (*arrowhead*) reaching the diaphragm.

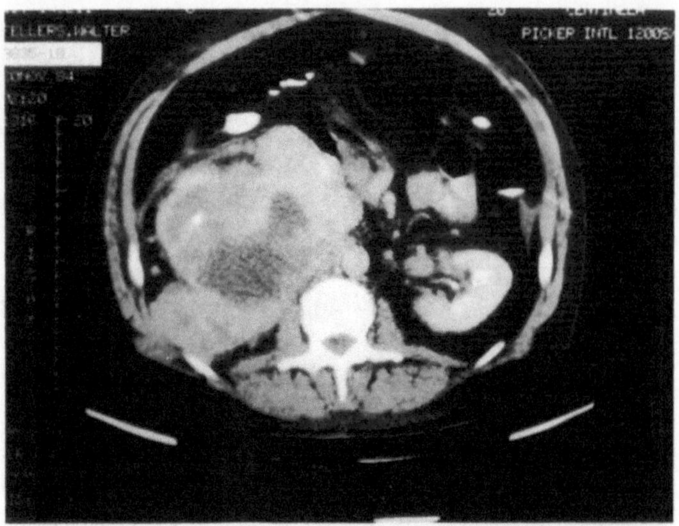

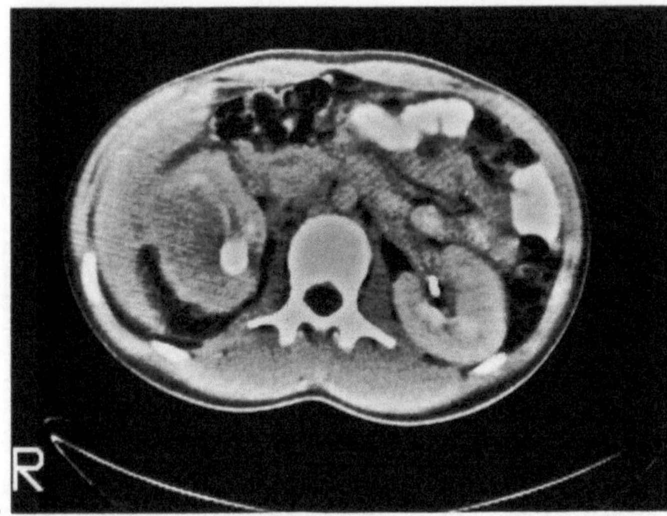

b

Fig. 32.7. **a** CT scan of a right renal cell carcinoma suggesting invasion of the posterior abdominal wall. **b** CT scan of a right renal mass suggesting invasion of the liver by the tumour. In both cases no tumour extension beyond the renal capsule was found during surgery.

tion test and serum calcium. If all are negative for metastases we do not perform radionuclide liver/ spleen scanning or bone scanning.

References

1. Robson CJ, Churchill BM, Anderson W (1968) The results of radical nephrectomy for renal cell carcinoma. Trans Am Assoc Genitourin. Surg 60:122
2. Beahrs OH, Myers MH (eds) (1983) American Joint Committee on Cancer: Manual for staging cancer, 2nd edn. Lippincott, Philadelphia, p 178
3. Holland JM (1973) Cancer of the kidney: natural history and staging. Cancer 32:1030
4. Weyman PJ, McClennan BL, Standley RJ, Levitt RG, Sagee SS (1980) Comparison of computed tomography and angiography in the evaluation of renal cell carcinoma. Radiology 137:417–424
5. Richie JR, Garnick MB, Seltzer S, Bettman MA (1983) Computerized tomography scan for diagnosis and staging of renal cell carcinoma. J Urol 129:1114–1116
6. Cronan JT, Zeman RK, Rosenfield AT (1982) Comparison of computerized tomography, ultrasound and angiography in staging renal cell carcinoma. J Urol 127:712–714
7. Hricak H, Crooks L, Sheloon P, Kaufman L (1983) Nuclear magnetic resonance of the kidney. Radiology 146:425–432
8. Kulkarni MV, Shaff MI, Sanoler MP et al. (1984) Evaluation of renal masses by MR imaging. J Comput Assist Tomogr 8:861–865
9. Lindner A, Goldman DG, deKernion JB (1983) Cost effective analysis of prenephrectomy radioisotope scans in renal cell carcinoma. Urology 22:127–129

Chapter 33

The Diagnosis of Solid Renal and Adrenal Masses by Aspiration Cytology

S. R. Orell

The first series of ultrasonographically guided percutaneous fine needle aspiration biopsies (FNAB) of solid renal masses was reported by Kristensen et al. In 1972 [1]. Technical improvements and wider access to ultrasonography (US) and computed tomography (CT) have made needle biopsy of deep lesions relatively easy. This is reflected by the large number of series of FNAB of renal and other deep masses reported in the literature in recent years (see Table 33.2). At the same time, it has been argued that modern radiological techniques, particularly CT, have such a high diagnostic accuracy by themselves that a tissue diagnosis may not often be necessary. In this chapter, our local experience at the Flinders Medical Centre (FMC) and the experience reported in the literature are reviewed in order to evaluate the role of aspiration cytology in the routine preoperative investigation of suspected solid renal and adrenal masses.

Material

Between the years 1977 and 1985 percutaneous FNAB was performed on 101 patients with renal or adrenal masses which appeared to be solid by radiological and/or ultrasonographic investigations. Cases in which needle aspiration was done only to confirm simple cysts are not included. There were 60 male and 41 female patients. The age of the patients ranged from 13 to 82 years and the mean age was 52 years. There were 94 renal and 7 adrenal masses. The final diagnosis was a benign process in 18 cases and a malignant neoplasm in 83.

The size of the biopsied masses varied from 14 mm in diameter to over 200 mm. The exact size was recorded in 79 cases; 10 were less than 30 mm in diameter, 17 were up to 50 mm, 18 were up to 70 mm, 18 were up to 100 mm and 16 were over 100 mm. Most of the lesions for which an exact measurement was not recorded were large.

Biopsy Technique

Aspiration biopsy was guided by fluoroscopy in relation to excretory urography or to angiography, by US or by CT. During the first 3 years of our study half of the biopsies were guided by fluoroscopy and the other half by US, whereas since 1980 US has been the standard guidance method, and CT and fluoroscopy have been used only in selected cases (seven and four cases, respectively). Free-hand biopsy was performed in six patients with large, palpable tumours.

A 22-gauge, 90 mm, disposable lumbar puncture needle was used for most biopsies. This was fitted to a 10-ml plastic syringe mounted in a Cameco syringe pistol. If aspiration through a standard needle yielded an abundance of blood from which diagnostic cells were difficult to retrieve, a 22-gauge Rotex screw needle was sometimes tried and found to give a better yield of cells. The point of entry was usually from the back. Using fluoroscopy or CT, a vertical approach is preferable; with US the entry and direction are chosen to allow the shortest and most convenient pathway to the lesion. With real-time US control, the tip of the needle can often be seen within the lesion [2]. We have found no particular advantage in using guide needles or a tandem technique for renal masses. Depending on the cytologist's assessment of the adequacy of the aspirate, up to 5 passes from different parts of the mass were made in any one session. In most cases one or two passes proved sufficient. The biopsy was usually performed while the patient was in hospital for investigations. However, in about 25% of the cases, US guided biopsy was undertaken as an out-patient procedure.

Cytological and Histological Methods

Each aspirate was checked immediately in the radiology department for adequacy of cell material by staining one smear with the Diff-Quik stain (Harleco, Philadelphia). Whenever possible, aspirates were smeared on several slides; some were air-dried, some fixed in ethanol, so that both May–Grünwald–Giemsa (MGG) and Papanicolaou or haematoxylin and eosin (H & E) staining could be made. If the aspirate contained visible tissue fragments, these were fixed in formalin for paraffin embedding. In a few cases, determined by the preliminary examination of the Diff-Quik smear, an aspirate was fixed in glutaraldehyde for electron microscopy. Paraffin blocks and histological sections of surgically removed tissue were available in 71 cases, including 69 of the 86 tumours. In the remaining cases, the final diagnosis was based on the findings at exploratory laparatomy, further radiological investigations or on the subsequent clinical course.

Criteria for Cytological Diagnosis

Renal Masses

Normal cortical tissue is represented by whole isolated glomeruli and tubular epithelial cells from different levels of the nephron (Fig. 33.1). This mixture of several types of tubular epithelial cell, distinguished by cell size, nuclear/cytoplasmic ratio, presence or absence of distinct cell borders and size and shape of sheets of cells, is characteristic of normal renal tissue and should preclude a misdiagnosis of low-grade renal cell tumour of granular

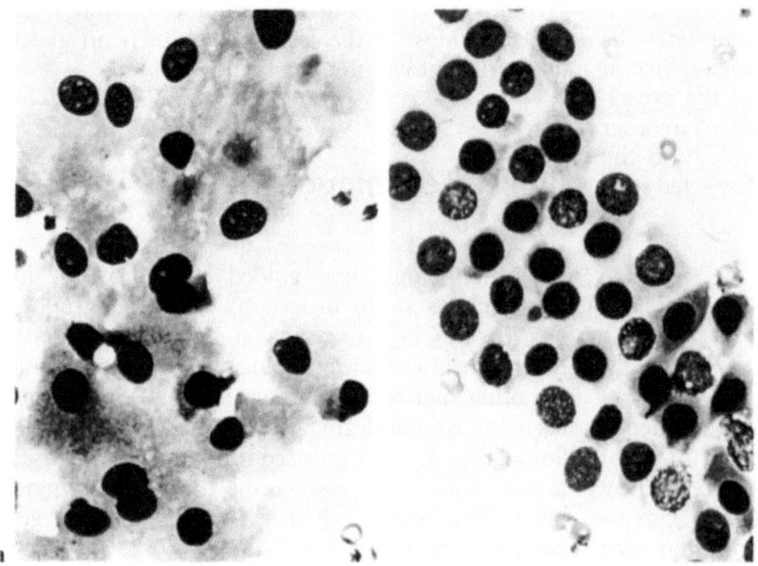

Fig. 33.1a,b. Aspirate of normal kidney. Tubular epithelial cells from different segments of the nephron are seen in the same smear. (MGG, × 610)

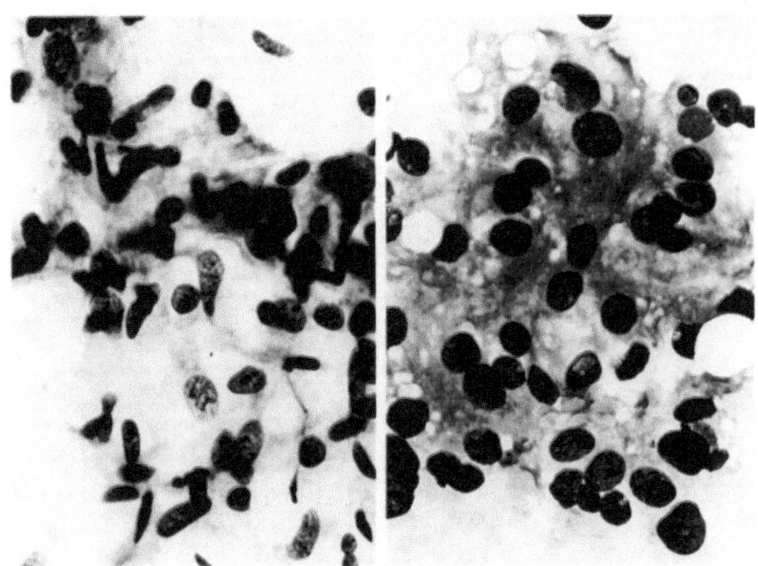

Fig. 33.2a,b. Angiomyolipoma. **a** Typical pattern of spindle cells of smooth muscle type and large fat vacuoles. (H & E, × 610). **b** Atypical round cell pattern. Note deep staining syncytial cytoplasm and large fat vacuoles. (MGG, × 610)

cell type, the cells of which may closely resemble the cells of proximal convoluted tubules. Also, normal tubular epithelial cells lack fine cytoplasmic vacuolation, nuclei are quite uniform in size and shape and lipofuscin granules are often conspicuous.

Normal cortical elements consistently obtained by repeat aspirations from a radiologically abnormal area indicate a cortical pseudo-tumour or a congenital deformity of the kidney, provided correct placement of the biopsy needle can be ascertained. The diagnosis may be confirmed by further radiological investigations such as angiography. Our series includes three cases in this category.

Only those *renal cysts* in which the US diagnosis was in any way in doubt are included in this series. A cyst aspirate consists of clear fluid which contains only a few macrophages and/or degenerating benign cuboidal epithelial cells. Occasionally, macrophages may be more numerous and appear atypical, but if the cells can be clearly identified as macrophages, this should not arouse a suspicion of malignancy. In doubtful cases, aspiration should be followed by contrast injection to ensure that the cyst has a smooth outline. Bleeding into the cyst caused by the needle puncture may render diagnosis difficult. A purulent aspirate indicates an *abscess*. The material should be used for staining and culture for micro-organisms.

Smears of an *angiomyolipoma* show clusters of bland spindle cells of smooth muscle type and a background of blood and numerous fat vacuoles (Fig.33.2a). Smooth muscle cells have an eosinophilic, non-granular, non-vacuolated attenuated cytoplasm which anastomoses with adjacent cells into a syncytial pattern. Smears from highly cellular areas with a round cell pattern may closely resemble a low-grade renal cell tumour of granular cell type (Fig. 33.2b). Close attention to cytoplasmic detail and the presence of some more characteristic spindle cells should indicate the correct diagnosis. As the smooth muscle cells of angiomyolipoma sometimes may appear severely atypical in histological sections, it seems possible that sarcoma could be suspected cytologically in some cases, but we have not yet encountered this particular problem.

Cells of *renal cell tumours* have rounded nuclei ranging from 6 to 40 μm (average 11 μm) in diameter, a low nuclear/cytoplasmic ratio and a well-defined, finely vacuolated or finely granular cytoplasm which usually stains pale amphophilic although the staining may be variable (Figs. 33.3a, 33.4, 33.5). In smears, the cytoplasm does not have the clear appearance familiar from histological sections [3]. The cells may form monolayered sheets, irregular multilayered aggregates, papillary structures or, less commonly, tubular or glandular patterns. Cells are often arranged along strands of basement membrane material (staining bright pink with MGG; Fig. 33.4). Cell cohesion, in other words the proportion of single cells, is quite variable. Overall nuclear size, nuclear pleomorphism, nucleolar size and loss of cell cohesiveness are in general proportional to tumour grade. Although the abundance of cytoplasm is a characteristic feature, sometimes clusters of small cells with round nuclei and scanty, indistinct cytoplasm which resemble thyroid follicular epithelial cells may be found (Fig.

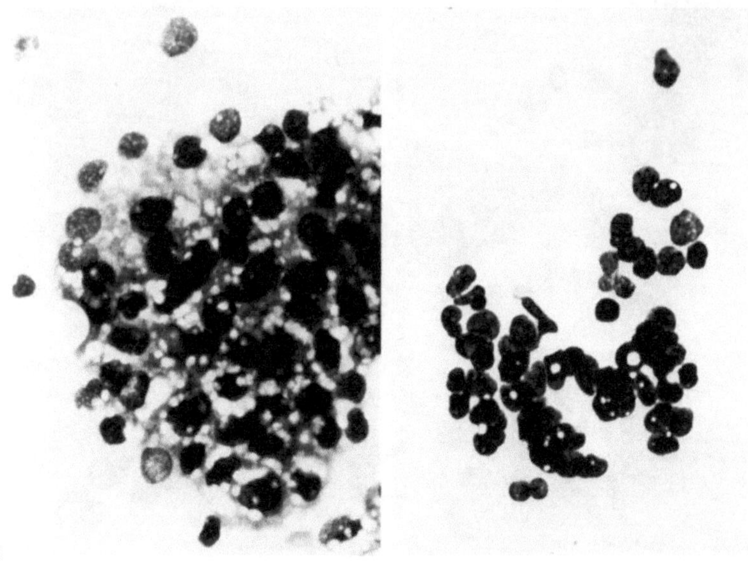

Fig. 33.3. a,b. Renal cell tumour grade I. Cohesive clusters of cells with small, round, uniform nuclei. **a** The cells have a well-defined cytoplasm with small vacuoles. **b** Nuclei are clustered, bare, resembling thyroid follicular epithelial cells. (MGG, × 610)

33.3b). Intranuclear cytoplasmic inclusions are common and are found in about one-third of renal cell tumours (Fig. 33.5).

It is not possible to distinguish clearly renal cortical adenoma and low-grade carcinoma cytologically. Such cases are reported as "low-grade renal cell tumours" and prediction of behaviour is based on size and clinical findings.

We have not seen an example of *renal oncocytoma* in our series. Since foci of bland oncocytic cells may be present in renal cell carcinoma, a cytological diagnosis should be made with caution and only if radiological appearances are typical [4].

Renal pelvic carcinoma invading the kidney is usually not well differentiated. Aspiration smears contain obviously malignant, pleomorphic epithelial cells, single or in small solid clusters. The cells have a moderate amount of dense cytoplasm, well-defined cell borders and excentric nuclei. Squamous differentiation may be evident. Nuclear chromatin is dense and irregular. Malignant cells are often mixed with normal tubular epithelial cells

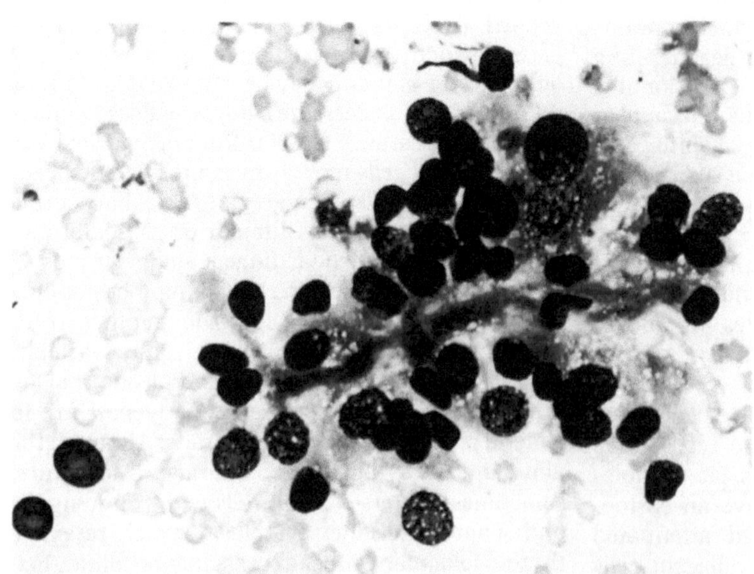

Fig. 33.4. Renal cell carcinoma grade II. Moderate nuclear pleomorphism, well-defined cytoplasm with small vacuoles, cells arranged along a strand of dense material, probably basement membrane. (MGG, × 610)

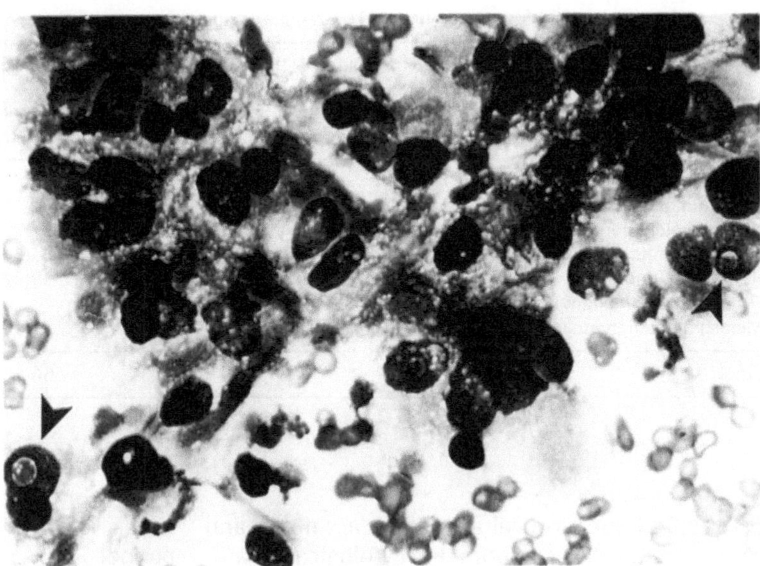

Fig. 33.5. Renal cell carcinoma grade III. Marked nuclear enlargement and pleomorphism. Lesser cell cohesion. Note intranuclear cytoplasmic inclusions (*arrowheads*). (MGG, × 610)

as the cancer may infiltrate renal parenchyma diffusely.

The cytological pattern of *Wilms' tumour* is that of a malignant small round cell tumour. Exact typing within this group of tumours is not often possible on routine cytological smears and, if possible, aspirated material should be fixed and processed for electron microscopy. Better differentiated tumours have cells arranged in tubular structures with a background of dispersed, small undifferentiated cells.

Adrenal Masses

The cytological pattern of *adrenal cortical adenoma and carcinoma* may closely resemble that of renal cell tumours. Radiological and clinical findings including hormone assays must be taken into account. As with most endocrine tumours, nuclear pleomorphism is not a reliable indicator of malignant behaviour. Smears of *phaeochromocytomas* show mainly dispersed cells and prominent anisokaryosis but a uniformly bland chromatin pattern. The cells have a fragile cytoplasm with red (MGG staining) granulation which may be coarse.

Metastatic carcinoma and *malignant lymphoma* of the kidney and of the adrenal gland are usually easily identified in aspiration smears as the cytological pattern clearly differs from that of primary tumours. However, there may be a problem if tumours are poorly differentiated. In practice, there is in most cases a history of a previous malignancy and histological material from this primary tumour may be available for comparison.

Diagnostic Accuracy

The FMC Experience

A definitive cytological diagnosis of malignancy was given in 74 of the 83 cases of malignant renal or adrenal tumour. Of these 74 tumours, there were 60 renal cell carcinomas, 1 adrenal cortical carcinoma, 2 renal pelvic carcinomas, 1 Wilms' tumour and 10 metastatic malignancies, including malignant lymphomas, 4 of which were in the adrenal gland, 6 in the kidney. In addition, a phaeochromocytoma which later metastasised was correctly diagnosed as a phaeochromocytoma with no attempt to predict behaviour. Thus, the sensitivity of FNAB in the diagnosis of renal and adrenal malignancy was 90%; it remains the same if only cases with histological confirmation are included (Table 33.1).

Specific typing within the group of malignant tumours is usually possible. Distinction between primary tumours, metastatic malignancies and malignant lymphoma is not often difficult except in highly anaplastic tumours. Immunocytochemistry and electron microscopy may be helpful, particularly in Wilms' tumour and in identifying the origin of metastatic tumours. For example, unsuspected metastasis of amelanotic melanoma was diagnosed by electron microscopy in one case. However, as has already been pointed out, we do not feel it is possible to distinguish between renal cell adenoma and well-differentiated carcinoma or

Table 33.1. Results of aspiration cytology of 101 renal and adrenal masses at the Flinders Medical Centre

Cytological diagnosis	Final diagnosis (histologically confirmed in brackets)				
	Non-neoplastic process	Angiomyolipoma	Renal cell carcinoma	Other primary malignancy	Metastatic malignancy
Unsatisfactory, no yield	2		2(2)		
Cyst fluid, pus or normal cells	13(2)				
Angiomyolipoma		2(1)			
Blood, necrotic debris only			4(3)	1(1)	1(1)
Renal cell tumour		1(1)	59(48)		
Other primary malignancy				5(3)	
Metastatic malignancy			1(1)		10(6)
Total	15	3	66	6	11

between adrenal cortical adenoma/carcinoma and similar renal cell tumours by cytological examination alone in many cases.

FNAB was reported as unsatisfactory in two cases of benign tumour, both cysts with heavily calcified walls which could not be penetrated by the needle. There were two unsatisfactory biopsies in malignant tumours; in one, only blood and a few tubular epithelial cells were obtained in spite of four passes and clear radiological evidence of tumour; in the other, aspiration yielded clear fluid. In the latter case, subsequent injection of contrast revealed a solid mass in the cyst wall. Attempts to biopsy this area gave a poor yield reported as unsatisfactory; however, on reviewing these slides a small number of cells diagnostic of low-grade renal cell tumour were found.

In six cases, which included four renal cell carcinomas, one adrenal cortical carcinoma and one metastatic carcinoma in an adrenal, multiple FNABs yielded only old blood and necrotic debris. As no preserved neoplastic cells were found, a definitive diagnosis could not be made. However, in our experience the presence of altered blood and necrosis strongly supports a clinical/radiological suspicion of malignancy.

One false positive diagnosis was made: a large angiomyolipoma was reported cytologically as a low-grade renal cell tumour. This tumour contained areas of a highly cellular, round cell, epithelioid pattern. Awareness of this variant of angiomyolipoma, close attention to the cytoplasm which is densely eosinophilic, non-granular and non-vacuolated and a search for more typical spindle cells, should prevent misdiagnosis. Diagnostic specificity of FNAB in our series is thus 95%. This figure should be seen in relation to the small number of benign lesions included in the series. Cytological findings were consistently negative in fluid aspirated from 85 simple cysts seen during the same period.

Review of the Literature

The results of 18 series of FNAB of renal and sometimes adrenal masses reported in the literature, including our own, are listed in Table 33.2. Unfortunately, these series are rather heterogeneous, which makes comparison of results difficult. Some include simple renal cysts, most only cysts in which the radiological diagnosis was doubtful. Some series include only avascular solid renal masses, others include all solid renal masses and a few also adrenal tumours. This heterogeneity to some extent explains the differences in diagnostic accuracy.

The reported series include a total of 1585 lesions investigated by FNAB. There were 603 malignant tumours, mainly renal cell carcinomas, but also a few adrenal tumours, renal pelvic carcinomas, Wilms' tumours and metastatic malignancies. Diagnostic sensitivity by FNAB was 86%, specificity was 98% with a false positive rate of 2.3%. The predictive value of a positive result was 96%. The main factor which limits diagnostic sensitivity is the difficulty in obtaining preserved neoplastic cells from large, extensively necrotic and haemorrhagic tumours. Although aspiration of necrotic debris and old blood without preserved neoplastic cells is recorded as a false-negative cytological diagnosis, such findings indirectly support a clinical/radiological diagnosis of malignancy. The disturbingly high false positive rate in this review is generated mainly by two of the series. One is the series of Beyer and Fiedler [5]; these authors found atypical cells in fluid aspirated from renal cysts which were reported as "suspicious of intracystic papilloma". We feel that if radiological appearances

Table 33.2. Aspiration cytology of renal and adrenal masses—review of the literature

Reference	Total cases	Malignant tumours	True positive cytology	False positive cytology	Significant complications
von Schreeb et al. [11]	12	12	12	0	0
Jeans et al. [9]	80	14	14	1	0
Lang [13]	388	19	17	0	1
Edgren et al. [7]	55	42	32	2	0
Holm et al. [20]	110	49	43	0	0
Thommesen and Nielsen [21]	49	18	18	1	—
Sherwood and Trott [22]	88	7	5	0	0
Beyer and Fiedler [5]	185	13	13	6	0
Vilaplana [23]	66	34	33	0	0
Phillips and Schneider [24]	8	6	6	0	0
Porter et al. [25]	28	28	19	0	0
Sundaram et al. [26]	20	5	5	0	0
Nosher et al. [27]	21	7	6	0	0
Triller et al. [28]	25	12	8	0	0
Helm et al. [8]	31	23	19	1	0
Droese et al. [29]	17	13	9	0	0
Juul et al. [6]	301	218	185	11	0
FMC series	101	83	75	1	1
Total	1585	603	519 (86%)	23	2

are typical of a cyst, the aspirated fluid is clear and cytological atypia is mild, such cases may be managed conservatively. Fourteen false positive diagnoses were recorded in the large series reported by Juul et al. [6]. Three of these were adenomas which we feel cannot be distinguished from well-differentiated carcinomas cytologically and therefore should be reported as low-grade renal cell tumours to be managed on the basis of, for example, size. The other false positive diagnoses were made in six cases of an inflammatory condition, three benign cysts and two benign non-renal cell tumours. Apart from these two series, only occasional false positive diagnoses have occurred; two in cases of chronic infection [7, 8], one in a case of polycystic degeneration [7], one in a case of a cyst with massive haemorrhage [9] and in our case of angiomyolipoma with a prominent round cell pattern.

Cytological Grading of Renal Cell Carcinoma

It is well known that the histological pattern often varies considerably within a given renal cell carcinoma. Nurmi et al. [10] found that the histological grade was variable in 24.2% of 150 renal cell tumours examined after nephrectomy. It is to be expected that cytological grading by FNAB, which only samples minimal parts of a large tumour, is less reliable and of limited value in clinical practice. Only a few authors have taken an interest in this aspect of cytological diagnosis (von Schreeb et al. [11], Edgren et al. [7], Nurmi et al. [10]). These studies show that a high grading by cytology is fairly reliable, whereas a significant proportion of high-grade tumours are undergraded by cytology. Cytological grading was based on the subjective evaluation of a combination of nuclear, cytoplasmic and architectural features, the latter mainly an evaluation of cell cohesiveness. We feel that dissociation of cells from aspirated tumour fragments may be influenced by biopsy and smearing technique and that cytoplasmic features are less important. Our grading is therefore based entirely on nuclear size and pleomorphism and follows the nuclear grading proposed by Fuhrman et al. [12] for histological diagnosis, except that we combine grade IV with grade III (Figs 33.3–33.6). The result is similar to that reported by others; only two tumours were overgraded cytologically, whereas 8 of 20 grade III tumours were undergraded (Table 33.3).

Complications of FNAB

Complications have been nearly absent in the reported series of FNAB of solid renal masses. Tran-

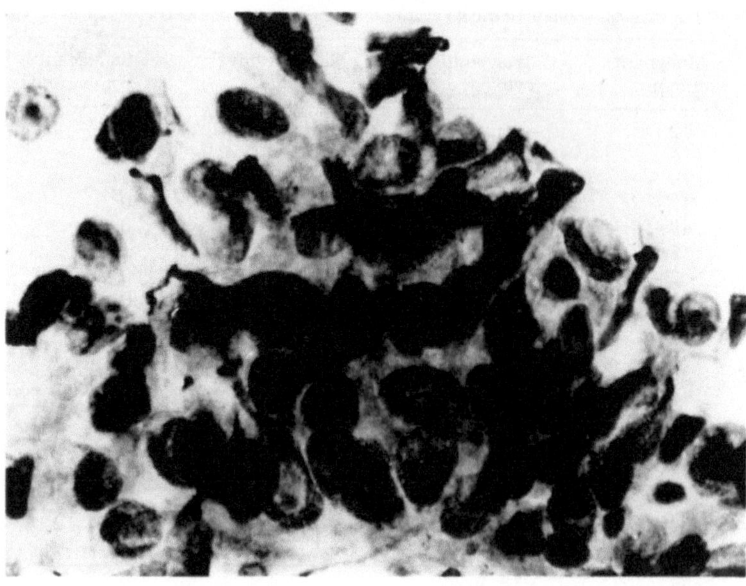

Fig. 33.6. Renal cell carcinoma grade III. Anaplastic carcinoma pattern. Difficult to recognise renal cell origin. (MGG, × 610)

Table 33.3 Grading of renal cell tumours

Cytological grade	Histological grade		
	I	II	III
I	3	1	
II	2	20	8
III			12

sient pain and minor pneumothorax have been recorded occasionally, but significant complications were only one case of infection of a large renal cyst [13] and one case in our series of major post-biopsy haemorrhage into an extensively necrotic renal cell carcinoma causing severe pain. Two cases of tumour implantation in the needle track have been reported. In one, an 18-gauge needle was used [14]; in the other, the size of the needle was not mentioned [15]. This complication has not occurred in our experience nor in any of the series reviewed [16].

Conclusions

It would thus appear that FNAB of solid renal masses has a diagnostic sensitivity of 85%–90% and a specificity of 95%–98%, and that complications are extremely rare. Although in most published series false positive diagnoses have been rare, a couple of centres have had a less favourable experience in this respect. There appears to be a risk of misdiagnosis in chronic inflammatory conditions, probably resulting from the alarming atypia that macrophages may sometimes show in smears [17]. Extensive experience with fine needle aspiration cytology of all tissues and optimal biopsy and processing techniques are essential to maintain a high diagnostic accuracy. It is equally essential to interpret cytological findings in relation to clinical and radiological findings. Thus, it must be recognised that there is no clear distinction cytologically between renal cortical adenoma and low-grade renal cell carcinoma, and that it may be difficult or impossible to decide if a tumour is of renal or adrenal cortical origin by cytological findings alone.

The obvious indications for FNAB of solid renal masses are as follows:

1. Tumours in which there is evidence of dissemination
2. Tumours which are considered to be inoperable for other reasons
3. Tumours in which preoperative treatment such as embolisation or irradiation is considered
4. Tumours in which the results of radiological and ultrasonographic investigations are equivocal.

The initial detection of a renal mass lesion by intravenous pyelography and/or US is usually followed by a CT examination. In most cases, CT provides not only a diagnosis but also information regarding tumour stage [18, 19] and the function of the contralateral kidney, and a preoperative tissue diagnosis by FNAB is unnecessary. However, if the results of the CT examination are equivocal, FNAB is usually preferable to angiography as the next investigation. Also, if staging information is not of paramount importance, or if CT is not readily available, FNAB in association with ultrasonographic examination can be a valuable alternative.

References

1. Kristensen JK, Holm HH, Rasmussen SN, Barlebo H (1972) Ultrasonically guided percutaneous puncture of renal masses. Scand J Urol Nephrol (Suppl) 15:49–56
2. Orell SR, Langlois SLP, Marshall VR (1985) Fine needle aspiration cytology in the diagnosis of solid renal and adrenal masses. Scand J Urol Nephrol 19:211–217
3. Linsk JA, Franzen S (1984) Aspiration cytology of metastatic hypernephroma. Acta Cytol 28:256–260
4. Rodriguez CA, Buskop A, Johnson J, Fromowitz F, Koss LG (1980) Renal oncocytoma. Preoperative diagnosis by aspiration biopsy. Acta Cytol 24:355–359
5. Beyer D, Fiedler V (1977) Ist die Nierenzystenpunktion eine brauchbare Methode zur Differentialdiagnostik gefassarmer raumfordernder Nierenprozesse? Urologe [Ausg A] 16:339–345
6. Juul N, Torp-Pedersen S, Gronvall S, Holm HH, Koch F, Larsen S (1985) Ultrasonically guided fine needle aspiration biopsy of renal masses. J Urol 133:579–581
7. Edgren J, Taskinen E, Alfthan O, Makinen J, Juusela H (1975) Radiology and fine needle aspiration biopsy in the diagnosis of tumours of the kidney. Ann Chir Gynaecol 64:209–216
8. Helm CW, Burwood RJ, Harrison NW, Melcher DH (1983) Aspiration cytology of solid renal tumours. Br J Urol 55:249–253
9. Jeans WD, Penry JB, Roylance J (1972) Renal puncture. Clin Radiol 23:298–311
10. Nurmi M, Tyrkko J, Puntala P, Sotarauta M, Antila L (1984) Reliability of aspiration biopsy cytology in the grading of renal adenocarcinoma. Scand J Urol Nephrol 18:151–156
11. von Schreeb T, Franzen S, Ljungqvist A (1967) Renal adenocarcinoma. Evaluation of malignancy on a cytologic basis. A comparative cytologic and histologic study. Scand J Urol Nephrol 1:265–269
12. Fuhrman SA, Lasky LC, Limas C (1982) Prognostic significance of morphologic parameters in renal cell carcinoma. Am J Surg Pathol 6:655–663
13. Lang EK (1973) Roentgenographic assessment of asymptomatic renal lesions. Radiology 109:257–269
14. Gibbons RP, Bush WH, Burnett LL (1977) Needle tract seeding following aspiration of renal cell carcinoma. J Urol 118:865–867
15. Auvert J, Abbou CC, Lavarenne V (1982) Needle tract seeding following puncture of renal oncocytoma. Prog Clin Biol Res 100:597–598
16. von Schreeb T, Arner O, Skovsted G, Wilkstad N (1967) Renal adenocarcinoma. Is there a risk of spreading tumour cells in diagnostic puncture? Scand J Urol Nephrol 1:270–276
17. Zajicek J (1979) Aspiration biopsy cytology, part 2. Cytology of infradiaphragmatic organs. Karger, Basel (Monographs in clinical cytology, vol. 7)
18. Cronan JJ, Zeman RK, Rosenfield AT (1982) Comparison of computerized tomography, ultrasound and angiography in staging renal cell carcinoma. J Urol 127:712–714
19. Love L, Churchill R, Reynes C, Schuster GA, Moncada R, Berkow A (1979) Computed tomography staging of renal carcinoma. Urol Radiol 1:3–10
20. Holm HH, Pedersen JF, Kirstensen JK, Rasmussen SN, Hancke S, Jensen F (1975) Ultrasonically guided percutaneous puncture. Radiol Clin North Am 13:493–503
21. Thommesen P, Nielsen B (1975) The value of fine needle aspiration biopsy and intravenous pyelography in the diagnosis of renal masses. Fortschr Roentgenstr 122:248–251
22. Sherwood T, Trott PA (1975) Needling renal cysts and tumours: cytology and radiology. Br Med J 3:755–758
23. Vilaplana E (1980) Cytology of special sites: kidney. In: Proceedings of the 7th international congress of cytology, Munich, Federal Republic of Germany, 19–22 May 1980
24. Phillips G, Schneider M (1981) Ultrasonically guided percutaneous fine needle aspiration biopsy of solid masses. Cardiovasc Intervent Radiol 4:33–38
25. Porter B, Karp W, Forsberg L (1981) Percutaneous cytodiagnosis of abdominal masses by ultrasound guided fine needle aspiration biopsy. Acta Radiol [Diagn] (Stockh) 22:663–668
26. Sundaram M, Wolverson MK, Heiberg E, Pilla T, Vas WG, Shields JB (1982) Utility of CT-guided abdominal aspiration procedures. AJR 139:1111–1115
27. Nosher JL, Amorosa JK, Leiman S, Plafker J (1982) Fine needle aspiration of the kidney and adrenal gland. J Urol 128:895–899
28. Triller J, Schneekloth G, Marincek B, Kraft R (1982) Computertomographisch gezielte Feinnadelpunktion abdominaler Raumforderungen. Radiologe 22:484–492
29. Droese M, Altmannsberger M, Kehl A, Lankisch PG, Weiss R, Weber K, Osborn M (1984) Ultrasound-guided percutaneous fine needle aspiration biopsy of abdominal and retroperitoneal masses: accuracy of cytology in the diagnosis of malignancy, cytologic tumour typing and the use of antibodies to intermediate filaments in selected cases. Acta Cytol 28:368–384

Chapter 34

Computed Tomography-guided Fine Needle Biopsy and Cytological Diagnosis of Adrenal Masses

M. Lüning and E. Hoppe

Introduction

Radiological methods are used in the wake of clinical examination and determination of laboratory values for the morphological functional diagnosis of adrenal lesions. Conventional radiographic diagnosis, scintigraphy, sonography and angiography are additionally used and provide differentiated information. Computed tomography (CT) is a method by which visualisation of even clinically intact adrenals has become possible for the first time. Changes in size and expansive lesions can also be diagnostically established, the limit of detectability being 5–8 mm. Values around 95% have been quoted in the international literature for the accuracy of this approach in the diagnosis of adrenal masses (sensitivity 85%–95%; specificity 95%–98%), although any reference to accuracy should obviously be related to tumour size.

The type of an adrenal mass, however, can be safely established in only 70%–80% of all cases by characteristic CT-morphological criteria. Problems in delimitation for differential diagnosis have been primarily attributable to the appearance, quite often uniform, of metastases and adenomas and even of tumours, which are generally liable to present themselves in conjunction with necrosis, haemorrhage and calcification. In other words, the morphological criteria depend not only on the histological substrate but also on tumour size. While size, contrast enhancement and consistency are discriminators of malignant from benign adrenal masses [1], they are sometimes not sufficiently manifest for correct diagnosis in the individual patient.

The increasing use of CT scanning, as compared with the past, has also proved helpful in the positive diagnosis of hormone-inactive expansive adrenal lesions. Inactive adenomas, for example, have accounted for something between 2% and 9% of histological findings and thus are no rare, accidental results [2, 3]. In clinical practice, a high percentage of patients are admitted for CT scanning for the detection of metastases of known malignant processes, and adrenal metastases have been recorded from 8% to 9% of all tumour patients. Hence, radiologists are finding themselves quite often confronted with the problem of differential diagnosis between metastasis and hormone-inactive adenoma.

Assessment of tissue in terms of differentiation between benign and malignant adrenal masses as well as typing of an adrenal mass have decisive impact upon therapeutic decision-making and on prognostication of the disease. CT-guided percutaneous fine needle biopsy (FNB), followed by cytological diagnosis, is an approach by which to meet the demand for unambiguous and highly non-invasive determination of tumour type.

Methods

The technique of percutaneous CT-guided adrenal biopsy is largely in agreement with recommendations generally made for coping with abdominal lesions [4, 5].

Patients are examined in the supine or prone position, depending on the given topographic situation. A grid is fixed to the patient's skin to localise the target of biopsy. The grid consists of polyethylene catheter pieces arranged in the fashion of organ pipes and spaced horizontally and vertically 1 cm from each other. The fine needles with mandrin are between 0.6 and 0.7 mm in outer diameter and between 150 mm and 200 mm in length. The route of biopsy should be chosen with the view to causing the least possible alteration to abdominal structures. However, there are cases in which passage through other organs cannot be avoided. Yet, the transhepatic route has been explicitly recommended by Price et al. [6] for biopsy of the right adrenal. The triangle technique [7] should be used whenever possible for target localisation, in order to avoid injury to the lung in the posterior costophrenic sulcus.

On removal of the stylet, a 10-ml syringe is attached to the needle. With suction being applied, the tip of the needle is rotated and moved back and forth within the target object in increments of 0.5 cm–1 cm. Cytological specimens collected are fixed for at least 30 min in 95% unmethylated alcohol and are subsequently stained by standard procedures: haemotoxylin and eosin, Papanicolaou, PAS reaction. The order of staining methods is an indication of their value. Haematoxylin and eosin staining is primarily used for orientation but may, partially, contribute to diagnosis. Papanicolaou's stain is suitable for more precise specification of cells from metastases of a squamous cell carcinoma in the process of hornification, since cellular hornification is properly visualised. The PAS reaction, finally, has proved to be useful in the detection of mucopolysaccharides in metastases of adenocarcinomas, or of glycogen-containing tumour cells in metastases of carcinomas of renal epithelial cells. Additional air-dried smears are prepared for Giemsa stain or for combined May–Grünwald–Giemsa stain, provided that sufficient amounts of material have been sampled. Actual use of the latter stain, however, will depend on the discretion and specialisation of the cytodiagnostician involved. Cytopathologists generally prefer the staining methods common in histopathology.

Our own experience is likely to suggest that more information can be obtained from representation of lipoids by Sudan-III stain. This is particularly true for differential diagnosis, since lipoids are usually present in an adenoma. Sometimes, only a small quantity of material is available. Under such circumstances, specimens may have to be decolourised in selected cases, taking also into consideration the fact that the smear in which informative cell complexes are present can be microscopically determined only from the stained specimen. Such decoloration will have to be followed by special purpose-oriented staining according to the problem at hand, i.e. either to expand and support a diagnosis or to rule out suspicious diagnosis.

Complications

Clinically irrelevant pneumothorax was recorded from one of the patients in the series reported here. Only rarely have complications been described in the literature, with pneumothorax being established. A somewhat increased rate of complications of 12% has been reported only by Bernardino et al. [8]. Most of the complications were asymptomatic haemorrhage which did not require treatment. However, therapeutic transfusion had to be undertaken in two cases. Bacteraemia which persisted for 24 h was additionally recorded from one of these two patients. Biopsy can be safely applied on an outpatient basis, provided that these outpatients are monitored for 3 h.

Material and results

A total of 55 CT-guided FNBs were applied in 5 years to 52 patients from whom suspicion of expansive adrenal lesions had been recorded by CT scanning. There were 24 males and 28 females, aged between 24 and 73 years (mean 51 years); 40 tumours were recorded from the left adrenal and 8 from the right. Tumour diameters were between 0.5 cm and 15 cm, with 29 tumours having diameters of less than 3 cm and 6 of these having diameters of less than 1 cm.

The results obtained from retrograde analysis of 42 FNBs in 40 patients will be reported in the following paragraphs. Cytological findings were verified by histological examinations in 15 cases and by clinical and CT follow-up checks in 25 cases [9].

Adenoma

Adenoma was definitely diagnosed in 25 patients and was histologically secured in 8 cases. Of these patients 16 were admitted for CT scanning for oncological problems, 7 for hypertension with suspicion of phaeochromocytoma (3 patients) and of adenoma (4 patients) and 1 patient each for clinical manifestation of Cushing's syndrome and of hyperadrenalism. Differential diagnosis to distinguish between metastasis and adenoma was considered only for the latter patients, with one phaeochromocytoma and one metastasis being falsely diagnosed. Malignoma was cytologically ruled out from 17 patients, and the diagnosis of adenoma was

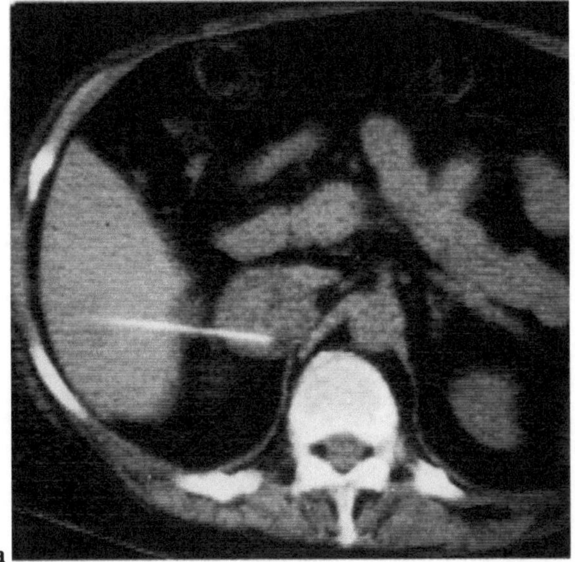

a

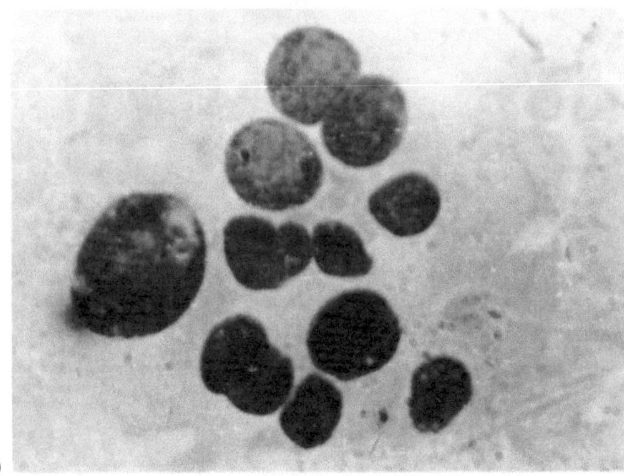

b

Fig. 34.1a,b. Cushing's syndrome. **a** Biopsy of a hypodense tumour of the right adrenal. **b** Cytological findings: primarily round cell nuclei, different in size, with scattered nuclear vacuoles; differently granulated cellular structure; marked adenoid cell grouping. Diagnosis: Pap. II K; adenoma more probable than not. Histologically confirmed. (Lüning et al. [9]; reproduced by kind permission of the Editor of *Fortschritte auf dem Gebiete der Röntgenstrahlen und der Nuclearmedizin. Ergänzungsband*)

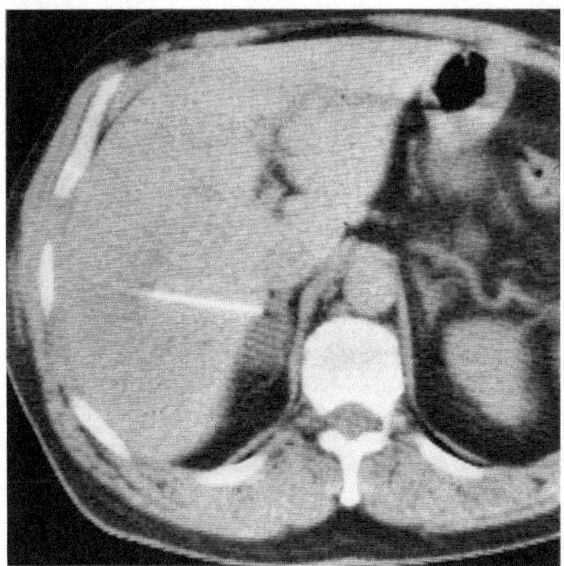

a

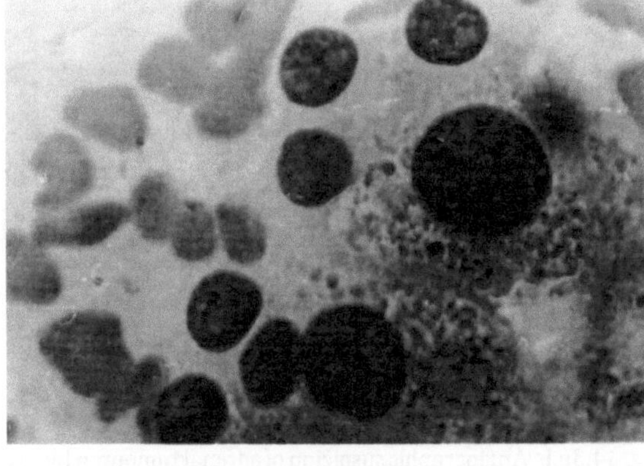

b

Fig. 34.2a,b. Unconfirmed sonographic suspicion of large abdominal tumour. **a** Biopsy of an oval hypodense tumour in the right adrenal. Suspicion of metastasis. **b** Cytological findings: spherical cell nuclei, different in size, with irregular chromatin structure. Diagnosis: Pap. III, abnormal cells; adenoma more probable than not. Histologically confirmed.

supported (Figs. 34.1, 34.2). Phaeochromocytoma was diagnosed in one case, while no differentiation was possible between adenoma and highly differentiated carcinoma in another. Repeat biopsies were applied to two patients for insufficient material or equivocal first findings. The diagnosis of adenoma was thus reaffirmed. In four additional patients with insufficient material (no tumour first established) tumour sizes were between 0.7 cm and 1.4 cm.

Primary and Secondary Malignancies (Figs. 34.3–34.6)

In seven patients with a definite diagnosis of adrenal metastases CT was applied for staging. Metastasisation was diagnosed by CT in four cases. No differentiation between adenoma and metastases was possible in three cases. Histological verification was made on two patients, while enlargement in

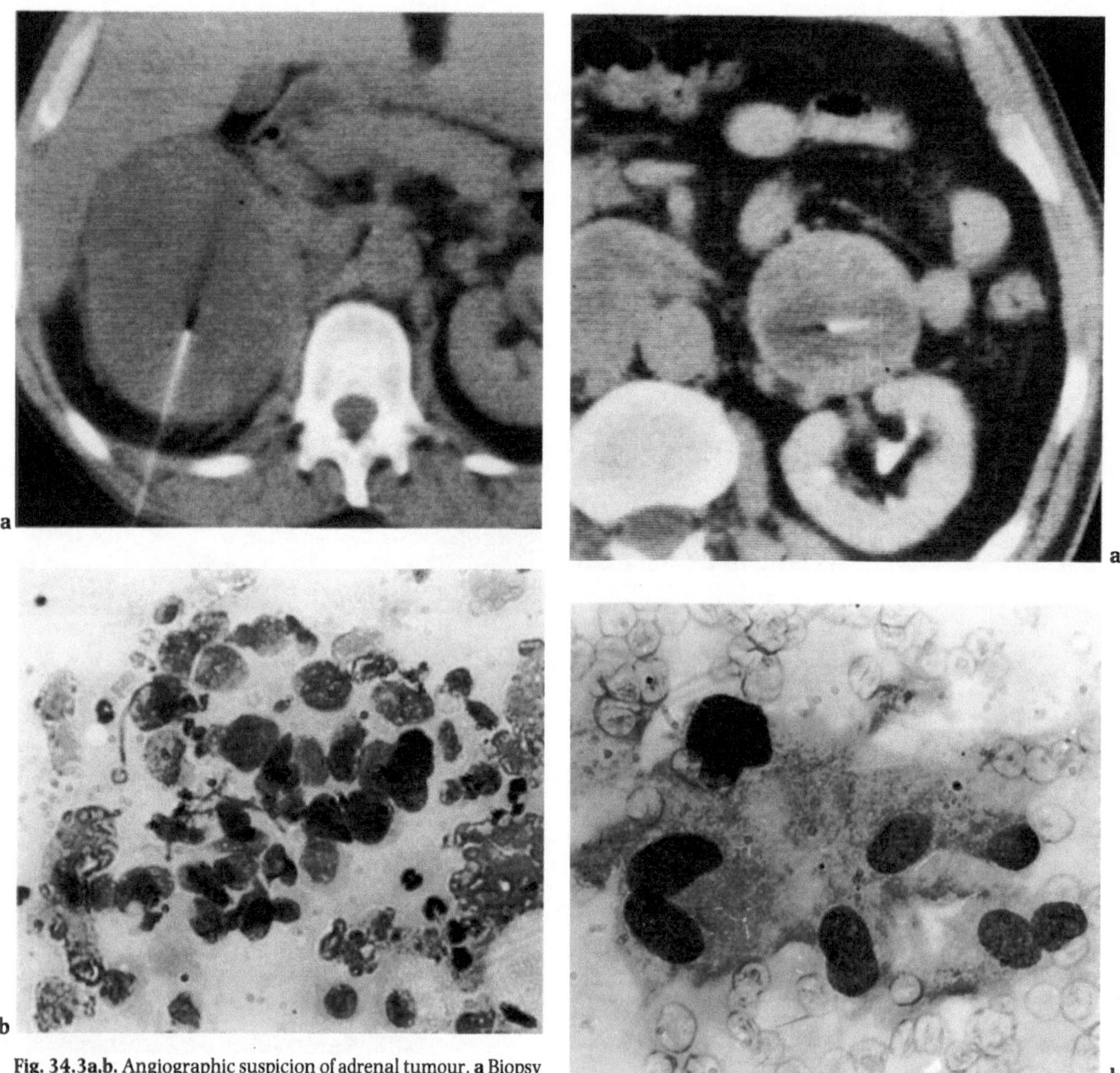

Fig. 34.3a,b. Angiographic suspicion of adrenal tumour. **a** Biopsy of a monstrous hypodense tumour of the right adrenal. **b** Cytological findings: concentration of atypical, polymorphous cells with honeycomb loosening of nuclei. Diagnosis: Pap V, tumour cells. No other primary tumour diagnosed in the process of examinations; liver metastases; therefore, assumption of adrenal carcinoma. Chemotherapy.

Fig. 34.4a,b. Hypernephroma. **a** Biopsy of a round tumour of inhomogeneous structure in the left adrenal. **b** Cytological findings: enlarged cells, some of them oval-shaped, with fine-grained chromatic structure. Histological confirmation of metastasis of hypernephroma.

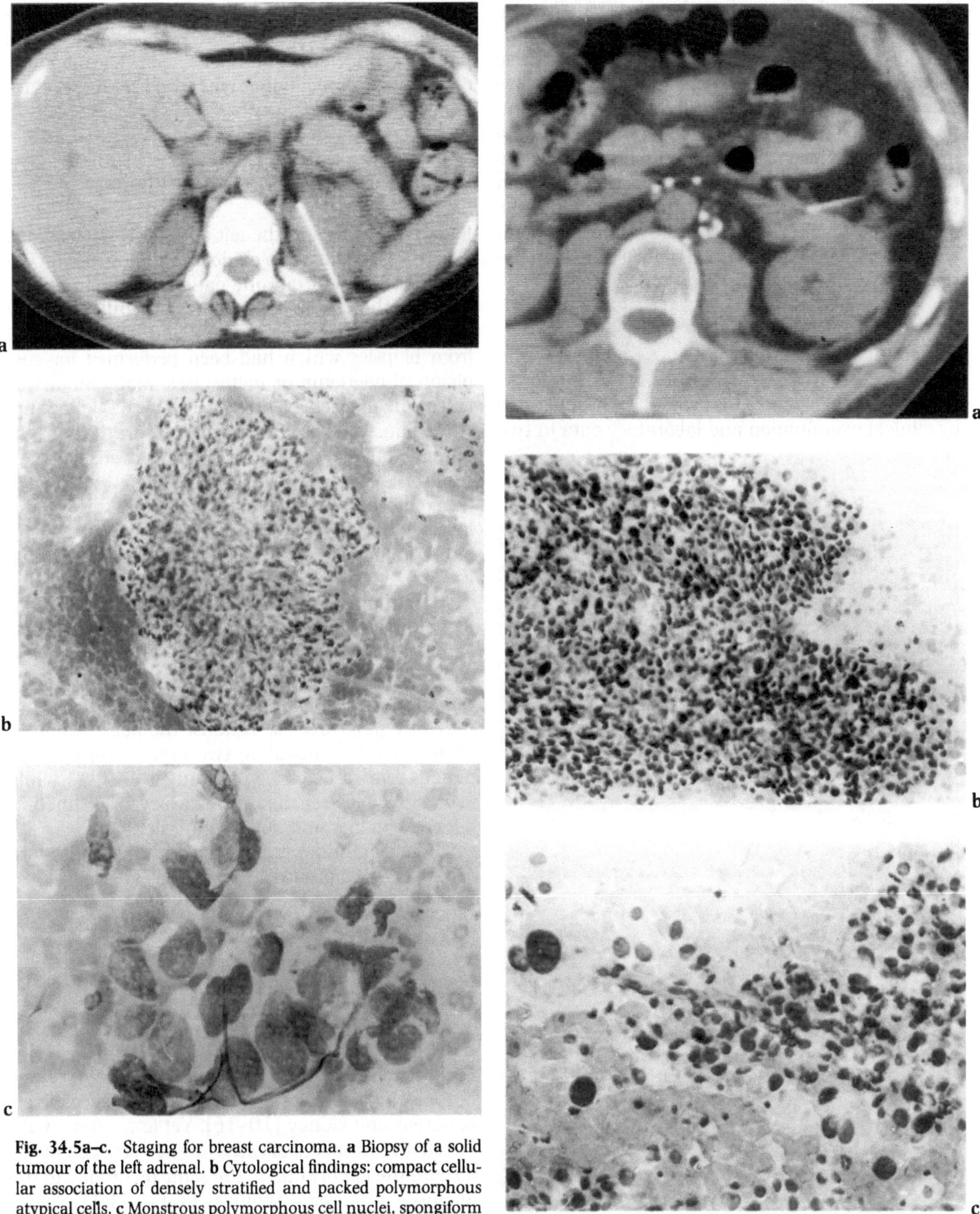

Fig. 34.5a–c. Staging for breast carcinoma. **a** Biopsy of a solid tumour of the left adrenal. **b** Cytological findings: compact cellular association of densely stratified and packed polymorphous atypical cells. **c** Monstrous polymorphous cell nuclei, spongiform loosening, transparent. Diagnosis: Pap. V, cells of solid carcinoma more probably than not. Histological confirmation of metastasis of adenocarcinoma.

Fig. 34.6a–c. Condition after surgery for adrenal carcinoma. **a** Biopsy of an isodense, irregular substrate in front of the left kidney. Differential diagnosis: recurrence—scarred lesions. **b,c** Cytological findings: densely packed cells with pronounced nuclear polymorphism and hyperchromatosis. Diagnosis: Pap. V, unambiguous tumour cells of carcinoma. Histological confirmation of peritoneal metastasis of adrenal carcinoma.

size was recorded from the others by CT follow-up check.

Metastases were cytologically diagnosed in five cases, while in two cases no distinction was possible between metastases and primary adrenal carcinoma. Histological diagnosis of adrenal carcinoma in one patient was confirmed by CT scanning. No cytological distinction had been possible between adenoma and carcinoma. On the other hand, the cytological result recorded from another patient with adrenal carcinoma, not included in this analysis, was (without type specification).

Phaeochromocytoma

Benign phaeochromocytoma had been suggested by clinical examination and laboratory data in two patients and was histologically confirmed. No distinction had been possible between adenoma and phaeochromocytoma on the CT scan. The diagnosis of phaeochromocytoma was confirmed by cytological testing (Fig. 34.7).

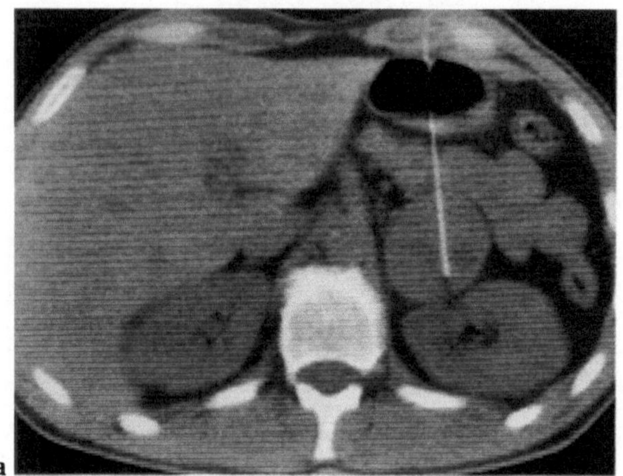

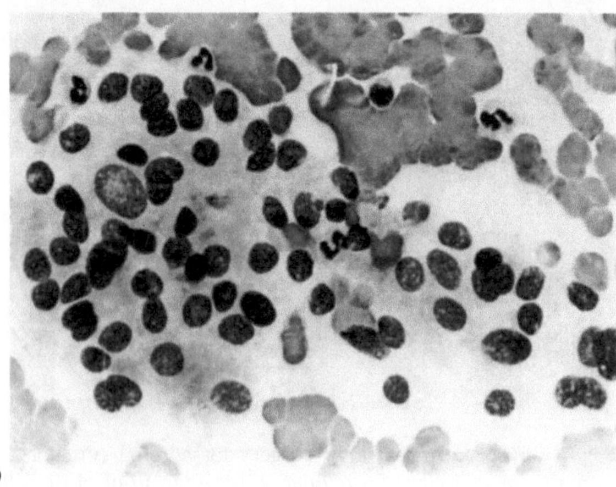

Fig. 34.7a,b. Suspicion of phaeochromocytoma. a Biopsy of a round tumour of the left adrenal. b Cytological findings: differentiated sizes of round or oval cell nuclei; loosened, granulated nuclear structure. Diagnosis: Pap. V, atypical cells: phaeochromocytoma highly probable. Histologically confirmed

Inflammatory and Other Alterations

Expansive lesions to the left adrenal were recorded from three patients in a clinical condition after acute pancreatitis and from one patient who had undergone nephrectomy for carcinoma. Inflammatory changes or inconspicuous epithelia were recorded from biopsies which had been performed for suspicion of adenoma or metastases. Malignoma was ruled out in each of these cases by CT follow-up. An expansive process was recorded from the right adrenal of one patient with clinical suspicion of Addison's disease (with no previous history of tuberculosis). Opaque fluid was aspirated by biopsy in seven cases. Tuberculosis was bacteriologically and cytologically established (Fig. 34.8) and was confirmed by surgical intervention.

A summary review of the results obtained from FNB and cytological testing is likely to suggest that cytodiagnostic material was gained from initial biopsy on 35 of 40 patients (88%). One false diagnosis and five dubious findings brought the accuracy of typing to 73%. Against the background of all 42 biopsies, including two repeat biopsies, the patient-related accuracy amounted to 78%, while even 85% was reached in accuracy of distinction between benign and malignant lesions.

Three aspirates were collected in the majority of cases (18/40 biopsies).

Discussion

Comprehensive and informative results have become available from CT-guided biopsy of liver, pancreas and kidney [10–16]. Yet few reports with conclusive numbers of cases have so far been available on the application of the same technique to the diagnosis of adrenal tumours (Table 34.1). Therefore, only limited assessment of its diagnostic accuracy is possible. The result of percutaneous biopsy of the adrenal was found to depend largely on the size of the tumour. Hence, accuracy data between 90% and 100%, given elsewhere in the literature, must primarily refer to larger tumours. Bernardino et al. [8] attributed their failure rate of 17% to

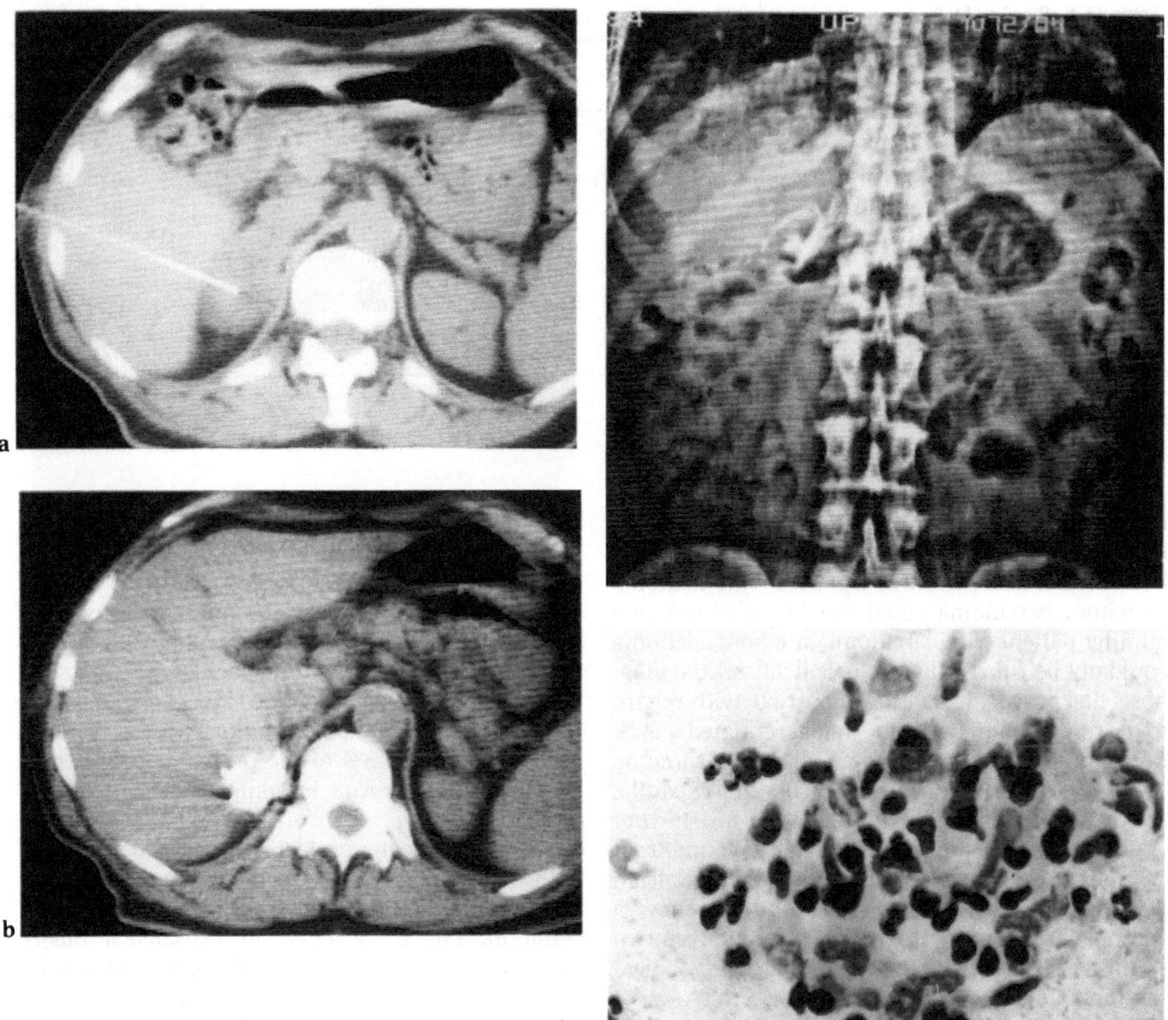

Fig. 34.8a–d. Clinical symptoms of Addison's disease; no known history of tuberculosis. **a** Biopsy of an oval, cystic tumour in the right adrenal; aspiration of 6 ml of yellowish liquid. **b** Condition after instillation of 3 ml of contrast medium. **c** Cavity represented by scout-view scan. **d** Cytological findings; focal accumulation of epithelia with reactive nuclear changes; lymphocytes and epithelioid cells interspersed. Diagnosis: Pap. II K: tuberculosis. Histologically confirmed. (Lüning et al. [9]; reproduced by kind permission of the Editor of *Fortschritte auf dem Gebiete der Röntgenstrahlen und der Nuklearmedizin. Ergänzungsband*)

inadequate technique, small size of lesions and adiposity of patients.

We also analysed our 42 cases of FNB of adrenals for the necessary number of aspirates per biopsy and found that number to be of relevance to the diagnostic result [9]. Only one or two aspirates were taken from three patients from whose biopsies no diagnostically useful material was obtainable. On the other hand, two aspirates were sufficient to gain valuable material from the other seven patients. Sufficient material for adequate cytological diagnosis was also obtained from each of 15 patients with three aspirates per biopsy and from all patients with four or five aspirates. Therefore, three aspirates should be gained by FNB in the context of this application.

Diagnostic accuracy in the evaluation of aspirates from adrenal tumours has been proved to depend on the experience of the cytologist involved and on the objectively existing cytological problems (Fig. 34.9). Such objective problems have been thoroughly described by Heaston et al. [18]. Some of them are relating to the possible difficulty of distinguishing adenomas from highly differentiated carcinomas of the adenoid or papillary type (according to Kettler [22]). Montali et al. [19], for example,

Table 34.1. Results obtained from percutaneous biopsies of adrenal tumours (literature sources with highest number of cases)

Reference	No. of biopsies	Technique	Tumour size (cm)	Diagnostic material (%)	Accuracy (%)
Berkmann et al. [17]	16	CT	1.7–6	94	94
Bernardino et al. [8]	58 (53 patients)	CT	1.5–9	83	83
Heaston et al. [18]	14	CT 13, TV 1	1.5–10	93	93
Lüning et al. [9]	42 (40 patients)	CT	0.5–15	88	85
Montali et al. [19]	18	Sono	2.5–10	100	80–90
Pagani [20]	46	CT	normal adrenals in 32 patients	[a]	[b]
Oliver et al. [21]	15	CT	mostly meta., 3 cm, adenomas 2.4 cm	[a]	100
Zornoza et al. [16]	21	CT 4, TV 17	2–17	100	82

[a] No data.
[b] 90.6%, including four repeat biopsies.

reported a case of highly differentiated carcinoma in which efforts to make that distinction had failed. In our series also there was a patient with adenoma in whom carcinoma could not be ruled out and another patient with carcinoma in whom adenoma could not be ruled out. Cytological differential diagnosis has proved difficult or impossible with regard to distinguishing between undifferentiated carcinoma, on the one hand, and metastases of sarcoma or of a neurogenic tumour of the adrenal medulla, on the other. Problems of this kind are further aggravated by the impossibility of discovering from aspirates whether the cells have been samples from the cortical or medullary zone. The presence of adrenal carcinoma could not be ruled out from two of our cases in which metastases of gynaecological carcinomas had been sampled. Generally, cytodiagnosis is not too problematic in the context of metastases, particularly in those cases in which the histological type of the primary tumour is known. Specialised staining techniques may provide an orientation in cases in which the site of the primary tumour is unknown. Metastases are usually characterised by conspicuous strange cells or cellular complexes according to tumour type, so that diagnosis becomes less difficult.

There are no unambiguous cytomorphological characteristics for diagnosis of adenoma but only certain clues that may be contributory to diagnosis. According to Lopes Cardozo [23], Noltenius [24] and Katz et al. [25], these clues are discrete dimensional differences among cellular nuclei with relatively homogeneous fine-grained structure, pronounced nucleoli, and, occasionally, nuclear vacuoles as well as rather marked adenoid cell formations. Such alterations tend to permit diagnoses based on mere suspicion which would be better made in conjunction with radiological examinations and with due consideration of clinical and paraclinical findings. However, in the context of cytological testing of oncological patients, the examiner's interest is concentrated on CT-established hormone-inactive expansive lesions, for the purpose of confirming or ruling out malignoma. This major demand appears to be achievable, provided the needle is correctly positioned in the tumour. Such determination was accurately accomplished in 85% of the probands involved, a fairly acceptable result in comparison with literature data.

Somewhat greater problems may also be involved in the diagnosis of phaeochromocytoma. Difficulties have been experienced even in histological differentiation, but more often in cytological differentiation, between benign and malignant forms. Should positive chromaffin reaction be successfully established from a smear, this may be helpful in diagnosis under certain circumstances. Different cellular patterns are provided by phaeochromocytoma, in particular an irregular chromatic structure and granulation of the tumour cells. Other manifestations are similar to those of adenoma [26].

Tuberculous processes can be cytomorphologically diagnosed, provided that epithelioid cells or Langerhans' giant cells are present. The detection of acid-proof rods in the smear, rarely accomplished, is not absolutely decisive, provided that such cells are present and typically stratified. Differential diagnosis will have to be attempted with emphasis on other types of granulomatous, non-tuberculous processes [24] with reactive atypical cells which may render exploration extremely difficult.

In conclusion, cytodiagnosis of primary tumours of the adrenals should be regarded as diagnosis of suspicion or probability. This should be its status in the spectrum of diagnostic variants. Its limitations

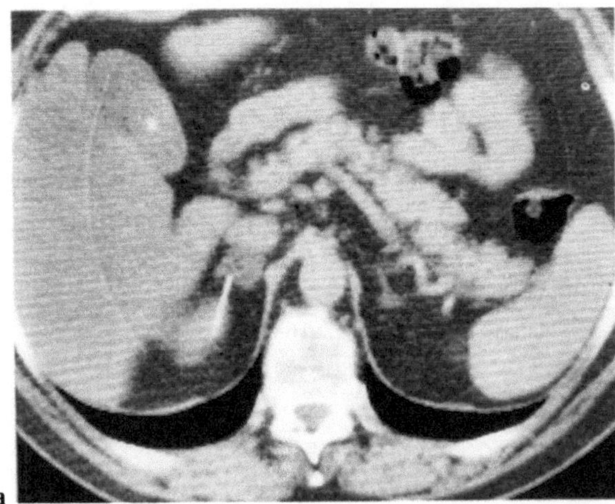

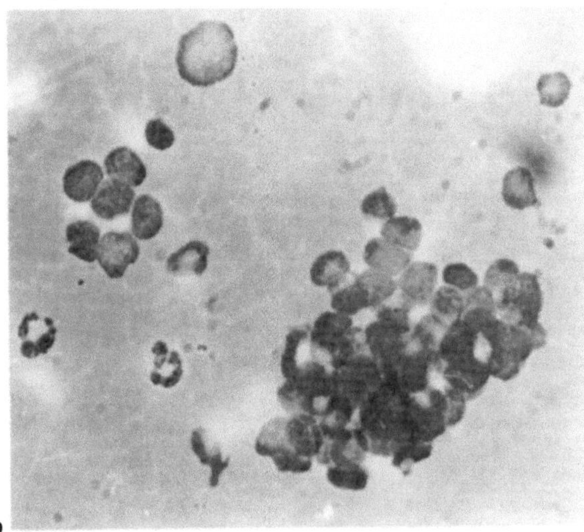

Fig. 34.9a,b. Staging for carcinoma of the urinary bladder. **a** Biopsy of a rounded, hypodense tumour in the right adrenal. **b** Cytological findings: group of "atypical" cells of moderate nuclear polymorphism, with spongy nuclear structure: cells relatively undifferentiated. Diagnosis: Pap. V; tumour cells of undifferentiated carcinoma. Histological confirmation of severe adenomatous hyperplasia of adrenal cortex.

have to be accepted. Clinical, paraclinical, and radiological parameters should be taken into due consideration primarily for purposes of differential diagnosis, so that, with all its restrictions, cytodiagnosis may helpfully contribute to a definite diagnosis.

The use of CT-guided FNB in the diagnosis of adrenal masses is important to the purpose of reducing more invasive tests and unnecessary therapeutic interventions. It is indicated in cases in which CT-recorded expansive processes are to be elucidated, after attempts have failed to distinguish between benign and malignant forms and to determine the type of tumour. The cytological result can be also used for the necessary documentation of inoperable primary malignancies of the adrenal as well as of metastases, resulting from primary tumours on various sites. In other words, it may be relevant to the assessment of operability and to prognostication.

References

1. Hussain S, Belldegrun A, Seltzer SE, Richie JP, Gitties RF, Abrams HL (1985) Differentiation of malignant from benign adrenal masses: predictive indices on computed tomography. AJR 144:61–65
2. Richter HJ (1980) Zur pathologischen Anatomie der Nebennierentumoren. Radiologe 20:149–157
3. Sommers SC (1977) Adrenal glands. In: Anderson WAP, Kisseane JM (eds) Pathology. Mosby, St Louis, pp 1671–1674
4. Ferrucci JT, Wittenberg J, Mueller PR, Simeone JF, Harbin WP, Kirkpatrick RH, Taft PD (1980) Diagnosis of abdominal malignancy by radiological fine-needle aspiration biopsy. AJR 134:323–330
5. Haaga JR, Alfidi RJ (1976) Precise biopsy localization by computed tomography. Radiology 118:603–607
6. Price RB, Bernardino ME, Berkman WA, Sones PJ, Torres WE (1983) Biopsy of the right adrenal gland by the transhepatic approach. Radiology 148:566–568
7. Van Sonnenberg E, Wittenberg J, Ferrucci JT, Mueller PR, Simeone JF (1981) Triangulation method for percutaneous needle guidance: the angled approach to upper abdominal masses. AJR 137:757–761
8. Bernardino ME, Walther MM, Phillips VM, Graham SD, Sewell CW, Gedgaudos-McClees K, Baumgartner BR, Torres WE, Erwin BC (1985) CT-guided adrenal biopsy: accuracy, safety, and indications. AJR 144:67–69
9. Lüning M, Hoppe E, Schöpke W (1986) Ergebnisse der Diagnostik von Nebennierenraumforderungen durch perkutane CT-gestützte Feinnadelbiopsien. Fortschr Geb Rontgenstr Nuklearmed Erganzungsband 144:135–254
10. Harter LP, Moss AA, Goldberg HJ, Gross BH (1983) CT-guided fine needle aspirations for the diagnosis of malignant and infectious disease. AJR 140:363–367
11. Lüning M, Schmeißer B, Wolff H, Schöpke W, Hoppe E, Meyer R (1984) Ergebnisanalyse 96 CT-gestützter Feinnadelbiopsien bei Raumforderungen der Leber. Fortschr Geb Rontgenstr Nuklearmed Erganzungsband 141:267–275
12. Lüning M, Kursawe R, Schöpke W, Lorenz D, Menzel A, Hoppe E, Meyer R (1985) CT guided percutaneous fine-needle biopsy of the pancreas. Eur J Radiol 5:104–108
13. Martino CR, Haaga JR, Bryan PJ, LiPuma JP, El Youself SJ, Alfidi RJ (1984) CT-guided liver biopsies: eight years' experience. Radiology 152:755–757
14. Mitty HA, Efremedis SC, Yeh HC (1981) Impact of fine-needle biopsy on management of patients with carcinoma of the pancreas. AJR 137:1119–1121
15. Pagani JJ (1983) Biopsy of focal hepatic lesions. Radiology 147:673–675
16. Zornoza J, Ordonez N, Bernardino ME, Cohen MA (1981) Percutaneous biopsy of adrenal tumors. Urology 18:412–416
17. Berkman WA, Bernardino ME, Sewell CW, Price RB, Sones

PJ (1984) The computed tomography-guided adrenal biopsy: an alternative to surgery in adrenal mass diagnosis. Cancer 53:2098–2013
18. Heaston DK, Handel DB, Ashton PR, Korobkin M (1982) Narrow gauge needle aspiration of solid adrenal masses. AJR 138:1143–1148
19. Montali G, Solbiati L, Bossi MC, DePra L, DiDonna A, Ravetto C (1984) Sonographically guided fine-needle aspiration biopsy of adrenal masses. AJR 143:1081–1084
20. Pagani JJ (1984) Non-small cell lung carcinoma adrenal metastases. Computed tomography and percutaneous needle biopsy in their diagnosis. Cancer 53:1058–1060
21. Oliver TW, Bernardino ME, Miller JI, Mansour K, Greene D, Davis WA (1984) Isolated adrenal masses in nonsmall-cell bronchogenic carcinoma. Radiology 153:217–218
22. Kettler LH (1976) Lehrbuch der speziellen Pathologie. Kap IV. Fischer, Jena
23. Lopes Cardozo P (1975) Atlas of clinical cytology. Hertogenbosch,
24. Noltenius H (1981) Systematik der Onkologie, Bd 2. Urban und Schwarzenberg, Munich
25. Katz RL, Patel S, Mackay B, Zornoza J (1984) Fine needle aspiration cytology of the adrenal gland. Acta Cytol (Baltimore) 28:269–282
26. Streicher H-J, Sandkühler St (1953) Klinische Zytologie. Thieme, Stuttgart, pp 162–163

Chapter 35

Percutaneous Biopsy of Adrenal Tumours

J. Zornoza

Although adrenal lesions are not common in the routine practice of diagnostic radiology, they are at present being identified more frequently because of the availability of new imaging modalities, ultrasound and computed tomography (CT). The incidence of adrenal metastases is surprisingly high in certain types of malignant tumour, such as carcinoma of the breast, lung and melanoma [1–4]. However, in general they are clinically silent because they fail to destroy enough tissue to cause clinical symptoms of adrenal insufficiency.

Adrenal cysts [5] and myelolipoma [6–8] are rare benign lesions involving the adrenal gland. Previously these tumours were surgically resected to obtain a diagnosis. Scheible et al. [9] described the value of percutaneous adrenal cyst aspiration obviating the need for laparotomy. We report our experience with percutaneous biopsy of adrenal masses in 21 patients, demonstrating its value, risk and technique.

Material and Methods

Percutaneous needle biopsy specimens of the adrenal gland in 21 patients were examined pathologically. The biopsies were performed to assess the presence of metastatic disease or to confirm radiological findings. Of the 21 patients, 11 had a known tumour and nine had an unknown primary malignancy. One patient had double primary malignancies (melanoma and squamous cell carcinoma of the nasopharynx).

For mass localisation a combination of intravenous pyelography (IVP) and ultrasonography was used in 9 cases, ultrasound and CT in 8, angiography and CT in 2, and ultrasound alone in 2. The adrenal lesions were unilateral in 19 cases and bilateral in 2. Biopsies were done on right adrenal masses in 9 cases and on left adrenal masses in 12 cases. The lesions ranged from 2 to 17 cm in diameter.

Biopsy specimens were obtained with 18- to 22-gauge spinal-type and Tru-Cut needles. A posterior puncture site was chosen for all the biopsies. Fluoroscopic guidance after injection of intravenous contrast medium was used in 17 cases and CT guidance in 4. The biopsy technique has been described previously [10, 11]. After aspiration of a cystic lesion, contrast material (meglumine diatrizoate) and air were injected.

The aspirate was placed in a preservative solution (Mucolex), centrifuged, smeared on glass slides and post-fixed in 95% ethanol. Cores of tissue were placed in 10% formalin for light microscopy, and material for electron microscopy was fixed in 2% glutaraldehyde in phosphate buffer. Papanicolaou stain was used for cytological material and haematoxylin and eosin for histological specimens.

The biopsies were performed on both outpatients and inpatients with no fasting required. No premedication was administered unless required by the patient. The only relative contraindication was a history of abnormal bleeding.

Results

Of the 21 patients undergoing percutaneous needle biopsy of the adrenal gland, true positive diagnoses were obtained from 12 biopsies, 5 yielded true negatives and 4 were false negative. No false positive diagnoses were made. The success rate was 81% (Table 35.1).

Table 35.1. Results of percutaneous needle biopsy of the adrenal gland in 21 patients

Lesions	No. of patients	Diagnosis by puncture/pathology
Malignant		
Adenocarcinoma	9	7
Adrenal cortical carcinoma	3	2
Neuroblastoma	3	2
Melanoma	1	1
Benign		
Cyst	4	4
Myelolipoma	1	1
TOTAL	21	17

Seven of nine patients with metastatic adenocarcinoma from the lung were correctly diagnosed (Figs. 35.1, 35.2). Of the nine patients with an unknown primary tumour, a definite diagnosis was obtained in seven. Three proved to be simple adrenal cysts (Fig. 35.3), two adrenal cortical carcinomas (Fig. 35.4) and two neuroblastomas. The fluid aspirated from the cysts was clear, and no malignant cells were found. Clinical follow-up in four cases and surgery in one (myelolipoma) confirmed the true negative results (Fig. 35.5). No complications related to the procedure were encountered.

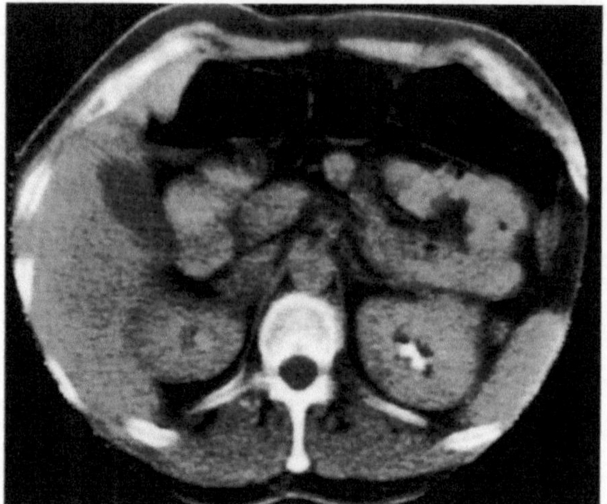

a

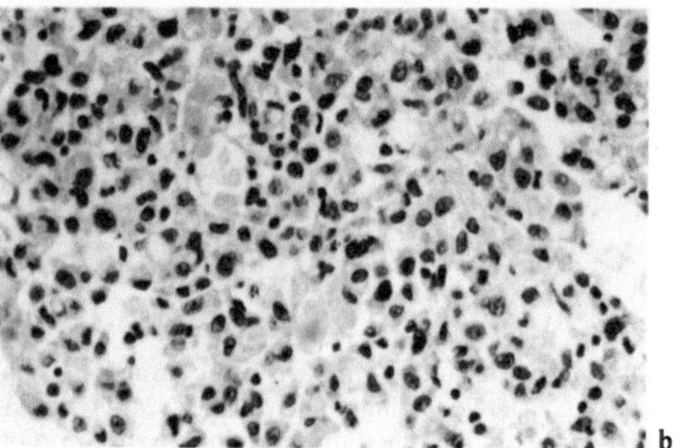

b

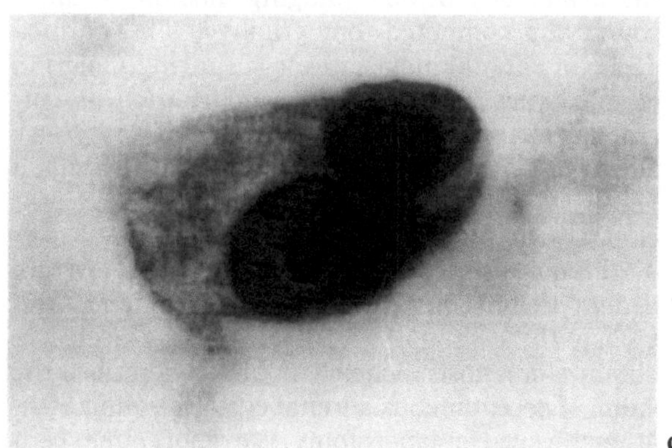

c

Fig. 35.1a–c. A 58-year-old man with carcinoma of the lung. **a** CT scan of abdomen shows 2-cm mass in right adrenal. **b** Percutaneous needle biopsy revealed adenocarcinoma; pleomorphic cells with pyknotic irregular nuclei. (H & E, ×134). **c** Cytological specimen showing two overlapping, finely vacuolated cells with moderately irregular nuclei and prominent nucleoli. (Papanicolaou stain, ×1000)

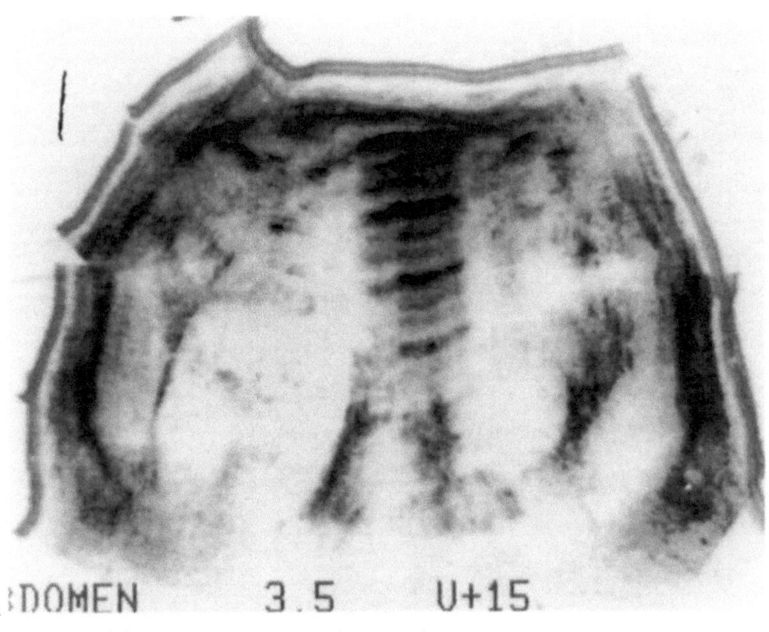

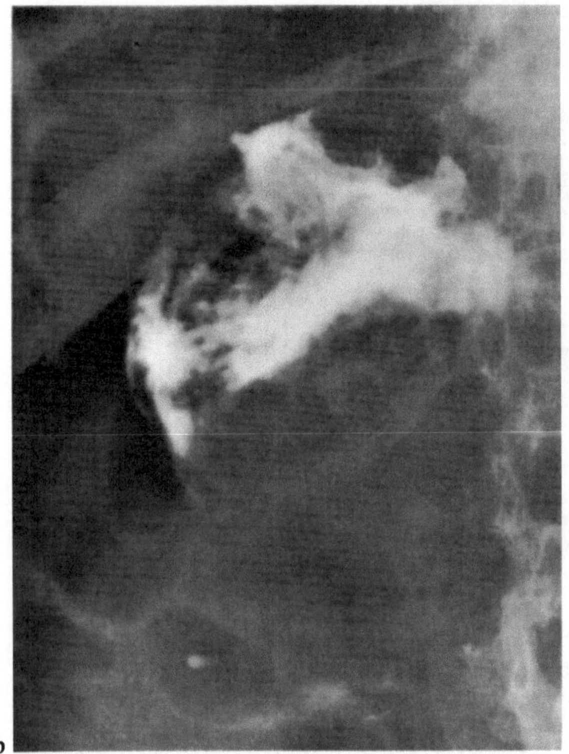

Fig. 35.2a,b. A 45-year-old man with melanoma. **a** Sonogram shows hypoechoic bilateral adrenal masses. **b** Injection of contrast demonstrates an irregular cavity lesion. Metastatic melanoma.

Discussion

The identification of adrenal pathology has improved remarkably with the advent of ultrasound and CT [12, 13]. Until the advent of these imaging modalities, the diagnosis of adrenal lesions required more invasive procedures such as nephrotomography, angiography, venography and retroperitoneal pneumonography [14]. Since only a small amount of functioning adrenal tissue is required to cover the daily body requirements of adrenal hormones, many non-functional adrenal lesions remain silent and are discovered accidently or at autopsy. Adrenal cysts, metastases, mesenchymal tumours and some adrenal cortical carcinomas are the most common types of lesion presenting as a mass without clinical signs or symptoms of hyperfunction or insufficiency. The frequency of adrenal metastases depends on the primary site, most commonly occurring in patients with carcinoma of the breast, lung and melanoma as determined from autopsy series [1–4].

Because of the lack of symptoms, the demonstration of adrenal metastases depends on radiological recognition. Ultrasound [15] and CT are more sensitive than previous radiological tests [16] and consequently more lesions are now detected. Lesions of the adrenal glands as small as 1 cm can be identified by CT, since they distort the normal outline of the gland.

Adrenal metastasis generally appears late in the course of the disease and is normally a mani-

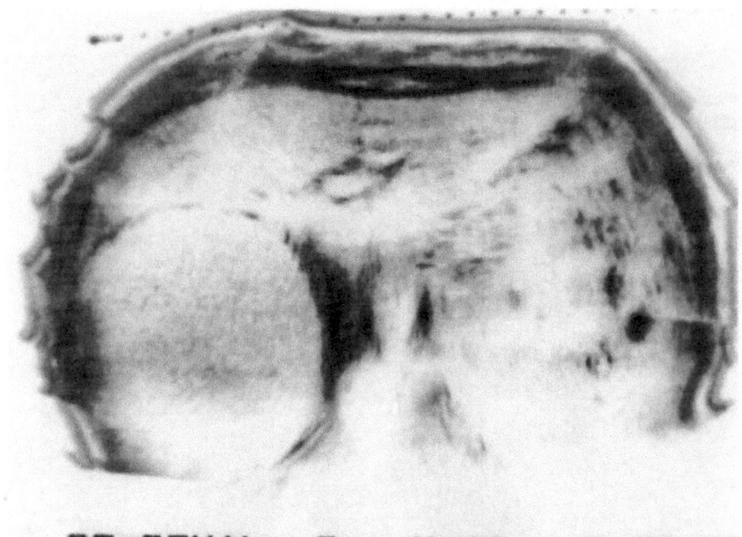

a

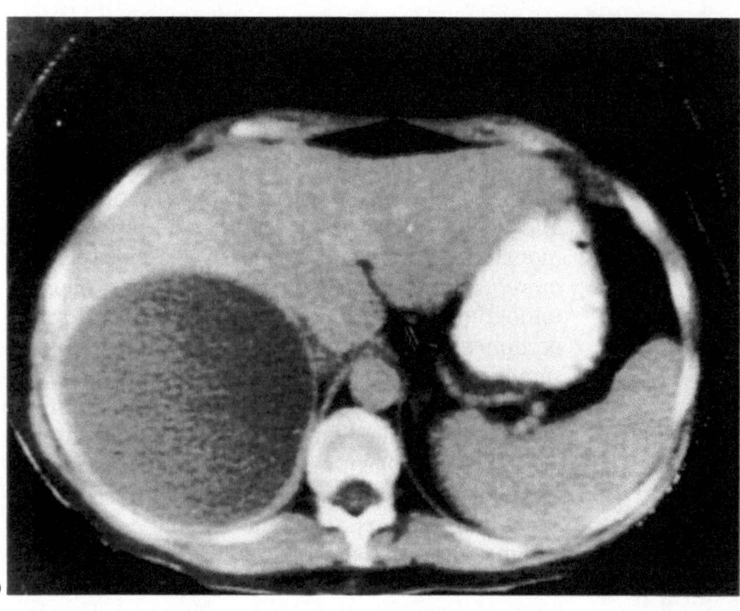

b

Fig. 35.3a,b. A 65-year-old woman with abdominal mass. **a** Transverse sonogram shows mass with cleavage plane. **b** CT scan shows large mass behind liver with smooth wall and water attenuation. Adrenal cyst.

festation of diffuse disease. However, not all adrenal masses identified by CT and ultrasound in patients with other primary neoplasms are necessarily metastatic deposits. In such cases, a tissue analysis is needed to determine the diagnosis and appropriate treatment. Percutaneous needle biopsy may spare selected patients the higher risk of exploratory surgery, prolonged hospitalisation and delay of treatment.

The same principle applies to patients with an unknown primary tumour and a non-functioning adrenal mass. The percutaneous biopsy can provide a pathological diagnosis, guiding future treatment of the patient. Only a few cases of percutaneous adrenal biopsy have been reported [17–20] and to our knowledge none includes a case of myelolipoma or adrenal cortical carcinoma. The use of small calibre spinal-type needles, 20–22 gauge, provides enough material for a cytological diagnosis of malignancy. However, larger needles (Tru-cut) are necessary to obtain enough histological material for the diagnosis of mesenchymal tumour or myelolipoma.

Light microscopic examination of cancer cells in

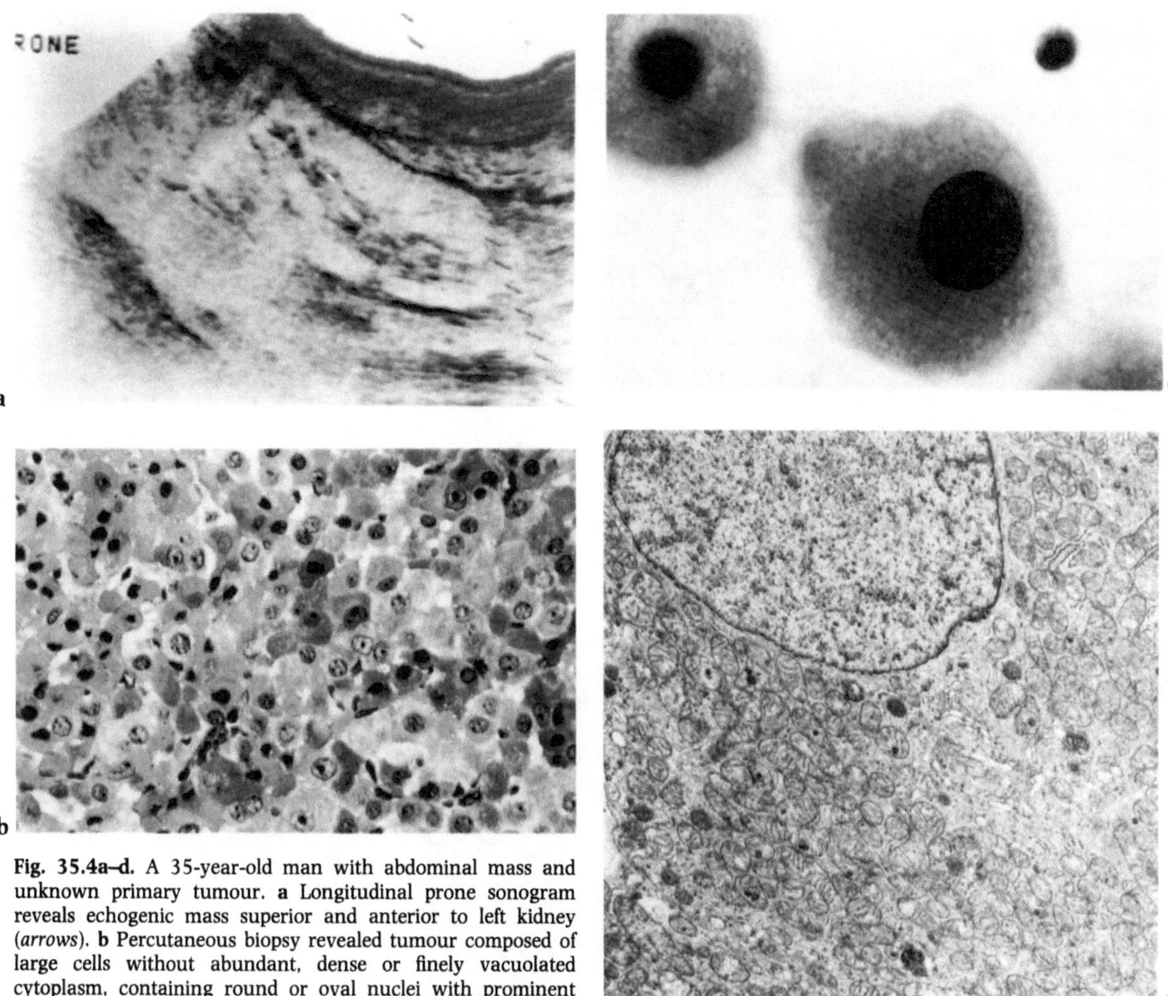

Fig. 35.4a–d. A 35-year-old man with abdominal mass and unknown primary tumour. **a** Longitudinal prone sonogram reveals echogenic mass superior and anterior to left kidney (*arrows*). **b** Percutaneous biopsy revealed tumour composed of large cells without abundant, dense or finely vacuolated cytoplasm, containing round or oval nuclei with prominent single or multiple nuclei. (H & E, × 136). **c** Cytological specimen showing cells with dense eosinophilic cytoplasm and round nuclei with large nucleoli. (Papanicolaou stain, × 1000). **d** Electron microscopy: cell cytoplasm well-developed, endoplasmic reticulum, prominent Golgi complexes and numerous mitochondria with tubular cristae and dense intramatrical condensations. (× 3380)

cytological smears often does not allow us to determine the specific type of neoplasm. Since one of the advantages of electron microscopy is the use of very small amounts of tissue, material obtained by needle aspiration biopsy is suitable for diagnostic purposes. The high magnification and resolution provided by the electron microscope permits the diagnosis of tumours derived from cells showing ultrastructural differential features, such as characteristic membrane-bound cytoplasmic granules and inclusions, specific intracytoplasmic filaments and cell junctions. The case of adrenal cortical carcinoma illustrates how electron microscopy can be an aid to establishing a specific diagnosis. The finding of large amounts of endoplasmic reticulum and the identification of numerous mitochondria with tubular cristae confirms the adrenal origin of this tumour and at the same time allows us to rule out other neoplasms which may be indistinguishable from adrenal cortical tumours by light microscopy alone [21].

The success rate of 81% coupled with the lack of complications underscores the value of percutaneous needle biopsy as a diagnostic procedure in evaluating adrenal masses of unknown origin.

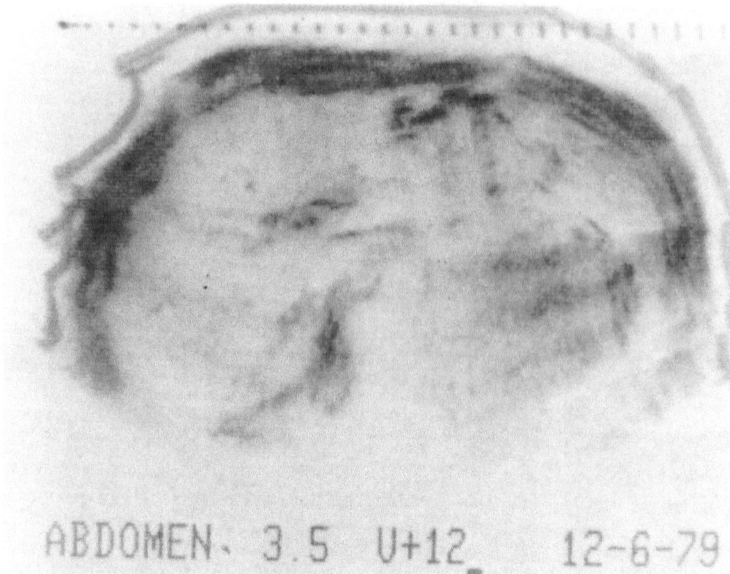

a

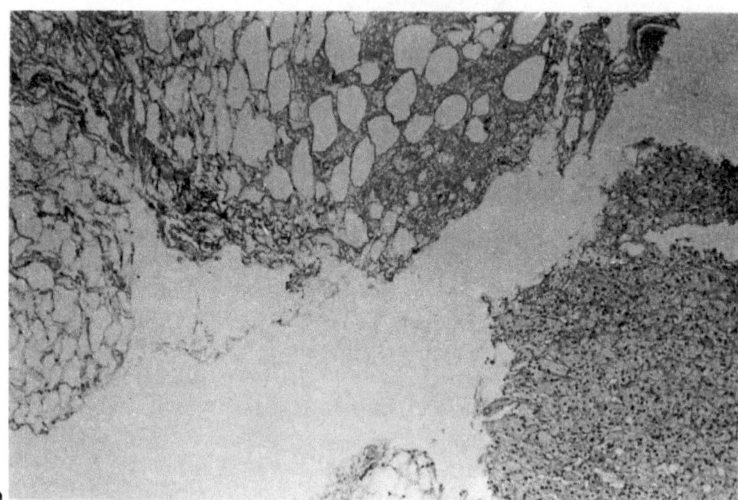

b

Fig. 35.5a,b. A 39-year-old woman with carcinoma of the cervix. **a** Transverse sonogram shows echogenic left adrenal mass. **b** Percutaneous biopsy revealed fragment of adrenal cortex and adipose tissue. (H & E, × 100)

References

1. Glomset DA (1938) The incidence of metastasis of malignant tumors to the adrenals. Am J Cancer 32:57–61
2. Abrams HL, Spiro R, Goldstein N (1950) Metastases in carcinoma analysis of 1,000 autopsied cases. Cancer 3:74–85
3. Lumb G, Mackenzie DH (1959) The incidence of metastases in adrenal glands and ovaries removed for carcinoma of the breast. Cancer 12:521–526
4. Aldrete JS, Bohrod MG (1967) Adrenal metastases in cancer of the breast, their prognostic significance when found at adrenalectomy. Am Surg 33:174–180
5. Foster DG (1966) Adrenal cysts: a review of the literature and report of a case. Arch Surg 92:131–143
6. Gee WF, Chikos PM, Greaves JP, Ikemoto H, Tremann JA (1975) Adrenal myelolipoma. Urology 5:562–566
7. Rubin HB, Hirose F, Benfield JR (1975) Myelolipoma of the adrenal gland. Am J Surg 130:354
8. Behan M, Martin EC, Muecke EC, Kazam E (1977) Myelolipoma of the adrenal: two cases with ultrasound and CT findings. AJR 129:993–996
9. Scheible W, Coel M, Siemers PT, Siegel H (1977) Percutaneous aspiration of adrenal cysts. AJR 128:1013–1016
10. Zornoza J, Wallace S, Goldstein HM, Lukeman JM, Jing BS (1977) Transperitoneal percutaneous retroperitoneal lymph node aspiration biopsy. Radiology 112:111–115
11. Zornoza J (1980) Abdomen. In: Zornoza J (ed) Percutaneous needle biopsy. Williams and Wilkins, Baltimore, pp 102–140
12. Bernardino ME, Goldstein HM, Green B (1978) Gray scale ultrasonography of adrenal neoplasms. AJR 130:741–744
13. Karstaedy N, Sagel SS, Stanley RJ, Melson GL, Levitt RG (1978) Computed tomography of the adrenal gland. Radiology 129: 723–736

14. McAlister WH, Koehler PR (1967) Diseases of the adrenal. Radiol Clin North Am 5:205
15. Zornoza J, Barnardino ME (1980) Bilateral adrenal metastasis: "head light" sign. Urology 15:91–92
16. Zornoza J, Bracken R, Wallace S (1976) Radiologic features of adrenal metastases. Urology 8:295–299
17. Pereiras R, Meyers W, Kunhardt B, Troner M, Huston D, Barkin JS, Viamonte M (1978) Fluoroscopically guided thin-needle aspiration biopsy of the abdomen and a retroperitoneum. AJR 131:197–202
18. Wilson SR, Rosen JE (1979) Abdominal biopsy with ultrasound guidance. J Can Assoc Radiol 30:138 – 140
19. Haaga JR (1979) New techniques for CT-guided biopsies. AJR 133:633–641
20. Ferrucci JT, Wittenberg J, Mueller PR, Simeone JF, Harbin WP, Kirkpatrick RH, Taft PD (1980) Diagnosis of abdominal malignancy by radiologic fine-needle aspiration biopsy. AJR 134:323–330
21. Silva EG et al. (1979) The fine structure of adrenocortical carcinoma. In: Proceedings of 37th annual meeting of the Electron Microscopy Society of North America. Claitor's, Baton Rouge, Lousiana, pp 230–231

SECTION VII

Testicular Cancer

Chapter 36

Radiological Evaluation of Retroperitoneal Nodes in Testicular Tumours

R. Musumeci, J. D. Tesoro Tess, L. Balzarini, E. Ceglia and R. Petrillo

Introduction

For the evaluation of the retroperitoneal lymph nodes in cancer of the testis, a number of radiological investigations can be used. These diagnostic tools include urography, phlebography, lymphography, ultrasound, computed tomography (CT) and magnetic resonance imaging (MRI). In some cases, fine needle aspiration cytology under fluoroscopic, ultrasound or CT guidance can be used. Our experience in this field is limited. Each of these procedures has different accuracy, sensitivity and specificity. However, as a general rule, the diagnostic value of each radiological procedure, either direct or indirect, is much lower when the diagnostic tool is used alone than when a combination of two or more procedures is used.

Urography

Urography is the most widely used indirect procedure. The diagnostic possibilities depend upon the site and size of the abnormal lymph nodes and upon the possibility that enlarged nodes will displace the kidney and/or the ureters. For this reason urography is mainly useful in tumours of the left testis in which the presence of para-aortic metastases can be disclosed by a lateral displacement of the proximal ureter. The test is less significant in tumours of the right side, in which metastatic adenopathies are usually discovered anteriorly to and/or in between the inferior vena cava and the aorta. Only when the involvement is huge will the lateral displacement of the right ureter be seen (Fig. 36.1).

Vascular Investigations

Vascular investigations include opacification of the inferior vena cava, the left renal and the left testicular veins. The intra-arterial route of administration of the contrast medium is seldom used and only in selected cases. The diagnostic accuracy of these procedures correlates with the size and the site of lymph node metastases (Table 36.1). When the metastatic node is in a favourable position, it is possible to see an impression on the walls, a dislocation, or a partial or total obstruction of the opacified veins, with collateral circulation. Only in a few patients will phlebography be the only exam-

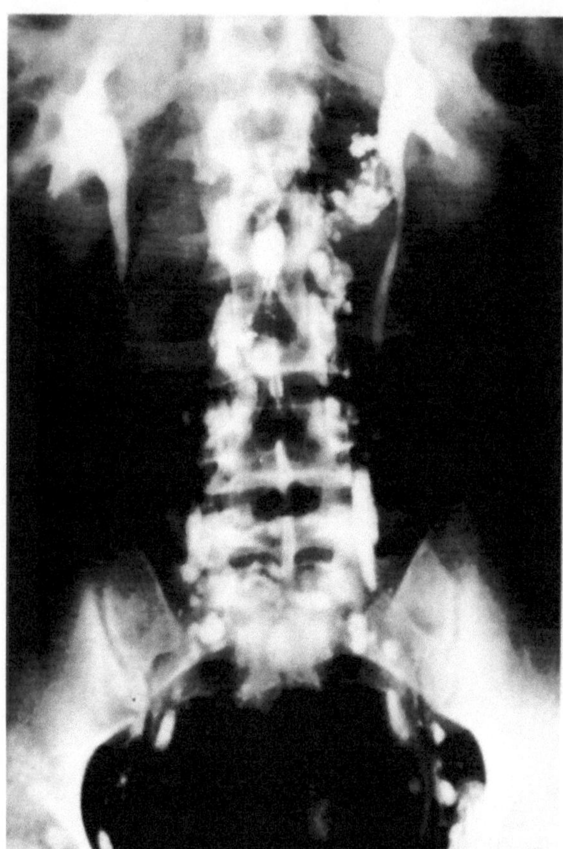

Fig. 36.1. Para-aortic adenopathies, opacified by bipedal lymphangiography, causing dislocation of the upper third of the left ureter.

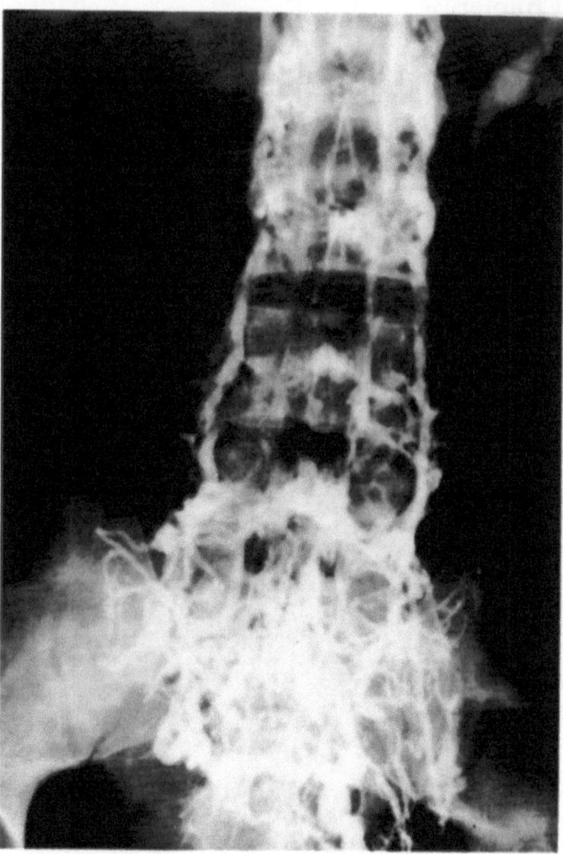

Fig. 36.2. Inferior vena cavography in a patient with right testicular seminoma. The inferior vena cava is completely occluded and collateral paralumbar veins are opacified. The right kidney is also excluded.

Table 36.1. Inferior vena cavographic interpretation[a]

Histology	Inferior vena cavography	
	Positive	Negative
Positive	37	19
Negative	2	17
Sensitivity	37/56	66%
Specificity	17/19	89%
Accuracy	54/75	72%

[a] Based on data from Dunnick and Javadpour [1] plus personal data.

ination indicating metastases. For this reason, because of its invasiveness and the availability of more sophisticated techniques, its routine use is no longer justified. However, the procedure may still be useful for demonstrating direct vascular invasion or anomalies of the inferior vena cava and left renal vein prior to retroperitoneal surgery (Fig. 36.2).

Lymphography

At present, lymphography is one of the best known diagnostic tools for the direct evaluation of retroperitoneal nodes in cancer of the testis. The published material is abundant but the results are often controversial (Table 36.2). The percentage of lymphograms which are abnormal for metastases ranges from 28.9% [2] to 57.5% [3]. According to the literature, the mean value is about 40%–42%. Moreover, the diagnostic accuracy of lym-

Table 36.2. Results of lymphography in testicular tumours[a]

No. of cases	3110
% positive	43.7 (range 28.9–57.5)
% accuracy	84.6 (range 62.5–100)

[a] Data from [1, 5–11].

Table 36.3. Results of lymphography in tumours of the testis

Histology	No. of cases	Positive (%)	Involvement		
			Unilateral (%)	Bilateral (%)	Para-aortic (%)
Seminoma	494	36	45	34	96
Carcinoma	872	59	44	36	99
	Total 1366	51	45	36	98
Non-germinal	50	30	61	30	100
	Total 1416	50	45	36	98

phography is controversial, and values ranging from 62.5% [4] to 100% have been presented [3]. The mean value according to the published data is about 85%.

Our experience concerns 1416 patients with histologically proven malignant testicular tumours collected between 1965 and 1985 (Table 36.3). The overall incidence of metastases was 50%; the incidence was higher in germinal (51%) than in miscellaneous non-germinal cases (30%). This value was also higher in carcinomas (59%) than in seminomas (36%). According to our knowledge of the lymphatic drainage of the normal testis, the ipsilateral para-aortic region is the first site of involvement. This finding is confirmed in our case material, with 98% of para-aortic metastatic diffusion. In about a half of the pathololological cases, the nodal involvement appeared to be limited to a single area, usually the ipsilateral para-aortic site. This occurred most frequently in non-germinal tumours (61%). Only one-third of the pathological lymphograms showed bilateral retroperitoneal node involvement (Figs. 36.3, 36.4).

A total of 454 patients (Table 36.4) underwent retroperitoneal lymphadenectomy or biopsy with an overall diagnostic accuracy for lymphography of 82.8%, sensitivity of 82.0% and specificity of 83.7%. There were 34 false positive and 44 false negative results. These data correlate well with those previously referred to in the literature. The diagnostic contribution of lymphography can be more adequately assessed in carcinomas where modified ipsilateral or radical retroperitoneal dissection was the treatment of choice. From the group of patients with carcinoma we selected 371 cases (Table 36.5) in whom the initial physical evaluation of lymph node areas was negative (NO). The patients then underwent lymphography followed by extended or modified retroperitoneal lymphadenectomy. More than half the cases were reclassified after lymphography as negative for retroperitoneal metastases, but in this group the

Table 36.4. Lymphographic/histological correlation in 454 patients following surgery

Histology	No. of patients	Lymphography/Histology				Accuracy (%)
		+/+	+/−	−/+	−/−	
Seminoma	21	18	1	1	1	90.5
Carcinoma	422	179	33	42	168	82.2
Non-germinal	11	4	—	1	6	90.9
Total	454	201	34	44	175	82.8
Sensitivity 82.0% (201/245)		Specificity 83.7% (175/209)				

Table 36.5. Results of lymphography in 371 selected cases of clinical NO carcinoma

	No. of cases	Lymphography/Histology				Accuracy (%)
		+/+	+/−	−/+	−/−	
Stage I	207	—	—	41	166	80.2
Stage II (A+B)	164	133	31	—	—	81.1
Total	371	133	31	41	166	80.6

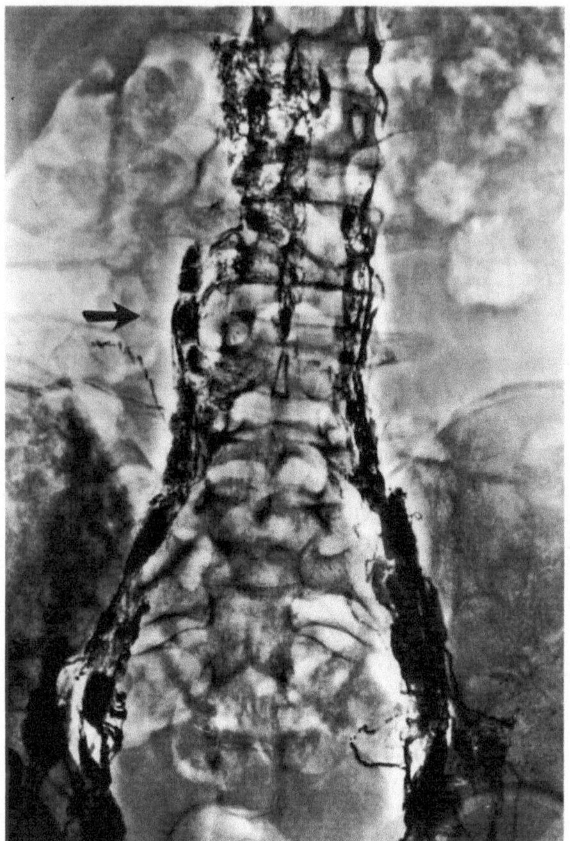

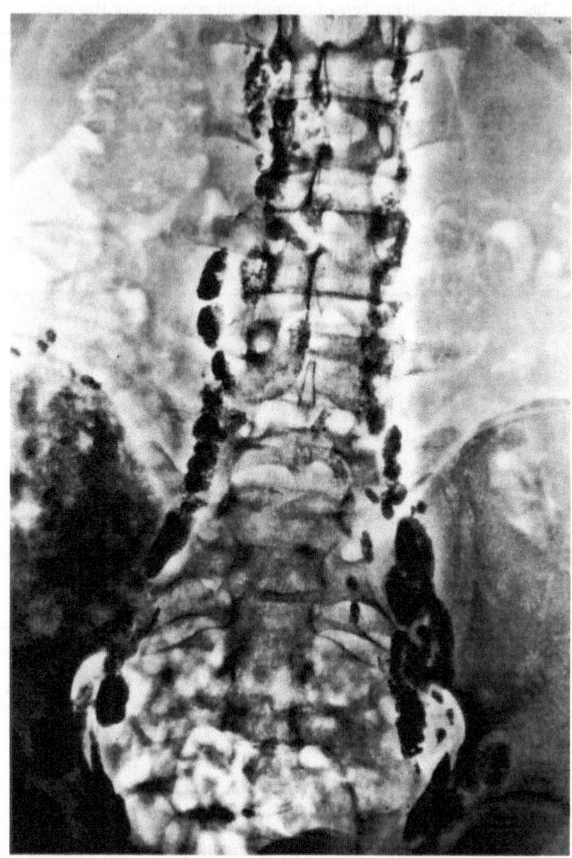

Fig. 36.3a,b. Bipedal lymphography in a patient with right testis carcinoma: lymphograms (**a**) demonstrate displacement of the right para-aortic lymphatic trunks and nodal chains. The lymphoadenographic phase (**b**) shows displacement of lymph nodes and a large filling defect at the level of the space L–3 to L–4 (*arrow*).

histological correlation was only 80.2%, with 41/207 false negative readings. Also, in the group with positive lymphography, a number of false positive readings occurred, mainly in patients with disease limited to a single ipsilateral lymph node. This is probably due to the objective difficulty in assessing the true nature of filling defects in a single node and to the subjective knowledge of the high rate of diffusion to the lymphatics of testicular carcinoma. Furthermore, one must remember that lymphography can understage the disease. This means that, even though only one or two radiologically involved nodes are seen, histological examination discloses more extensive diffusion. At the same time we found that, usually in false negative results, metastases are so large that these nodes are completely bypassed by the contrast and thus impossible to demonstrate. The last important finding to explain the false negative results is the location of the involved nodes. There are, in fact, node areas impossible to evaluate with lymphography, such as those in the high para-aortic area up to the diaphragmatic crura. The nodes of first drainage of the testis, i.e. the superficial intercavoaortic for right-sided tumours and those near the renal vein on the left side, are difficult or even impossible to visualise with standard lymphography. For this reason, when possible, the use of intraoperative funicular lymphography is suggested.

Echography

Echography is recommended among the imaging procedures in abdominal staging as the first diagnostic step since it is inexpensive, widely available and avoids ionising radiation. However, it must be stressed that lymph nodes with a diameter less than 1.5 cm are almost completely missed or cannot be interpreted with this procedure. A metastatic lymph

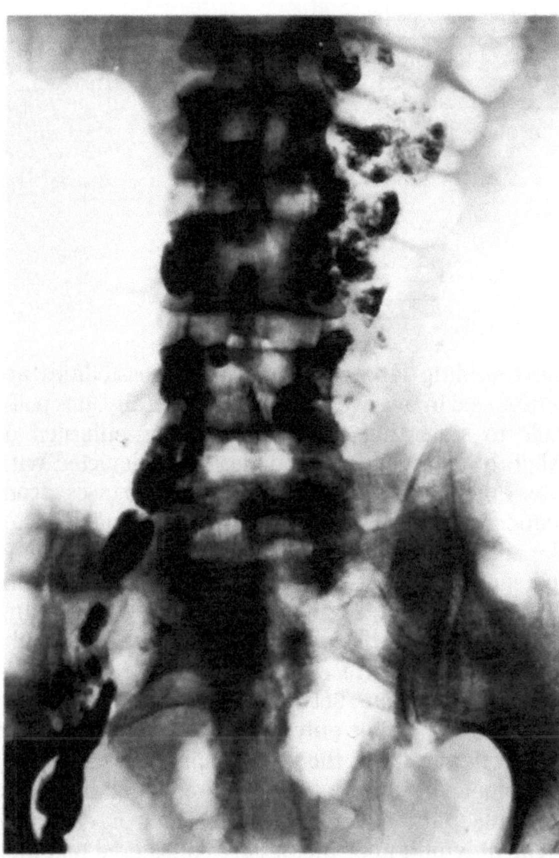

Fig. 36.4. Lymphogram of a patient with left testicular carcinoma showing enlarged para-aortic lymph nodes with multiple filling defects.

node is echographically homogeneous, i.e. transonic or only slightly echogenic, without posterior enhancement if completely solid. Not infrequently a cystic pattern is found if colliquative necrosis has occurred or in the presence of mature teratoma (Fig. 36.5).

The diagnostic value of echography is restricted by the size of the patient. Obese or moderately obese patients are poor candidates for examination because of the physical properties of ultrasound. Furthermore, meteorism influences the picture quality negatively in at least 20% of the patients. In Table 36.6 published results of echography in the evaluation of the retroperitoneal node chains are shown. The values of overall diagnostic accuracy range from 64% to 90% of the cases in whom the examination was technically possible. Specificity was high (95%) but sensitivity was low. On the basis of the data reported in the literature, only a positive result can be considered of value, while the importance of a negative result appears to be questionable [13, 18] (Fig. 36.6).

Moreover, one must consider that every kind of patient is included in reports, i.e. patients without disease, those with minimal disease and those with extensive disease. To assess the true possibilities of echography we performed a retrospective evaluation of our results in a group of 48 patients who underwent retroperitoneal lymphadenectomy. The patients had early stage disease; lymph node metastases, if present, measured less than 5 cm and in half of the cases less than 2 cm in diameter. The results of the review in this group of selected cases are reported in Table 36.7. Considering the various parameters, our results in the early stages of disease were disappointing.

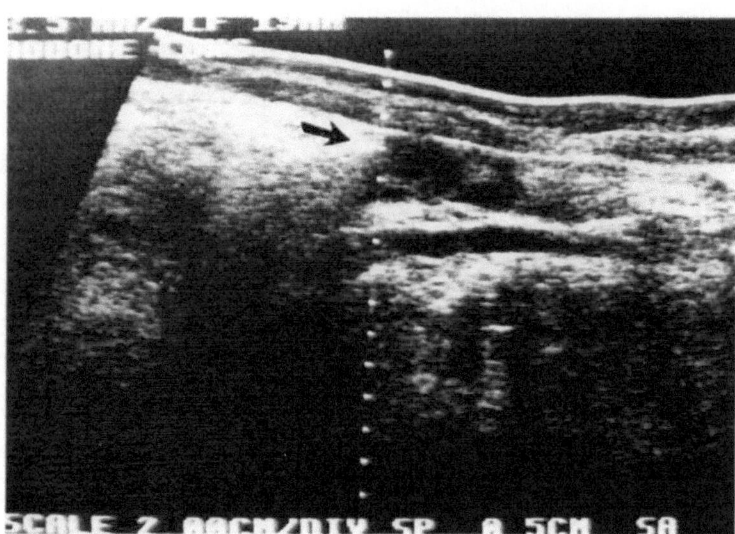

Fig. 36.5. Echography of precaval relatively transonic adenopathies (*arrow*) without involvement of the contiguous structures (i.e. inferior vena cava).

Table 36.6. Results of echography in the evaluation of retroperitoneal lymphadenopathy

Reference	No. of cases	Sensitivity (%)	Specificity (%)	Accuracy (%)
Winterberger [12]	46	—	—	90
Rochester et al. [13]	17	64	96	—
Brascho et al. [14]	56	—	—	87
Burney and Klatte [15]	39	—	—	75
David et al. [16]	62	79	95	89
Williams et al. [17]	21	93	57	81
Rowland et al. [18]	53	32	93	64

Table 36.7. Results of echography in early pathological stage testicular carcinoma

Sensitivity	8/19	42.1%
Specificity	23/29	79.3%
Overall accuracy	31/48	64.6%
Accuracy of positive result	8/14	57.1%
Accuracy of negative result	23/34	67.6%

Computed Tomography

Computed tomography (CT), performed with machines that permit high resolution and fast scanning times, is considered to be the best non-invasive imaging modality for precise identification of a number of abnormal conditions, including diseases of the lymph nodes. The CT criteria for determining nodal involvement are based solely upon identification of enlarged lymph nodes, which are considered abnormal if greater than 1.5 cm in diameter. Since lymph nodes are visualised by the presence of surrounding fat, the visualisation possibilities are enhanced in overweight adults. Generally, it is possible to state that metastases in non-enlarged or slightly enlarged lymph nodes go undetected with this procedure; however, the metastases from cancer of the testis are often large and CT can be considered suitable for the evaluation of the N category in this disease. According to the literature Table 36.8), the diagnostic results of CT scanning appear comparable with those reported for lymphography (Fig. 36.7).

A comparison between different radiological methods is possible only when patients are selected correctly. In fact, the evaluation is hampered if histologically different tumours are included, or if patients with advanced disease and/or clinically evident lymph node masses are considered together with those having small volume disease. Obviously, in these cases the values of accuracy are strictly dependent on the percentage of more advanced cases presented in the material (Fig. 36.8).

To evaluate this problem we selected from our case material a group of patients who had under-

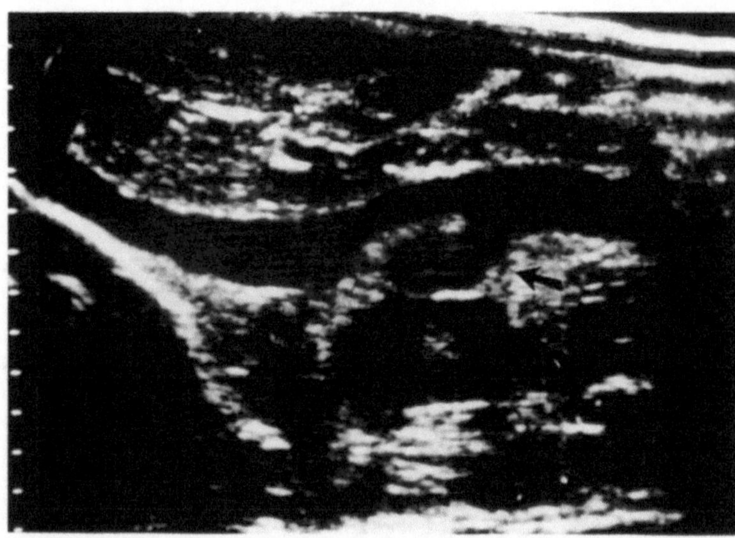

Fig. 36.6. Echography of retrocaval inhomogeneous echogenic adenopathies, which compress the posterior wall of the inferior vena cava (*arrow*).

Table 36.8. Results of CT in identification of lymph node metastases from cancer of the testis

Reference	No. of cases	Sensitivity (%)	Specificity (%)	Accuracy (%)
Burney [15]	39	—	—	76
Lackner [19]	64	80	79	80
Williams [17]	32	93	82	88
Dunnick [1]	50	66	100	74
Thomas [10]	27	90	83	89
Lien [20]	51	56	96	76
Ehrlichman [5]	17	60	50	59
Jing [6]	37	83	86	84
Richie [21]	30	65	90	73
Marincek [7]	30	44	81	70
Tesoro Tess [9]	35	74	75	74

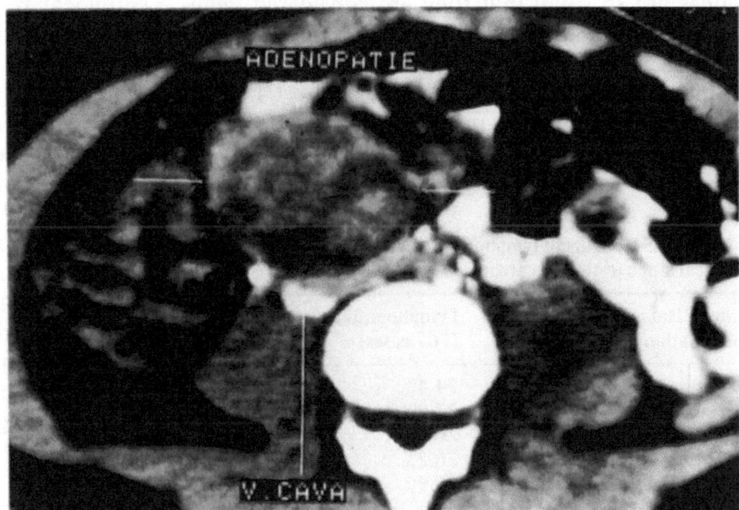

Fig. 36.7. CT scan after intravenous and oral contrast medium administration in a patient with right testis carcinoma: large mass of precaval nodes compressing and dislocating the inferior vena cava and bowel (*arrows*).

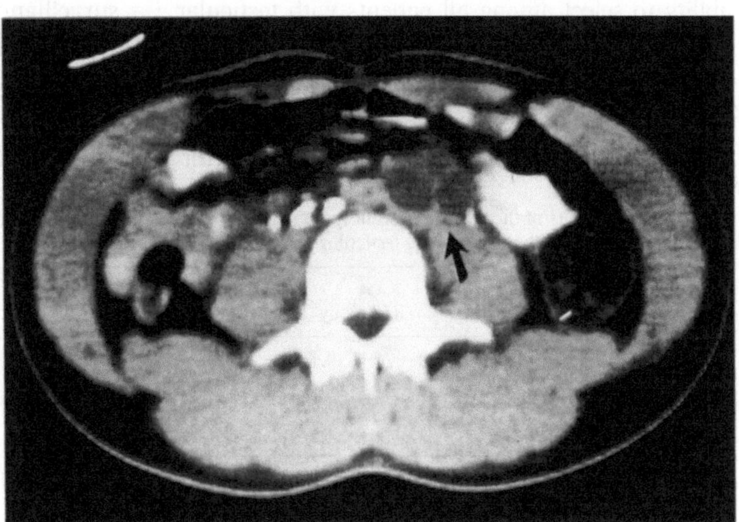

Fig. 36.8. CT scan showing low-density adenopathy from left testicular carcinoma in the paralumbar area (*arrow*).

gone retroperitoneal lymphadenectomy and who were pathologically classified as having testicular carcinoma at stage I, IIA or IIB, i.e. without lymph node metastases or with metastatic nodes less than 5 cm in diameter; in half of the cases the node diameter was less than 2 cm. The group included 167 patients, and the diagnostic results of this evaluation are reported in Table 36.9. It must further be stressed that lymphography was performed in all patients, whereas CT scanning was done in only 20% of the same cases. In this case material the overall diagnostic accuracies were 76.0% and 74.3% for lymphography and CT scanning, respectively. From a statistical viewpoint, lymphography shows an overall accuracy slightly superior to CT scanning, with the same sensitivity and a somewhat higher specificity, but the differences are so small that they are considered negligible. At the same time, a positive radiological report seems to have a better chance of being confirmed histologically when obtained by CT scanning (77.8%), while a negative report is more frequently accurate with lymphangiography (75.9%).

Table 36.9. Results of lymphography vs CT in 167 early pathological stage testicular carcinoma

Statistical evaluation	Lymphography (167 cases)	CT (35 cases)
Sensitivity	74.4%	73.7%
Specificity	77.6%	75.0%
Overall accuracy	76.0%	74.3%
Accuracy of positive result	76.2%	77.8%
Accuracy of negative result	75.9%	70.6%

The goal of each procedure can be identified as the ability to select among all patients with testicular carcinoma those with disease involving the retroperitoneal nodes. Our results indicate that selection of the true negative cases as well as those with involvement limited in size or extension, independent of the diagnostic tool used, is still difficult. In fact, the rate of false negative findings still reaches 25%–30% if different procedures are considered separately. Theoretically, the more adequate screening technique could be considered the combined use of lymphography and CT. Of our patients, 35 underwent both examinations before lymph node dissection (Table 36.10). Among the patients with limited retroperitoneal metastatic disease this combination was diagnostic of abnormal nodes in 89.5%, with good improvement of the diagnostic possibilities of the two separately evaluated tests. On the other hand, when patients with histologically negative nodes are considered, the combination of lymphography and CT decreased the overall accuracy to 62.5%, with 37.5% false positive readings (31.2% due to lymphography and 24.9% to CT, respectively). It must be stressed that we considered equivocal results of radiological examinations as positive for metastases, thus increasing the false positive rate. Furthermore, during lymphography there is an objective difficulty in assessing the true nature of a small filling defect in a single node, and experience can influence the subjective criteria for the determination of a positive or negative lymphangiography study. By contrast, with CT the difficulty of an evaluation of the intrinsic structure of the lymph node, and the 1.5 cm limit in diameter between normal and abnormal nodes, make it possible to consider a nodal swelling as pathological, resulting from non-neoplastic conditions.

On the assumption that approximately 60%–80% of clinical stage I patients could be cured with orchiectomy alone, and relying on the availability of an effective chemotherapy in the salvage of patients who developed metastases, several surveillance programmes in clinical stage I disease have been suggested [22–25]. The experience gained in our Institute with this combination allowed us to start a surveillance study in August 1981 when the availability of CT scanning and the half-life kinetics of serum tumour markers for alpha fetoprotein (AFP) and the β-subunit of human chorionic gonadotrophin (HCG) improved the accuracy of clinical staging [26, 27]. A total of 86 consecutive patients

Table 36.10. Results of the combination of lymphography (L) + CT (C) in a group of patients with early pathological stage testicular carcinoma

Histology	Radiology				
	L−/C− (%)	L+/C+ (%)	L−/C+ (%)	L+/C− (%)	Correct interpretation (%)
N−	62.5	18.7	6.2	12.5	62.5
N+	10.5	57.9	15.8	15.8	89.5

were entered into the study. Prerequisites were no clinical evidence of spread and no prior chemotherapy and/or radiotherapy following orchiectomy. The standard staging procedure included chest radiograph, intravenous pyelography, bipedal lymphography, and CT scanning of the abdomen, which were to be unequivocally negative. Furthermore, the measurement of serum tumour markers had to be normal or normalised following orchiectomy. Patients have been followed for 15–63 months after orchiectomy (median 32, mean 34 months).

Metastases developed in 23 patients (26.7%) (Table 36.11). In nine cases relapse was limited to the chest, and in one patient raised marker levels were the only sign of recurrence. In 13 cases (15.0%) retroperitoneal adenopathies occurred. Time of relapse post-orchiectomy ranged from 2 to 36 months, with shorter intervals for chest (4 months) versus retroperitoneal diffusion (7 months) (Table 36.12). In all cases lung metastases measured at diagnosis were less than 2 cm, whereas retroperitoneal adenopathies were greater than 5 cm in 5 out of 13 patients (38.5%). Pulmonary lesions could be quite easily identified at an early stage (< 2 cm), and the percentage of patients who relapsed in the chest (10/86; 11.6%) is comparable with that noticed in patients treated with retroperitoneal lymphadenectomy (11.6% versus 11.1%) [28]. Using the combination of lymphography and CT in monitoring the retroperitoneal lymph nodes, the relapse rate was expected to not exceed 10%–15% and metastases were supposed to develop within a few months from orchiectomy. The results partially disappointed our expectations. Considering the whole population, only 15.0% of retroperitoneal relapses were encountered, but 38.5% of them measured at diagnosis 5 cm or more in the major diameter. Furthermore, metastases occurred late, even 3 years after orchiectomy (median 7 months), and in one patient with primary pulmonary metastases a second relapse occurred in the retroperitoneal lymph nodes. This is probably attributable both to the natural history of clinical stage I testicular carcinoma and to the difficulty in detecting small retroperitoneal lymph node metastases even when frequently using the most sophisticated tests.

Table 36.11. Site of metastatic involvement in 23/86 patients

Site	No. of cases	%
Retroperitoneal lymph nodes > 5 cm	5	12.8
Retroperitoneal lymph nodes < 5 cm	6	
Retroperitoneal lymph nodes + distant	2	2.2
Lung	9	10.6
Occult (markers)	1	1.1
Total	23	26.7

Table 36.12. Time of relapse

Site	No. of cases	Time (months)		
		min.	max.	median
Retroperitoneal lymph nodes	11	3	36	7
Retroperitoneal lymph nodes + distant	2	3	7	
Lung	9	2	10	4
Markers	1		22	

Magnetic Resonance Imaging

Magnetic resonance imaging (MRI) is a new cross-sectional technique which accelerated the revolution in diagnostic imaging introduced by ultrasound and CT scanning. As the MR images contain information dependent on multiple physical parameters (spin density, tissue relaxation times T1 and T2, chemical shift, etc.) many efforts have been made to enhance the signal intensity differences between various normal tissues and depict abnormalities arising in them. Furthermore, the possibility of obtaining images in any plane, the avoidance of invasive procedures or contrast media and the characteristic high soft tissue resolution made it possible to consider MRI as a valid substitute for CT in the delineation of metastatic retroperitoneal lymphadenopathy [29, 30] (Fig. 36.9). Furthermore, MRI proved to be superior to CT because of its intrinsic physical possibilities. In fact, besides the spatial effects of surrounding structures such as compression and/or dislocation, adenopathies demonstrate a remarkable difference in signal intensity, mainly in T2-weighted images, which permit the distinction of nodes from muscles, vessels, fat and bowel. Especially in large adenopathic masses, a great dishomogeneity in signal occurs, with areas of low signal contiguous to areas of high signal. These features are due to necrosis and haemorrhages, commonly present in adenopathies from testicular carcinoma. For the same reason, the calculated T2 values appear to be significantly different from patient to patient and even from node to node in the same patient [31, 32].

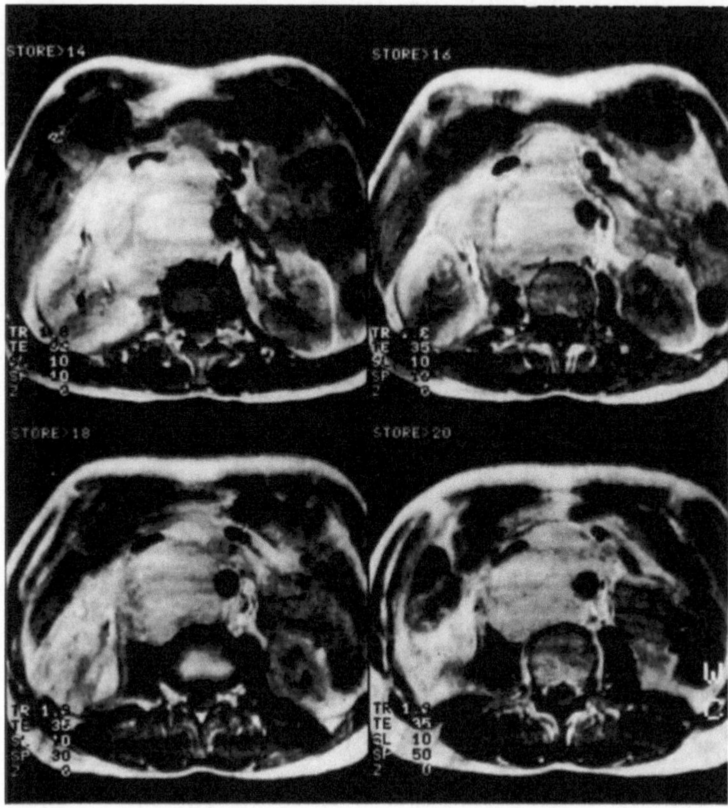

Fig. 36.9. MRI of a patient with right testis cancer: large retrocaval and aortocaval adenopathies displacing the inferior vena cava, aorta and superior mesenteric artery. The adenopathies show increased signal strength while vessels show signal void, indicating flow.

To assess the diagnostic possibilities of MRI in detecting retroperitoneal adenopathies we studied 60 patients with histologically proven testicular carcinoma. All the studies were performed on a 2 Tesla Magnetom machine, operating at 0.5 T during the first part of the study and actually at 1.5 T. Generally, two different Spin-Echo pulse sequences were used, using the multislice with both single and double-echo techniques (Table 36.13) (Fig. 36.10).

Obviously, and according to the previous reported results of CT, the main condition that influences the diagnostic accuracy is the size of metastases; therefore, patients with advanced disease are usually correctly diagnosed. If we consider patients with early stage disease (stages I, II A–B) who underwent radical bilateral or modified retroperitoneal lymphadenectomy (Table 36.14) and we compare the MRI results with those obtained by CT (Table 36.15), a good improvement in distinguishing the presence or absence of retroperitoneal metastases is achieved with MRI. Furthermore, in no patient with large adenopathic masses (diameter > 5 cm) did MRI understage the disease causing errors in treatment planning or patient management.

Table 36.13. Work sequences of MRI

Spin echo	Multislice		Multiecho
TR	500 ms		1200–3000 ms
TE	28–30		28–35, 70–120
No. of slices	5–6		8–15
SL thickness		10 mm	
SL gap		80%–100%	
Matrix		256 × 256	
Averages	2		1–2
Zoom factor		1.2–1.4	
SL orientation		X, Z ± Y	

TR, repetition time; TE, echo time; SL, slice.

Radiological Interpretation of the Evolution of Retroperitoneal Metastases

Whereas the diagnostic value of radiological workup in the staging of patients with testicular cancer has been well documented, there are only a few

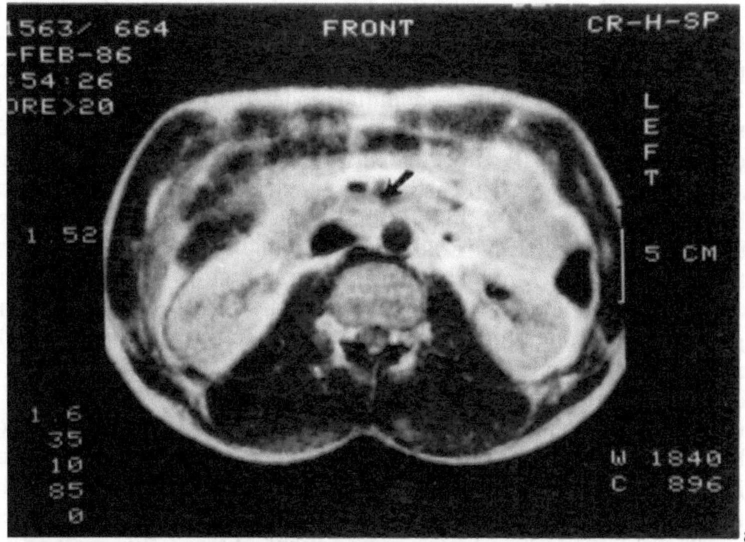

a

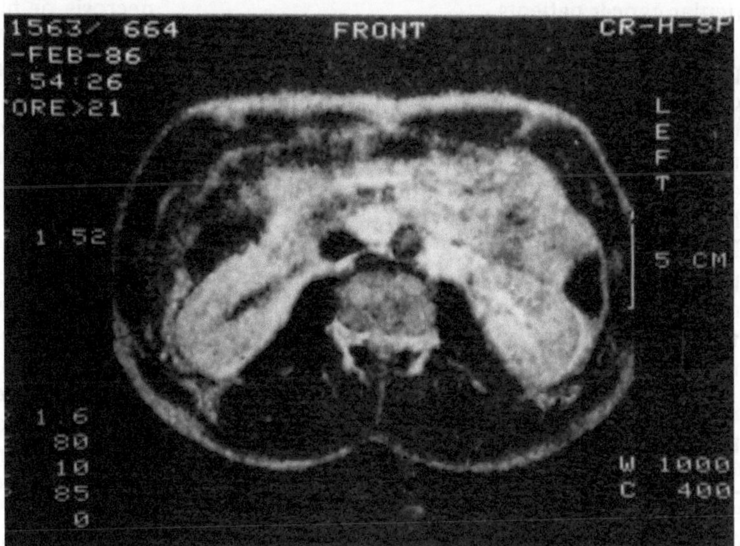

b

Fig. 36.10a,b. Small intercavoaortic adenopathy (*arrow*) compressing the inferior vena cava (**a**). With long TE (**b**), differentiation of the lymph node from vessels and bowel is better appreciable.

Table 36.14. Correlation of MRI and histology in 21 patients following surgery

MRI+/Histology+	MRI+/Histology−	MRI−/Histology+	MRI−/Histology−
10	1	1	9

Overall accuracy	19/21	90.5%
Sensitivity	10/11	90.9%
Specificity	9/10	90.0%
Predictive value of positive result	10/11	90.9%
Predictive value of negative result	9/10	90.0%

Table 36.15. Correlation of MRI and CT in 12 patients following surgery

Histology	MR+/CT+	MR+/CT−	MR−/CT+	MR−/CT−
N+	4	3	1	–
N−	1	–	–	3

reports on the accuracy of the different imaging techniques in interpreting the evolution of retroperitoneal metastases in the post-therapy cases [33–35]. Patients with bulky abdominal nodal metastases (stage II C–D) or with advanced testicular carcinoma (stage III) are usually treated with chemotherapy followed in partially responsive cases by radical lymphadenectomy. This approach is based on the knowledge that residual malignant tumour may be present following such a treatment, and the aim is the removal of bulky masses in those patients in whom mature teratoma is present in the retroperitoneal nodes. In most cases, however, no evidence of malignancy is recognised in the excised mass [36], and often surgical treatment could have been avoided. For these reasons, a more accurate interpretation of the evolution of retroperitoneal malignant masses could potentially have therapeutic value and allow better management of testicular cancer patients.

We reviewed a group of 38 patients with retroperitoneal lymph node metastases from testicular carcinoma who had been treated with chemotherapy followed by lymphadenectomy. All patients underwent bipedal lymphography at the initial diagnostic work-up and follow-up lymphograms both during chemotherapy and before surgery. Furthermore, CT scanning was performed in 16 patients. As expected (Table 36.16), in a majority of the patients (84.2%) no evidence of malignancy was found in the excised mass, as well as a high rate of mature teratoma (44.7%).

Table 36.16. Testicular carcinoma: clinical and pathological findings in 38 patients with advanced disease

Bulky abdominal mass (stages II and III)	34 cases
Advanced disease (stage III)	4 cases
Diameter of tumour at diagnosis:	
<5 cm	8 cases (21.1%)
5–10 cm	22 cases (57.8%)
>10 cm	8 cases (21.1%)
Histology after treatment:	
Necrosis/fibrosis	18 cases (47.4%)
Mature teratoma	14 cases (36.8%)
Mature teratoma + carcinoma	3 cases (7.9%)
Carcinoma	3 cases (7.9%)

Considering the different radiological investigations, regression in lymph node metastases was demonstrated in all patients independently of the diagnostic method used and the final histological diagnosis. Using follow-up lymphograms, we were unable to distinguish either persistent malignancies or patients with mature differentiated teratoma from those with tumour necrosis or fibrosis induced by therapy. In fact, the radiographic pattern of lymph nodes remains aspecifically "pathological" when necrosis or maturation of nodal metastases occurs, even though the therapy was successful. This makes follow-up lymphograms of little use in the management of these patients, since the estimation of tumour diameters and lymphographic characteristics of lymph nodes does not provide accurate information about the tumour's response to therapy. Similarly, even an enlargement of the adenopathic mass during intensive chemotherapy does not indicate a definite progression of the disease, since the enlargement may be due to the conversion of the solid tumour to a cystic mass which lymphographically simulates a therapeutic failure. CT scanning can undoubtedly be more accurate than lymphography because of its ability to detect gross changes in anatomy. In fact, by this method the conversion of malignant mass to necrosis or fibrosis, or to a benign tumour form, can be demonstrated, thus allowing a decision to remove the mass rather than to continue the chemotherapy as the sole treatment. In our series of patients, however, 25% of the cases considered to be in partial remission or cystic degeneration by CT, demonstrated residual viable malignant tumour at surgery.

Our results clearly indicate that serial CT examinations are extremely important in the monitoring of patients with metastatic testicular tumours, whereas other radiological examinations are revealed to be of doubtful utility. At present no diagnostic method alone or in combination can adequately exclude the presence of residual viable malignant tumour, even when favourable factors such as cystic changes or necrosis within metastases exist. It is possible that the newer imagining procedures such as MRI, allowing not only visual but also biochemical and metabolic assessments of tissues, will make the follow-up of these patients easier and more accurate. Until that time, however, our experience suggests that the combination of radiological evaluations with periodic measurement of serum tumour markers (HCG-AFP) is necessary to obtain more adequate parameters of the presence of active residual disease.

References

1. Dunnick NR, Javadpour N (1981) Value of CT and lymphography: distinguishing retroperitoneal metastases from nonseminomatous testicular tumors. AJR 136:1093–1099
2. Wallace S, Jing BS (1970) Lymphangiography: diagnosis of nodal metastases from testicular malignancies. JAMA 123:94–97

3. De Roo T, Van Minden SH (1973) Lymphographic findings in a series of 258 patients with tumors of the testis. Lymphology 6:97–100
4. Lien HH, Fossa SD, Ous S, Stenwig AE (1983) Lymphography in retroperitoneal metastases in non-seminoma testicular tumor patients with normal CT scan. Acta Radiol [Diagn] 24:319–322
5. Ehrlichman RJ, Kaufman SL, Siegelman SS, et al. (1981) Computerized tomography and lymphangiography in staging testis tumors. J Urol 126:179–181
6. Jing B, Wallace S, Zornoza J (1982) Metastases to retroperitoneal and pelvic lymph nodes: computed tomography and lymphangiography. Radiol Clin North Am 20:518–520
7. Marincek B, Brutschin P., Triller J, Fuchs WA (1983) Lymphography and computed tomography in staging nonseminomatous testicular cancer. Limited detection of early stage metastatic disease. Urol Radiol 5:243–246
8. Musumeci R, Mauri M (1980) La linfografia in oncologia. Ilford, Origgio
9. Tesoro Tess JD, Pizzocaro G, Zanoni F, Musumeci R (1985) Lymphangiography and computed tomography in testicular carcinoma: how accurate in early-stage disease? J Urol 133:967–970
10. Thomas JL, Bernardino ME, Bracken RB (1986) Staging of testicular carcinoma: comparison of CT and lymphangiography. AJR 137:991–996
11. Wobbes T, Blom JMH, Oldhoff J, Schraffordt Koops H (1982) Lymphography in the diagnosis of non-seminoma tumours of the testis. J Surg Oncol 19:1–4
12. Winterberger AR (1977) Correlation of lymphography findings in the abdomen with B-Scan ultrasonic laminography. In: Mayall RC, Witte MH (eds) Progress in lymphology. Plenum, New York, pp 173–178
13. Rochester D, Bouvie J, Kurzmann A, Lester E (1977) Ultrasound in the staging of lymphoma. Radiology 124: 483–487
14. Brascho D, Durant J, Green L (1977) The accuracy of retroperitoneal ultrasonography in Hodgkin's disease and non-Hodgkin's lymphoma. Radiology 125: 485–487
15. Burney B, Klatte E (1979) Ultrasound and CT in the staging of testicular carcinoma. Radiology 132: 415–419
16. David E, Van Kaick G, Ikinger U, Gerhardt P, Prager P (1982) Detection of neoplastic lymph node involvement in the retroperitoneal space. Eur J Radiol 2:277–280
17. Williams RD, Feinberg SB, Knight LC, Fraley EE (1980) Abdominal staging of testicular tumors using ultrasonography and computed tomography. J Urol 123: 872–875
18. Rowland RG, Weisman D, Williams SD, Einhorn LH, Klatte EC, Donohue GP (1982) Accuracy of preoperative staging in stages A and B non-seminomatous germ cell testis tumors. J Urol 127:718–720
19. Lackner K, Weisback L, Boldt I, Scherholz K, Brecht G (1979) Computer-tomographischer Nachweis von Lymphknotermetastasen bei malignen Hodentumoren. Fortschr Rontgenstr Nuklearmed Erganzungsband 130:636–643
20. Lien HH, Kolbenstvedt A, Talle K, Fossa SD, Klepp O, Ons S (1983) Comparison of computed tomography, lymphography and phlebography in 200 consecutive patients with regard to retroperitoneal metastases from testicular tumor. Radiology 146:129–132
21. Richie JP, Garnick MB, Finberg H (1980) Computerized tomography: how accurate for abdominal staging of testis tumours? J Urol 123:872–874
22. Jewett MAS, Herman JG, Sturgeon JFG Comisarow RH, Alison RE, Gospodarowicz MK (1984) Expectant therapy for clinical stage A nonseminomatous germ cell testicular cancer? Maybe. World J Urol 2:57–58
23. Johnson DE, Lo RK, von Eschenbach AC, Swanson DA (1984) Surveillance alone for patients with clinical stage I nonseminomatous germ cell tumors of the testis: preliminary results. J Urol 131:491–493
24. Peckham MJ, Barrett A, Husband JE, Hendry WF (1982) Orchidectomy alone in testicular stage I non-seminomatous germ-cell tumours. Lancet II:678–680
25. Peckham MJ (1984) Orchiectomy for clinical stage I testicular cancer. A progress report of the Royal Marsden Hospital Study. In: Proceedings of the first international symposium on testicular tumors. Paris, 8–10 October, 1984
26. Pizzocaro G, Zanoni F, Milani A, Salvioni R, Piva L, Pilotti S, Bombardieri E, Tesoro Tess JD, Musumeci R (1986) Orchiectomy alone in clinical stage I non-seminomatous testis cancer: a critical appraisal. J Clin Oncol 4: 35–40
27. Tesoro Tess JD, Balzarini L, Musumeci R (1985) Radiological follow-up in clinical stage I testicular carcinoma treated with orchidectomy alone. In: Casley Smith JR, Piller NB (eds) Progress in lymphology. University of Adelaide Press, Adelaide, pp 262–264
28. Pizzocaro G (1986) Retroperitoneal lymphadenectomy in clinical stage I non-seminomatous germinal testis cancer. Eur J Surg Oncol 12:25–28
29. Ellis JH, Bies JR, Kopecky KK, Klatte EC, Rowland RG, Donohue JP (1984) Comparison of NMR and CT imaging in the evaluation of metastatic retroperitoneal lymphadenopathy from testicular carcinoma. J Comput Assist Tomogr 8 (4):709–719
30. Lee JKT, Heiken JP, Ling D, Glazer HS, Balfe DH, Levitt RG, Dixon WT, Murphy WA Jr (1984) Magnetic resonance imaging of abdominal and pelvic lymphadenopathy. Radiology 153:181–188
31. Balzarini L, Petrillo R, Ceglia E, Tesoro Tess JD, Musumeci R (1986) Lo studio delle adenopatie: quale ruolo potrebbe competere oggi alla risonanza magnetica? Radiol Med (Torino) 72 (9):615–619
32. Musumeci R, Balzarini L, Petrillo R, Tesoro Tess JD, Ceglia E (1986) Possible role of MR in staging retroperitoneal nodes in testicular cancer. In: Proceedings of the fifth annual meeting of the Society of Magnetic Resonance in Medicine, Montreal, Canada, 18–22 August, 1986
33. Husband JE, Hawkes DJ, Peckham MJ (1982) CT estimations of mean attenuation value and volume in testicular tumors: a comparison with surgical and histologic findings. Radiology 144:553–558
34. Javadpour N, Anderson T, Doppman J (1979) Computed tomography in evolution of testicular cancer during intensive chemotherapy. J Urol 122: 565–567
35. Soo CS, Bernardino ME, Chuang VP, Ordonez N (1981) Pitfalls of CT findings in post-therapy testicular carcinoma. J Comput. Assist Tomogr 5: 39–41
36. Hong WK, Wittes RE, Hadju ST, Cvitkovic E., Whitmore WF, Golbey RB (1977) The evolution of mature teratoma from malignant testicular tumor. Cancer 40:2987–2992

Chapter 37

The Value of Fine Needle Aspiration Cytology in the Management of Metastatic Germ Cell Tumours

R. Scaletscky, C. Stephenson, R. T. D. Oliver, W. J. Highman and M. J. Kellett

Introduction

Testicular cancer is a rare malignancy, comprising only about 1% of all cancers in men. Significant advances have been made in recent years with a reduction in overall mortality. Cure of metastatic disease is dependent upon accurate clinical and surgical/pathological staging, appropriate use of tumour markers, intensive chemotherapy regimens and the correct timing of surgery to remove residual disease [1–4].

Treatment of metastatic disease is still dependent upon the histological type and the stage of the tumour. The accuracy of clinical staging of testicular neoplasms has improved as a result of the development of new technologies such as computed tomography (CT), ultrasound examination and sensitive radioimmune assays to measure tumour markers. The histological type of the primary tumour together with the local extent of the primary tumour have been shown to have significant prognostic implications in terms of risk of metastasis [5, 6].

The time interval between the onset of symptoms and initiation of treatment influences both the incidence of metastases and the outcome following chemotherapy [4, 7, 8]. At present, most of the deaths are in the group of patients who present with advanced disease as measured by tumour volume or tumour marker levels [4, 8, 9,10]. Both of these parameters increase with prolongation of delay following the first symptoms, and it is difficult to separate out the relative contribution of each of these factors (Table 37.1). As delay is potentially easier to reverse than the actual malignant potential, this is the more significant factor to address.

Germ cell tumours vary markedly in their growth rate. The occasional patients with seminoma may delay for up to 5 years without developing metastases, and it is more difficult to demonstrate the effect of delay on survival of patients with seminoma. For patients with malignant teratoma the correlation with delay is more significant, and patients with metastatic trophoblastic tumours may be dead in less than 6 weeks if left untreated, and the doubling of such tumours may be as short as 5–6 days [11].

Delay in starting appropriate chemotherapy is not always due to later presentation but may also be due to a surgical complication developing after laparotomy or thoracotomy, which may have been necessary in order to establish the diagnosis in those patients without an occult primary tumour. The morbidity and mortality associated with these procedures followed by cytotoxic chemotherapy

Table 37.1. Prognostic factors for disease-free survival of patients with malignant teratoma receiving platinum-containing chemotherapy

	No. of patients	% disease-free 2-year survival
Treatment prior to 1980	23	35
Treatment 1980–1982	42	77 (P=0.02)
Delay to first chemotherapy < 6 months	28	79
Delay to first chemotherapy > 6 months	37	51 (P=0.04)
AFP < 500 KU/litre and HCG < 1000 IU/litre	34	71
AFP > 500 KU/litre and HCG > 1000 IU/litre	32	47 (P=0.02)
Para-aortic mode < 2 cm and/or lung metastases < 2 cm	18	89
Para-aortic node 2–5 cm + lung metastases < 2 cm or less than 3 in number	12	58 (P=0.04)
Para-aortic node > 5 cm and/or lung metases > 3 in number and at least 1–2 cm in diameter	36	50

prompted an investigation into the role of fine needle aspiration cytology (FNA) in metastatic germ cell tumours. As a result of a preliminary indication of a better outcome in patients diagnosed by FNA than in those where the diagnosis was made by conventional techniques [12] an attempt has been made to evaluate the reliability of the FNA results in a larger series of patients.

Patients and Methods

Patients that were considered suitable for this study included:

1. Patients who presented with metastatic disease, whether or not there was an obvious tumour in the testicle. FNA of the mass was performed and serum levels of alpha fetoprotein (AFP) and beta human chorionic gonadotrophin (BHCG) were estimated urgently.
2. Patients under surveillance for stage I disease who presented with an obvious relapse on the CT scan, irrespective of tumour marker levels.
3. Patients who had a residual mass at the site of previous disease following completion of chemotherapy, in whom post-treatment surgical staging was indicated.
4. Patients who developed recurrent disease following completion of chemotherapy.

Technique

There are four main methods of FNA of intra-abdominal masses. The first is direct, when a mass is clinically palpable, and this technique can be performed on the ward. The second method is to use the ultrasound for direct guidance. This is only really accurate when puncturing cystic lesions as it is sometimes difficult to see the tip of the needle. The third method is standard fluoroscopy, which can be carried out in most hospitals using standard X-ray equipment. Retroperitoneal nodes can be opacified by lymphangiography, calcified lesions are easily identified and intravenous contrast may be given to relate a mass to the kidneys. Finally, over the past few years CT-guided aspiration biopsy has become very popular and has the obvious advantage that one can accurately locate the tip of the needle in very small retroperitoneal masses. Alternatively, CT can also be used to produce a map (scanogram) to define landmarks for fluoroscopy. All of these four methods of FNA were used in our study.

Fine needle aspiration was performed using a 23-gauge needle. Negative pressure was maintained whilst the needle was moved back and forth in the mass, and released before the needle was withdrawn. In an attempt to ensure adequate sampling, aspirations were made from several sites in large masses.

Results

Table 37.2 demonstrates the risks associated with open surgical biopsy in the diagnosis of metastatic germ cell tumour and gives a preliminary indication that the survival of those diagnosed by FNA cytology may be better, although as this is a historic comparison the groups may not be strictly comparable.

Review was undertaken of 43 cytology preparations from 34 patients, 9 of whom underwent repeat biopsies: 1 because of failure of the tumour to respond, 3 because of recurrence after chemotherapy and 5 in an attempt to confirm negative results at the first biopsy. Two of these repeat negative biopsies were positive, as were three of four biopsies for persistent or recurrent tumour.

It was possible to classify the cytological findings into four categories: definite malignant cells, no malignant cells, benign epithelial cells (possibly mature teratoma, possibly normal gut) and necrotic

Table 37.2. Impact of diagnostic procedure on results of treatment with large volume malignant teratoma metastases

	No.	Alive disease free	Early treatment-related death	Serious complications related to biopsy
Surgical excision biopsy	20	6	3[a]	3[b]
Positive FNA or positive tumour marker	10	8	—	—

[a]One patient with ascending renal vein thrombosis from an inguinal wound abscess. Two patients with leucopenic sepsis from unsuspected intra-abdominal infection following laparotomy, only manifested after beginning of chemotherapy.
[b]One accidental perforation of the duodenum in a patient with regressed primary tumour, leading to duodenal obstruction treated by intravenous feeding for 6 weeks. One patient with large bowel obstruction from postoperative adhesions. One patient with clostridial sepsis and renal failure from bowel perforation.

cells (Table 37.3). There were no apparent false positive results as all patients in whom malignant cells were detected underwent regression on chemotherapy, except one who subsequently died of drug-resistant tumour. In 7 of the 12 cases no malignant cells were detected, the results were subsequently proved to be false negative: Two patients had a repeat biopsy, four underwent regression on chemotherapy and one had a microscopic focus in a retroperitoneal mass excised at surgery after chemotherapy. The other four patients remain free of disease for more than 1 year at the time of writing. As expected, the incidence of viable tumour was higher in the biopsies performed prior to chemotherapy: 15/26 (53%) compared with 4/17 (26%). However, in the post-chemotherapy group the incidence of surgical verification was higher, as 8 of 13 cases were subsequently verified by surgical excision of the residual mass and the other 4 were carefully followed up for 8 and 18 months and 5 and 6 years without recurrence. As mentioned above, the one false negative result was in a patient

Table 37.3. Classification of cytological findings

	Positive	Benign epithelial cells	Necrotic cells	No malignant cells	
				True negative	False negative
Before chemotherapy ($n=26$)	15	4	–	1	6
After chemotherapy ($n=17$)	4	5	3	4	1

who had a small focus of degenerate malignant cells in a large necrotic mass. Apart from the high frequency of positives from direct aspiration of an obvious mass, there was no obvious difference between the other techniques (Table 37.4) but as the tumours biopsied under CT control were generally smaller, comparison is not really possible.

Table 37.4. FNA cytology results using different methods

Method	Positive result	Benign epithelium or necrotic cells	No malignant cells	
			True negative	False negative
Direct aspiration ($n=10$)	9	1	–	–
Fluoroscopy ($n=22$)	6	5	7	4
Ultrasound-guided aspiration ($n=3$)	1	2	–	–
CT scan-guided aspiration ($n=8$)	3	2	–	3

Table 37.5 correlates the cytological findings from the FNA biopsy of the metastasis with the histological findings of the primary tumour in those patients who underwent orchiectomy prior to chemotherapy. In 11 of 15 patients with adequate cells for evaluation, it was possible to demonstrate cells in the cytological preparation compatible with the histological findings in the primary (Figs 37.1–37.3). There were five discrepancies: Two were patients in whom it was not possible to be certain whether the cells were seminoma or undifferentiated; one was a patient with stage I seminoma who

Table 37.5. Primary histology and FNA findings

Cytological interpretation	Malignant teratoma undifferentiated	Malignant teratoma intermediate	Seminoma	Mixed seminoma
Seminoma cells	–	–	6	–
? Seminoma/ undifferentiated	–	–	1	1
Undifferentiated cells	–	–	1	1
Benign epithelial cells	1	3	1	–
False negative	1	1	2(+2)[a]	–

[a]Two cases had negative readings at first attempt, positive on repeat biopsy.

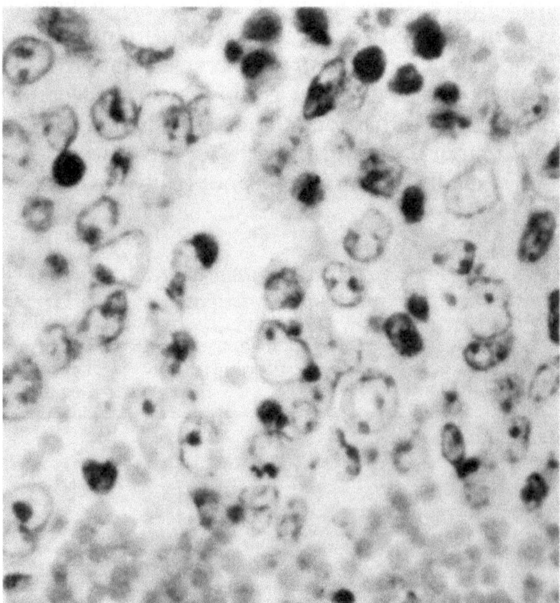

Fig. 37.1. Aspirate from para-aortic lymph node of patient with malignant teratoma undifferentiated.

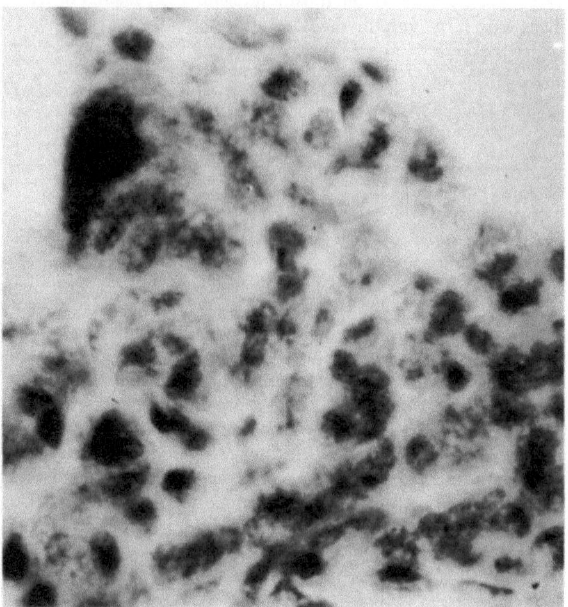

Fig. 37.3. Aspirate of lung metastasis of patient with malignant teratoma trophoblastic.

relapsed at 2 years, and the cytological appearances were more compatible with undifferentiated cells; one was a patient with stage I seminoma who relapsed at 10 years with probable malignant teratoma intermediate and only benign epithelial cells in the aspirate; and the final one was a patient with malignant teratoma undifferentiated who had benign epithelial cells detected in a mass which disappeared after chemotherapy. Three patients with drug-resistant tumour underwent biopsy before and after chemotherapy without obvious gross change in the morphology.

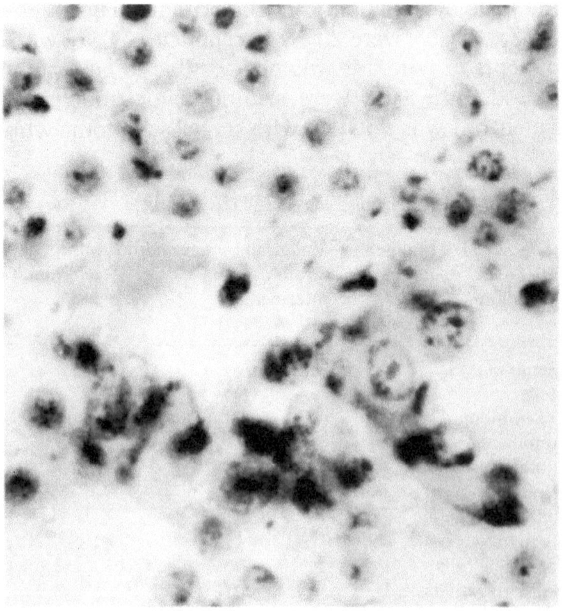

Fig. 37.2. Aspirate from supraclavicular lymph node of patient with seminoma.

Discussion

Delay is an important determinant of successful treatment of metastatic germ cell tumours. In addition, complications after major surgical procedures to establish the diagnosis in patients with widespread metastases can lead to the death of patients. These factors justify the simple approach of FNA rather than surgical biopsy. The possibility that this method involves risk of tumour recurrence in the needle tract has always been an anxiety. This has been extensively investigated by Esposti [13], who followed up 100 consecutive cases after aspiration of the primary testis tumours prior to orchiectomy over a period of up to 18 years. There was only one local recurrence (this was in a case where the primary tumour had extended in the scrotum prior to aspiration), and one metastasis to inguinal nodes.

As all the patients in our series who had positive cytological findings proceeded to chemotherapy, it is not possible to comment on the issue of needle tract recurrence. However, the most important conclusion is that for those patients who present with bulky metastases or an occult primary and have a positive FNA result, the result is sufficiently reliable to justify a trial of chemotherapy without the need to perform orchiectomy or laparotomy. This is important because most patients with bulky metastases will ultimately require post-treatment surgical staging and removal of residual masses after treatment to confirm the completeness of response to chemotherapy.

For the majority of patients who present without clinically obvious metastases, orchiectomy will remain the diagnostic procedure of choice because it is also therapeutic for patients who have no metastases [14]. However, as the majority of patients recover fertility after chemotherapy [15] and chemotherapy is known to produce complete remission of primary tumours [16], there is a case for considering the use of FNA as the diagnostic procedure of choice for patients with a suspicious mass in a single testis who wish to preserve fertility.

Although the results presented in this paper validate a positive FNA, the number of false negative results still leaves room for improvement in technique, though as reported by Esposti [13] it is likely that in patients with malignant teratoma intermediate, i.e. teratocarcinomatous elements, or patients with extensive malignant retroperitoneal fibrosis associated with seminoma it may not always be possible to get positive material. Nonetheless, the demonstration of malignancy in two patients who underwent a second biopsy when there was a strong suspicion of malignancy emphasises the need to persist if in doubt.

As far as correlation of cytological detail with histopathological subtype is concerned, it is obvious that FNA can never define all the refinements of histology, as histological subtype is not just dependent on the morphology of individual cells but also on how they relate to one another [17]. In addition, more than 50% of tumours consist of more than one component [18]. For pure seminoma and MTU the correlation between FNA and histology was good, though as with histology there is some uncertainty at the borderline between seminoma and malignant teratoma undifferentiated, which may be further evidence in support of the concept that seminoma could transform into malignant teratoma undifferentiated [19, 20].

As far as malignant teratoma intermediate is concerned, the cytology was less satisfactory because of the extensive variation in cellular content seen in these tumours and, as observed by Esposti [13], the greater adhesion between the mature components, making it difficult to determine whether the cells seen were part of the tumour or part of normal organs through which the aspiration needle passed. These problems in relation to diagnosing metastases with mature somatic tissue are even more pronounced after chemotherapy. However, given the increasing recognition that it is not safe to leave mature teratoma in situ after chemotherapy because of the risk of transformation into sarcoma and carcinoma [21], the discovery of benign epithelial cells in an aspirate from a residual mass after chemotherapy would be additional support for proceeding to excision after chemotherapy. In some instances it might justify earlier surgery if there was strong suspicion from the CT scan that there had been no actual shrinkage after chemotherapy.

Review of post-chemotherapy residual masses excised in our previously reported series of patients has demonstrated that in the minority who have persistent malignancy viable tumour often makes up less than 10% of the residual mass [12, 22]. Thus, unless there has been considerable shrinkage in comparison with pretreatment, it is not wise to postpone surgery for more than 4–6 weeks after chemotherapy, even if there is a negative post-treatment cytological report.

Conclusion

In addition to reducing the risk of serious post-operative complications when undergoing marrow suppressive treatment, FNA provides a rapid and less invasive procedure than surgery for establishing the diagnosis of germ cell tumours if the patient has extensive spread and needs treatment urgently; however, because of the problem of sampling, a negative result is of little value. A preliminary attempt to correlate cytological appearances with histological subtype shows good correlation for seminoma and malignant teratoma undifferentiated, though less precise definition of the variety of cell types seen in malignant teratoma intermediate.

References

1. Einhorn LH, Donohue JP (1977) Cis-diammine dichloroplatinum, vinblastine and bleomycin combination

chemotherapy in disseminated testicular cancer. Ann Intern Med 87:293–298
2. Newlands ES, Begant RHJ, Bagshaw KD (1980) Further advances in the management of malignant teratoma of the testis and other sites. Lancet I:948–951
3. Peckham MJ, Barrett A, Lias KH (1980) The treatment of metastatic germ cell testicular tumour with bleomycin, etoposide and cisplatin (BEP). Br J Cancer 47:613–619
4. Oliver RTD (1985) Testicular germ cell tumours—a model for a new approach to treatment of adult solid tumours. Postgrad Med J 61:123–131
5. DeWys WD, Green SB, William SD et al. (1985) Prediction of nodal involvement and of relapse after lymphadenectomy in early stage testicular cancer. In: Proceedings of the American Society of Clinical Oncology, p 134
6. Hoskin P, Dilly S, Easton D et al. (1986) Prognostic factors in stage I non-seminomatous germ-cell testicular tumours managed by orchiectomy and surveillance: implications for adjuvant chemotherapy. J Clin Oncol 4:1031–1036
7. Sher H, Bosl G, Gelier N et al. (1983) Impact of symptomatic intervals on prognosis of patients with Stage III testicular cancer. Urology 21:559–561
8. Medical Research Council Working Party on Testicular Tumours (1984) Prognostic factors in advanced non-seminomatous germ cell testicular tumours: results of a multicentre study. Lancet I:8–11
9. Germa-Lluch JR, Begent RMJ, Bagshaw KD (1980) Tumour marker levels and prognosis in malignant teratoma of the testis. Br J Cancer 42:850–855
10. Peckham MJ, Barrett A, McElwain TJ, Hendry WF, Raghavan D (1981) Non-seminoma germ cell tumours (malignant teratoma) of the testis. Results of treatment and an analysis of prognostic factors. Br J Urol 53:162–172
11. Garrera L, Debonniere C, Berigiron R, Cioppani F, Jofipouici JJ, Thomas JP (1970) Testicular tumour metastatic growth rate. Bull Soc Med-Chirurg Hopit Fond Sanit Armees 2:93–97
12. Oliver RTD, Highman WJ, Kellett MJ, Curling M, Dacie JE (1985) The value of fine needle aspiration cytology in the management of metastatic germ cell tumours. Br J Urol 57:200–203
13. Esposti PL (1979) Aspiration biopsy cytology, part 2. In: Zajicek J (ed) Monographs in clinical cytology. Karger, Basel, pp 113–123
14. Oliver RTD, Hope-Stone HF, Blandy JP (1983) A justification for the use of surveillance in the management of stage I germ cell tumours of the testis. J Urol 55:760–763
15. Oliver RTD (1985) Fertility of patients with germ cell tumours of the testis before and after treatment with platinum and etoposide containing combination chemotherapy regimens. In: Jones WG (ed), Germcell tumours II. Pergamon, London, p 467 (Advances in biosciences, vol 55)
16. Greist A, Einhorn LH, Williams SD et al. (1984) Pathological findings at orchidectomy following chemotherapy for disseminated testicular cancer. J Clin Oncol 2:1025–1027
17. Pugh RCB (1976) Testicular tumours—the panel classification. Blackwell Scientific, Oxford, pp 267–270
18. Mostofi FK (1984) Tumour markers and pathology of testicular tumours. In: Progress and controversies in oncological urology. Alan R Liss, New York, pp 69–87
19. Ragavan D, Heyderman E, Monoghan P et al. (1981) Hypothesis: when is a seminoma not a seminoma. J Clin Pathol 34:123–128
20. Oliver RTD (1987) HLA phenotype and clinico pathological behaviour of germcell tumours—possible evidence for clonal evolution from seminomas to non-seminomas. Andrologica (in press)
21. Ulbright TM, Lochrer PJ, Roth LM, Einhorn LH (1984) The development of non germ cell malignancies within germ cell tumours. Cancer 54:1824–1833
22. Oliver RTD, Blandy JP, Hendry WF, Pryor JP, Williams JP, Hope-Stone HF (1983) Evaluation of radiotherapy and/or surgico-pathological staging after chemotherapy in the management of metastatic germ cell tumours. Br J Urol 55:764–768

Section VIII

Urological and Non-urological Metastatic Disease

Chapter 38

Periureteral Aspiration Cytology in the Study of Ureteral Stenoses in Patients with Known Malignancy

L. Luciani, P. Scappini, T. Pusiol and F. Piscioli

Introduction

Disease involvement of the urinary collecting system in the course of neoplastic disease is a common occurrence, especially in patients with advanced pelvic malignancies [1]. Cancer of the cervix, prostate, breast, bladder, colon and rectum, and of the lymphoid system often cause urinary obstruction by direct extension or compression [2, 3]. On the other hand, post-therapeutic retroperitoneal fibrosis may be expected in patients undergoing surgery, radiation or combined surgery and radiation procedures on the pelvic area [4]. Therefore, when ureteral obstruction occurs in patients with known malignancy, it is important to establish the presence of pre-existing, recurrent or metastatic tumours before ascribing the effects to radiation or surgery. In such cases cytological diagnosis can be helpful since diagnostic imaging is often inconclusive [5]. In this chapter we describe our experience with aspiration biopsy cytology (ABC) in diagnosing the aetiology of the ureteral stenosis in 22 patients with known cancer. Of these cases, 15 have been reported elsewhere [6].

Materials and Methods

Our series consisted of 22 patients with known primary cancer who developed ureteral stenosis and who underwent fluoroscopy-guided ABC of the site of obstruction, using a 22- to 23-gauge modified Chiba needle[1]. There were 5 males and 17 females of age ranging from 35 to 75 years (mean 63 years).

The primary malignancies were adenocarcinoma of the uterine cervix (3), epidermoid carcinoma of the uterine cervix (6), endometrial adenocarcinoma (3), invasive urothelial carcinoma of the urinary bladder (4), prostatic adenocarcinoma (2), adenocarcinoma of the rectum (1), gastric cancer (1), pancreatic carcinoma (1) and pulmonary adenocarcinoma (1). All patients had undergone radical surgery and chemotherapy of the primary lesion, and irradiation was performed in 17 cases.

Visualisation of the site of aspiration was obtained by means of intravenous injection of contrast medium or an antegrade pyeloureterography through a previous transcutaneous nephrostomy (Fig. 38.1).

In the case of a retrograde catheterisation the opacification of the ureter was achieved by injection of contrast material through the ureteral stent under fluoroscopic control (Figs. 38.2, 38.3). Most of the patients had previous ultrasound and computed tomography (CT) study (Fig. 38.4). The anterior abdominal wall was prepared and draped in the usual manner and the needle inserted transperitoneally up to the point of obstruction or just distal to it. The correct placement of the needle was verified using fluoroscopy. Biplane fluoroscopy is

[1] Cyto-Aspir (patent pending). Cook Urological Inc., Spencer, Ind, USA.

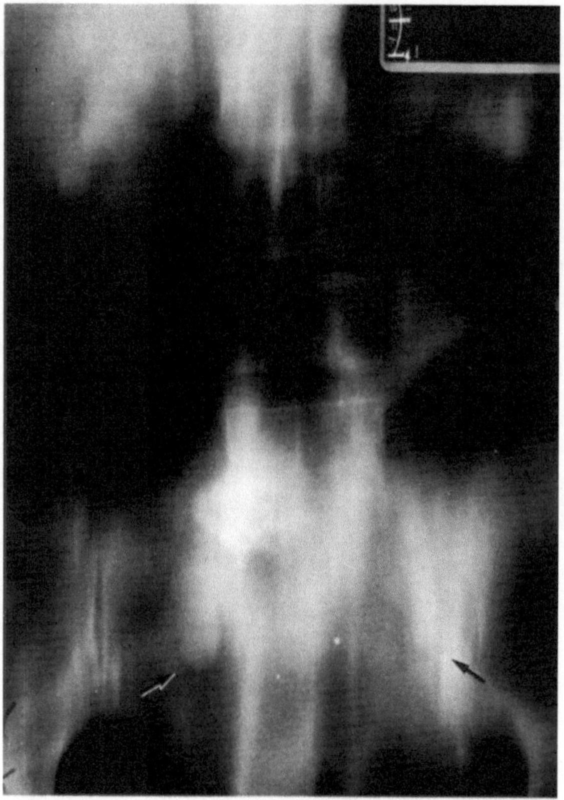
a

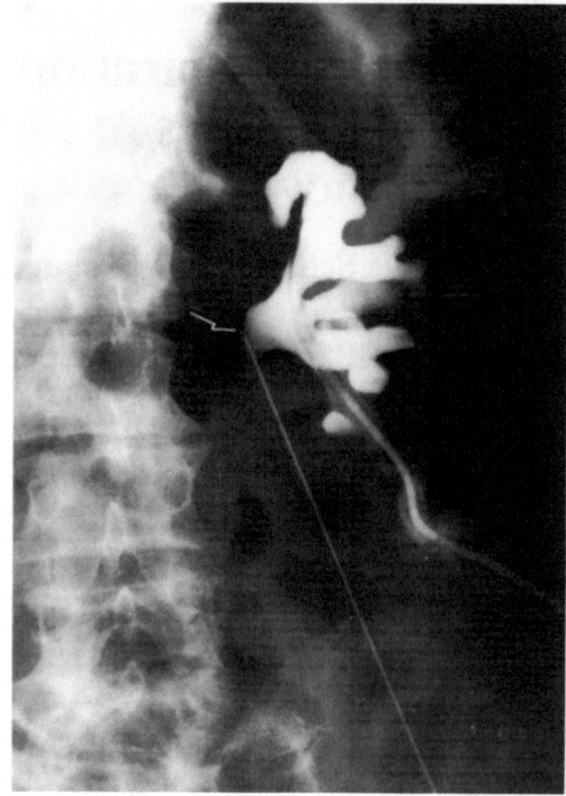
c

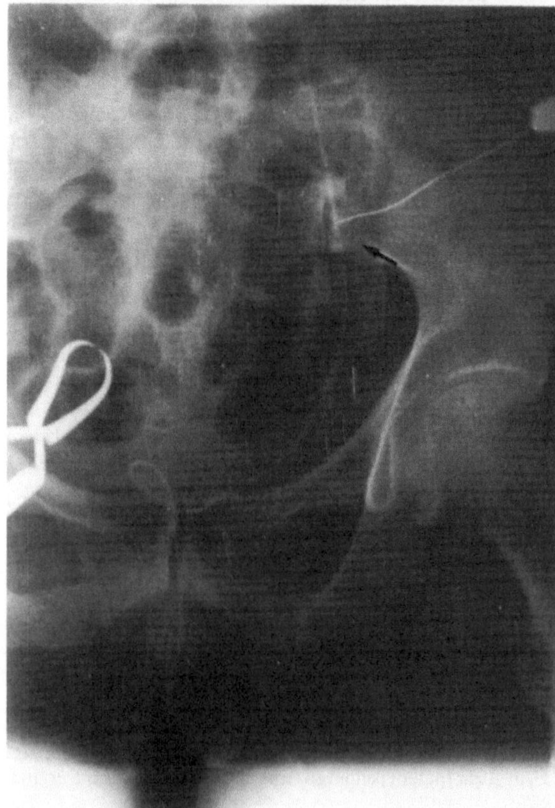
b

Fig. 38.1. (Case 4.) **a** Moderately differentiated endocervical adenocarcinoma with periureteral metastasis causing bilateral simultaneous narrowing of the ureters at pelvic brim (*arrows*) (February 28, 1984). **b** Transcutaneous aspiration biopsy was done with the help of a retrograde catheterisation which demonstrated incomplete obstruction of the left ureter (*arrow*). Malignant cells were found on the aspirate (March 2, 1984). **c** 11 days later, the patient developed complete proximal obstruction of the left ureter. Aspiration biopsy was repeated on the more proximal periureteral site (*arrow*) using an antegrade nephrostography under fluoroscopic control and showed malignant cells consistent with cervical carcinoma (March 13, 1984).

useful but not necessary. Subsequently, the stylet was removed and up and down movements were made with the needle before applying suction with a plastic syringe handle[2], which we recommend because it provides better connection and more forceful aspiration than the traditional metal-glass syringe handle.

The needle was then withdrawn and a small drop of the aspirate spread onto a sterile slide and smeared by pressing the slide on several others. Fixation in 95% ethyl alcohol for 30 min and

[2] Cameco, Sweden.

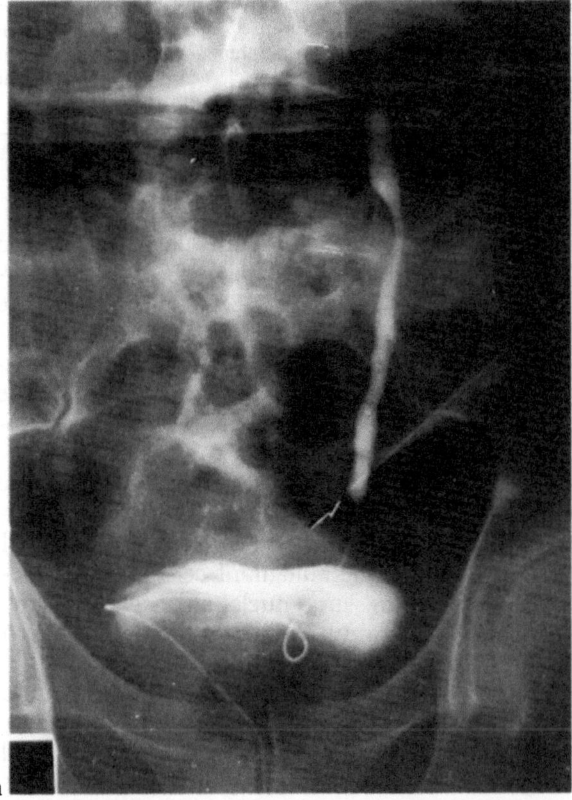
a

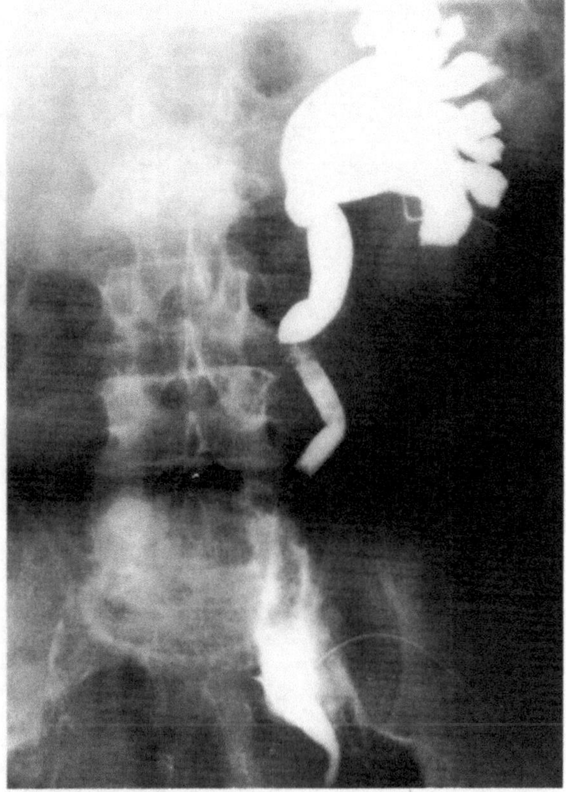

Fig. 38.3. (Case 11.) Antegrade pyelography showing left hydroureteronephrosis caused by obstruction of the distal ureter. Fluoroscopy-guided ABC of the site of stenosis was performed and the cytological findings were consistent with metastasis from prostatic adenocarcinoma. The extravasation of the contrast medium and the presence of air bubbles in the ureter confirms penetration of the needle tip through the ureteral wall. (Luciani et al. [6]; reproduced by kind permission of the Editor of *Cancer* and the publisher, J.B. Lippincott Co., Philadelphia)

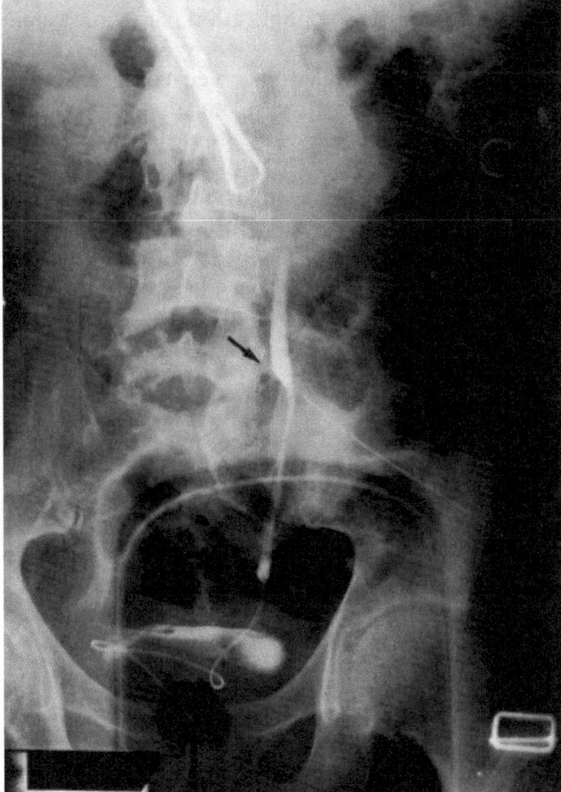
b

Papanicolaou stain were the last steps of the procedure.

Generally, four to six smears were obtained from a single aspiration, giving a total of 97 Papanicolaou-stained smears suitable for examination. In all cases the cytological diagnosis was verified by comparison with the histological findings of the surgical specimens.

Fig. 38.2a, b. (Case 8.) Distal incomplete obstruction of left ureter secondary to poorly differentiated gastric cancer. Transcutaneous aspiration biopsy was done under guidance of retrograde catheterisation and antegrade opacification of the ureter. The needle tip is seen localised in two different points (**a**, **b**) of the obstructed tract (*arrows*). Distal aspiration (**b**) showed metastasis (March 18, 1983).

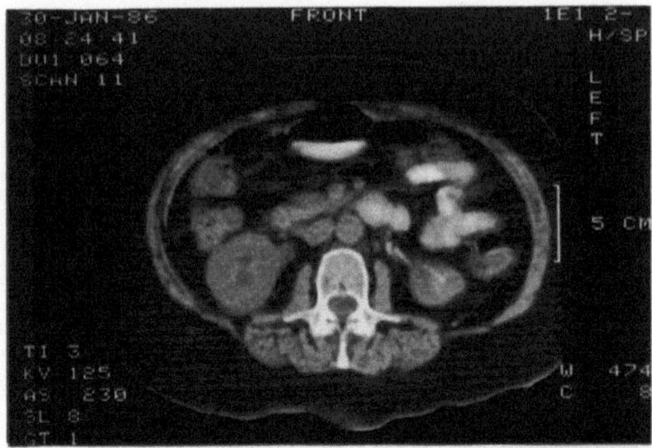

a

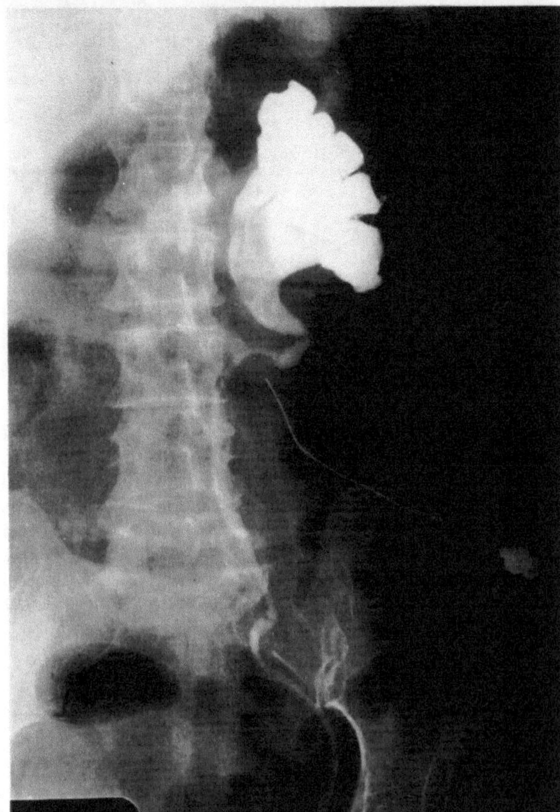

b

Fig. 38.4. (Case 16.) **a** CT scan showing dislocation of the upper third of the left ureter. **b** Left transcutaneous pyeloureterogram showing the needle placed in the lumbar periureteral tissue at the level of the dislocated ureteral tract. Aspiration was consistent with metastasis from bronchogenic adenocarcinoma.

Cytological and Histological Findings

The cytological and histological findings, along with the clinical and radiological findings, are summarised in Table 38.1.

Periureteral metastasis of poorly differentiated pancreatic carcinoma (Case 1) showed a dirty background with slight eosinophilic necrobiotic granular material, commixed with neoplastic naked nuclei and with loose syncytia or sheets of degenerated malignant cells (Fig. 38.5). The nuclei were characterised by high polymorphism and an irregular profile with folds and protrusions. The chromatin pattern progressed from fine to coarse. The cytoplasm had a polygonal, columnar configuration.

Columnar cells with acinic disposition were also observed in aspirate from poorly differentiated carcinoma of the rectum metastatic to periureteral tissue (Case 2). The malignant cells had well-presented cytoplasm, large nuclei of uneven size and configuration, granular hyperchromatic chromatin pattern and large nucleoli.

Haemorrhagic, necrobiotic background with high cellularity was found in aspirated specimens from metastatic adenocarcinoma of endometrial origin (Cases 3 and 10). Lymphocytes, neutrophils and diffuse nuclear debris were scattered throughout the smears. The malignant cells assumed mainly a glandular configuration, showing a good adherence to each other (Fig. 38.6). However, many naked nuclei were also observed commixed with the erythrocytes because of the tumour diathesis. The malignant elements sometimes presented their honeycomb disposition, but the cytoplasm was ill-defined for its degeneration. The nuclear shape was

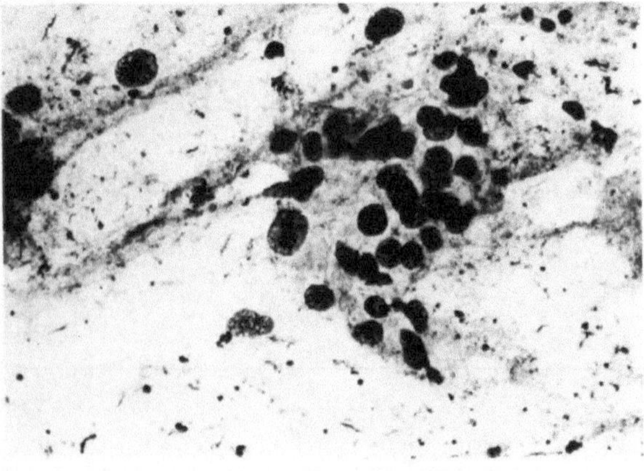

Fig. 38.5. Necrobiotic granular background containing a loose sheet of polymorphous cells with dark hyperchromatic nuclei. (H & E, × 76)

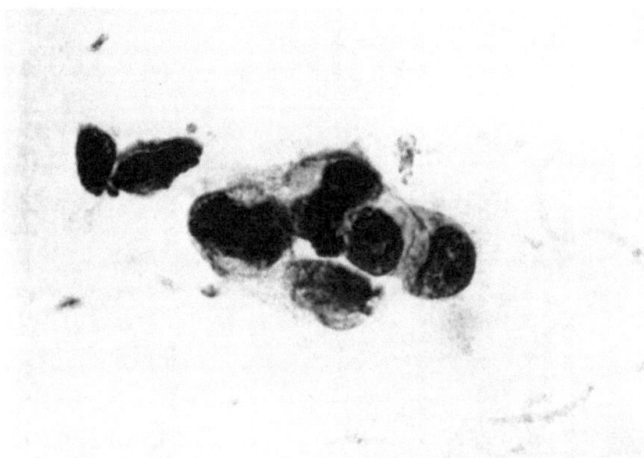

Fig. 38.6. A small group of columnar cells with eccentric nucleus and abundant frothy cytoplasm. Hyperchromasia is not striking. Irregular huge nucleoli are seen in some nuclei. (H & E, × 152.) (Luciani et al. [6]; reproduced by kind permission of the Editor of *Cancer* and the publisher, J.B. Lippincott Co., Philadelphia)

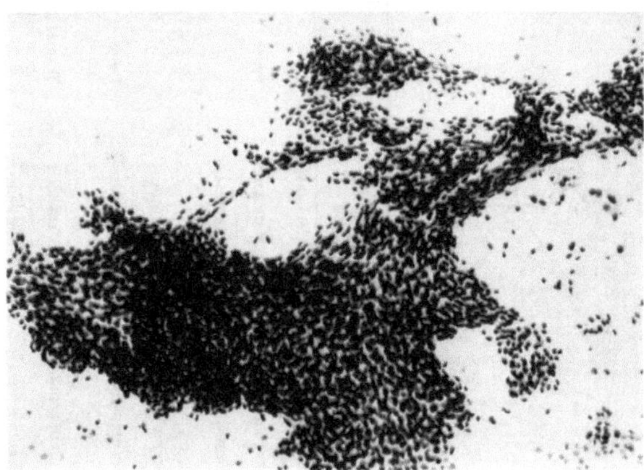

Fig. 38.8. High cellular sample of overlapping atypical elements characterised the periureteral metastases from poorly differentiated squamous cell carcinoma. (H & E, × 61)

polygonal or oval with smooth outline and opaque cribriform chromatin texture. Because of this intranuclear granulated structure the eosinophilic nucleoli were easy to recognise and showed a slight variation in size and position. Histological examination of the metastasis revealed a glandular proliferative pattern with cuboidal and columnar cells arranged in tubular formations of varied size and shape (Fig. 38.7).

Hyperchromatic, atypical cells arranged in small papillary clusters of four to eight elements characterised the aspiration biopsy of metastatic endocervical adenocarcinoma (Case 4). The abnormal cells were relatively uniform in configuration and small with chromocentres, but otherwise showed bland chromatin pattern on higher magnification.

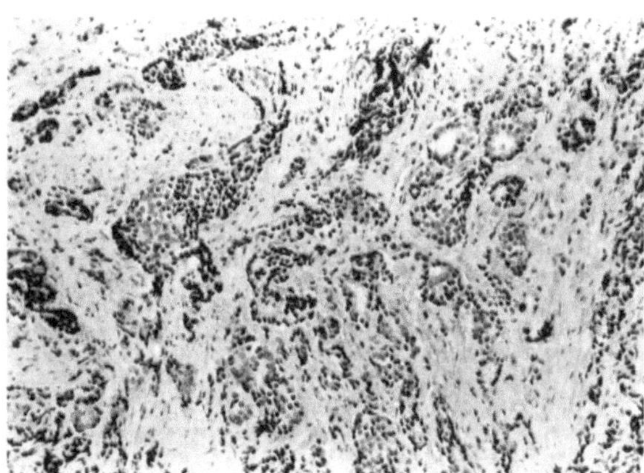

Fig. 38.7. Histological section of periureteral metastases from endometrial cancer. Islands and tubular formations of polygonal cell are evident. (H & E, × 61)

Abnormal nucleoli were disclosed in most nuclei. The cytoplasm had well-defined borders, stained eosinophilic and showed columnar configuration.

High cellular specimens were obtained from periureteral metastases of poorly differentiated squamous cell carcinoma of the cervix (Case 6). Marked pleomorphism and cellular sheets gave a tissue-like appearance to the smear and the clusters were typically branching (Fig. 38.8). The nuclei had the tendency to overlap and demonstrated a high degree of degeneration with smooth and thin borders and reticular chromatin provided by chromocentres, but they had only a few nucleoli. The cytoplasm was weakly eosinophilic. Many naked nuclei with remarkable polymorphism and sometimes central vacuoles were scattered on the background. Scanty inflammatory elements were present. The ureteral wall was infiltrated diffusely by atypical cells arranged in solid, compact nests (Fig. 38.9).

Well-differentiated cervical keratinising squamous carcinoma metastasised to periureteral tissue was easy to identify because of the histological finding of typical cells with dense orangeophilic cytoplasm and hyperchromatic picnotic nuclei (Case 7). The background was dirty for diffuse inflammatory cells and nuclear debris. Neoplastic cells were predominantly small in size and mainly isolated. The cytoplasm was keratinising, dense, with a tail or bizarre shape. The most important feature was the pearl disposition of the cancer cells indicating squamous origin (Fig. 38.10). The nuclei were dark and the nuclear structure was not easy to recognise. The nucleoli were difficult to identify due to nuclear hyperchromasia.

Table 38.1. Clinical, radiological, cytological and histological findings in 22 patients with known malignancy and ureteral obstruction

No.	Name	Age	Sex	Primary lesion	Previous radio-therapy	Excretory urography	CT scan	Guidance	Cytology	Treatment	Histology	Complication
1	G.M.	64	F	Poorly differentiated pancreatic carcinoma	NP	Non-functioning left kidney	Enlargement of the left renal pelvis	Retrograde pyelography under fluoroscopy	Positive	Left nephroureterectomy (pyonephrosis and urosepsis)	Metastasis	
2	G.M.	61	F	Poorly differentiated adenocarcinoma of the rectum	NP	Non-functioning right kidney, hydronephrosis		Antegrade pyelography under fluoroscopy	Positive	Left pyelostomy	Metastasis	Fever
3	D.E.	63	F	Endometrial adenocarcinoma	Yes	Delayed nephrogram and enlargement of the renal pelvis	Retroperitoneal lymph node enlargement. Right hydronephrosis	Antegrade pyelography under fluoroscopy	Positive	Pyelostomy	Metastasis	
4	F.T.	67	F	Moderately differentiated endocervical adenocarcinoma	Yes			Retrograde pyelography under fluoroscopy	Positive	Pyelostomy	Metastasis	
5	E.P.	35	F	Adenocarcinoma of the uterine cervix	Yes	Right hydroureteronephrosis due to a stenosis of the lower part of the right ureter	Right hydroureteronephrosis	Antegrade pyelography under fluoroscopy	Positive	Ureterocutaneostomy	Metastasis	Fever
6	G.E.	64	F	Poorly differentiated epidermoid carcinoma of the uterine cervix	Yes		Right hydroureteronephrosis	Antegrade pyelography under fluoroscopy	Positive	Right ureterocutaneostomy	Metastasis	Transient hypotension
7	G.A.	73	F	Well-differentiated epidermoid carcinoma of the uterine cervix	Yes	Bilateral hydronephrosis and non-functioning right kidney	Bilateral hydronephrosis and evidence of recurrence of malignant tumour infiltrating bladder and rectum	Antegrade pyelography under fluoroscopy	Positive	Ureterocutaneostomy	Metastasis	
8	G.S.	62	M	Poorly differentiated gastric cancer	NP	Hydronephrosis		Catheterisation under fluoroscopy	Positive	Ureterocutaneostomy	Metastasis	
9	T.M.	56	F	Adenocarcinoma of the uterine cervix	Yes	Non-functioning left kidney. Right hydronephrosis. Obstruction of distal right ureter		Catheterisation under fluoroscopy	Negative	1. Right ureterolysis with intraperitoneal omentoplasty 2. Right nephrostomy (ureteral obstruction after 6 months)	Fibrosis	
10	P.G.	64	F	Poorly differentiated endometrial adenocarcinoma	Yes	Signs of previous right nephrectomy. Hydronephrosis of the left kidney	Mass infiltrating the lower portion of the left ureter	Antegrade pyelography under fluoroscopy	Positive	Ureterocutaneostomy	Metastasis	
11	T.E.	73	M	Poorly differentiated prostatic adenocarcinoma	Yes	Bilateral hydronephrosis		Antegrade pyelography under fluoroscopy	Positive	Bilateral ureterocutaneostomy	Metastasis	Periureteral extravasation of contrast material. Incidental injection of air in the ureteral lumen
12	C.M.	64	M	Poorly differentiated prostatic adenocarcinoma	NP	Probable involvement of the bladder wall with obstruction of the meatus of the left ureter. Non-functioning left kidney	Left hydroureteronephrosis	Antegrade pyelography under fluoroscopy	Positive	Nephrostomy	Metastasis	
13	B.A.	69	F	Poorly differentiated epidermoid carcinoma of the uterine cervix	Yes	Not performed because the patient had already had a left pyelostomy and was readmitted for anuria		Antegrade pyelography under fluoroscopy	Positive	Bilateral nephrostomy	Metastasis	

Table 38.1 (*continued*)

No.	Name	Age	Sex	Primary lesion	Previous radio-therapy	Excretory urography	CT scan	Guidance	Cytology	Treatment	Histology	Complication
14	V.N.	75	F	Invasive urothelial carcinoma of the urinary bladder	Yes	Hydroureteronephrosis with no opacification of the bladder	Infiltrating tumour of the bladder with hepatic metastases	Antegrade pyelography under fluoroscopy	Positive	Pyelostomy	Metastasis	
15	B.L.	71	M	Invasive urothelial carcinoma of the urinary bladder	Yes		Infiltrating bladder tumour with adrenal metastasis	Intravenous pyelography under fluoroscopy	Negative	1. Ureterocutaneostomy 2. Left pyelostomy due to necrosis of distal ureteral stump after 1 month	Fibrosis	
16	G.G.	60	F	Undifferentiated pulmonary adenocarcinoma	NP	NP	Left hydroureteronephrosis	Antegrade pyelography under fluoroscopy	Positive	Pyelostomy	Metastasis	
17	Z.C.	39	F	Poorly differentiated squamous cell carcinoma of the uterine cervix	Yes	NP	Bilateral hydroureteronephrosis. Retroperitoneal mass involving aorta, cava, left common internal and external iliac nodal chains and ureters	Double J ureteral stent under fluoroscopy	Positive	Left ureterocutaneostomy. Catheterisation of the right ureter		
18	P.A.	61	F	Poorly differentiated squamous cell carcinoma of the uterine cervix	Yes	Bilateral hydroureteronephrosis from distal obstruction	Bilateral hydroureteronephrosis. Retrovesical mass	Antegrade pyelography under fluoroscopy	Positive	Bilateral ureterocutaneostomy		
19	M.M.	69	F	Infiltrating bladder carcinoma	NP	Distal obstruction of left ureter. Non-functioning right kidney	Bilateral hydroureteronephrosis. Mass involving uterus, bladder, ureters	Antegrade pyelography under fluoroscopy	Positive	Bilateral ureterocutaneostomy		
20	T.M.	54	F	Poorly differentiated squamous cell carcinoma of the uterine cervix	Yes	Non-functioning left kidney	Left hydroureteronephrosis. Mass involving left ureter bladder, vagina	CT	Positive	Left ureterocutaneostomy	Recurrence of poorly differentiated squamous cell carcinoma	
21	D.I.	69	F	Endometrial adenocarcinoma	Yes	Right hydroureteronephrosis, vesicovaginal fistula	NP	Antegrade pyelography under fluoroscopy	Inflammatory	Bilateral ureterocutaneostomy	Post-irradiation fibrosis	
22	M.E.	73	M	Infiltrating bladder carcinoma	Yes	Non-functioning right kidney. Left hydroureteronephrosis	Bilateral hydroureteronephrosis. Neoplasm infiltrating the right posterolateral bladder wall and extending to perivesical tissue	CT	Positive	Left ureterocutaneostomy		

NP, not performed.

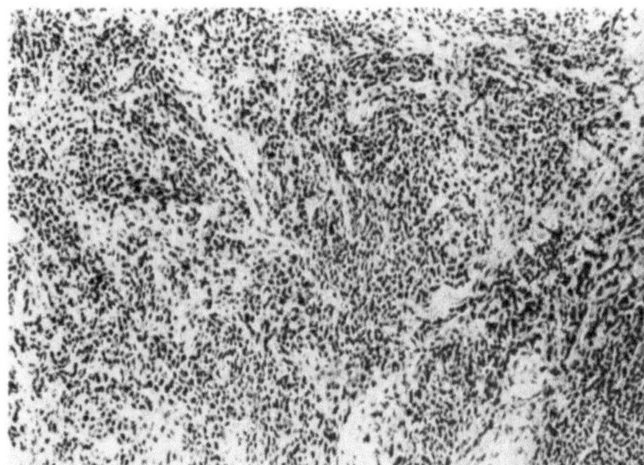

Fig. 38.9. Histological findings from periureteral metastatic poorly differentiated squamous cell carcinoma showing neoplastic solid nests without any tendency to differentiate, which repeat the cytological pattern of the aspirate. (H & E, × 61)

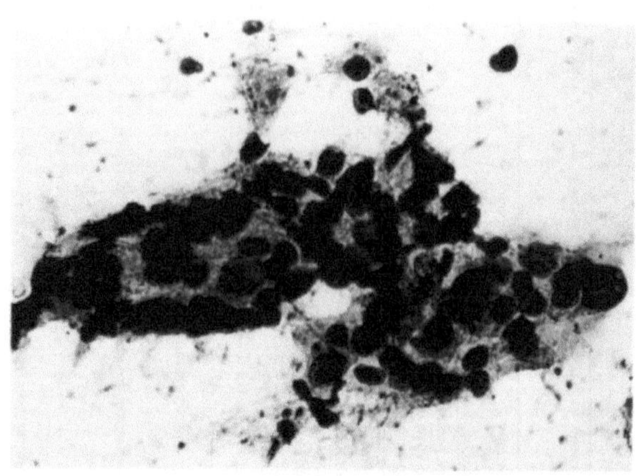

Fig. 38.11. Aspirate from periureteral involvement of prostatic adenocarcinoma shows glandular arrangement of neoplastic cells together with dirty background. (H & E, × 152)

The aspirate of gastric adenocarcinoma metastatic to periureteral tissue (Case 8) contained papillary clusters or single anaplastic cells with a high degree of degeneration. Irregular nuclear configuration, stripped nuclei and images of "cell in cell" arrangement were frequently seen.

Aspiration cytology from periureteral metastases of prostatic cancer (Cases 11 and 12) was made up of a mixture of necrotic material and damaged leucocytes with large, dense clusters of malignant cells. The wide glandular clusters and the sheets of polygonal cells showed high nuclear pleomorphism and pale eosinophilic cytoplasm with indistinct borders (Fig. 38.11). Generally, the nuclei showed a fine granular chromatin texture, slight hyperchromasia and prominent large single or multiple nucleoli (Fig. 38.12). In the histological sections these malignant elements assumed a tubular pattern with nuclear crowding (Fig. 38.13).

Scanty malignant cell sheets were obtained from bladder urothelial carcinoma metastatic to periureteral tissue (Cases 14, 19 and 22). Cancer cells showed a great variability in shape and size, particularly when they occurred singly. Dark, irregular hyperchromatic nuclei were found, and the cytoplasm was well preserved and stained eosinophilic.

Bronchogenic adenocarcinoma metastatic to periureteral tissue showed high cellular aspiration specimens (Case 16). The background was clear without tumour diathesis, and most malignant cells were loosely arranged. Poor cellular adherence, increased nuclear cytoplasm ratio and large irregular eosinophilic nucleoli were the most important features. Cellular signet-ring appearance may be

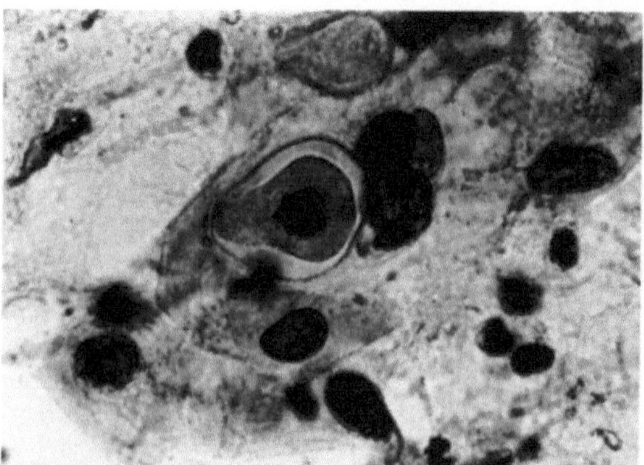

Fig. 38.10. Cannibalism and keratinising cytoplasm indicate well-differentiated squamous cell carcinoma in the aspirate. (H & E, × 244)

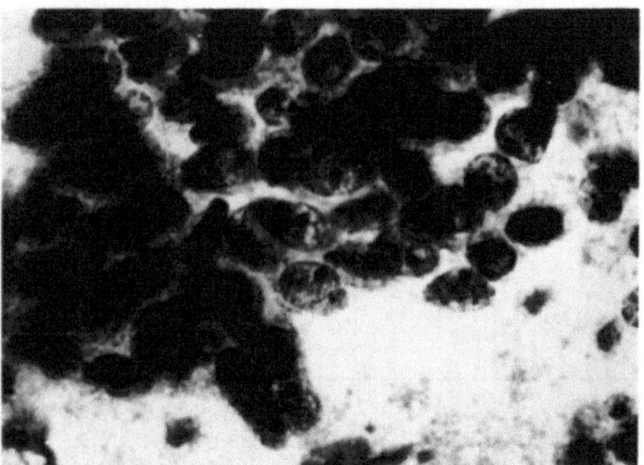

Fig. 38.12. The same sample on higher magnification reveals irregular chromatin texture and prominent single nucleoli. (H & E, × 244)

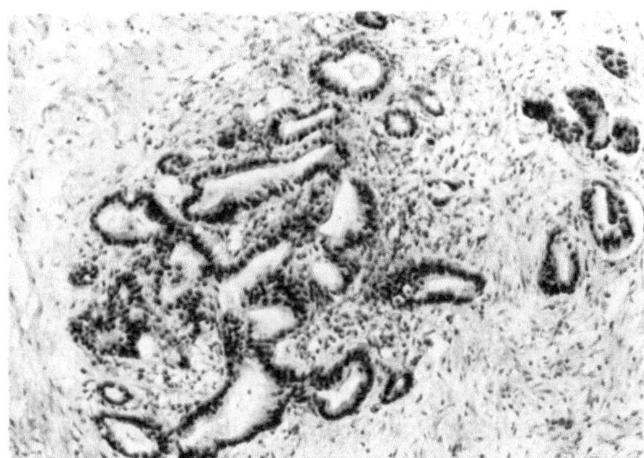

Fig. 38.13. Histological section of periureteral metastases of prostatic carcinoma. (H & E, × 61)

found. Very fine cytoplasmic vacuoles were more common than large ones. Atypical mitotic figures, polinucleated malignant elements and "intranuclear" vacuoles may also occur. The chromatin was finely granular without clumps along the nuclear borders.

Three patients affected by urothelial cancer of the urinary bladder and by endocervical carcinoma presented periureteral stricture caused by inflammatory reactive tissue.

The aspiration cytology smear obtained from periureteral benign stenosis showed a serous eosinophilic background crowded by lymphocytes, polymorphonuclear leucocytes, macrophages and few erythrocytes. No necrotic substance was observed. The identifiable fibroblastic cells appear in loose clusters as spindle, well-preserved elements characterised by a high number of mitotic figures but without atypical forms (Fig. 38.14a). These cells displayed an extremely thin, delicate, eosinophilic, elongated cytoplasm without any vacuoles that reached 150–200 μm in length and had the appearance of a tail or as a corkscrew (Fig. 38.14b). The cytoplasmic borders were sharp and sometimes exhibited small projections. The nuclear deposition in these cells was variable, but most frequently they were centrally placed and had a bland appearance. The nuclear texture was extremely fine, uniform and without nuclear border thickness (Fig. 38.14c).

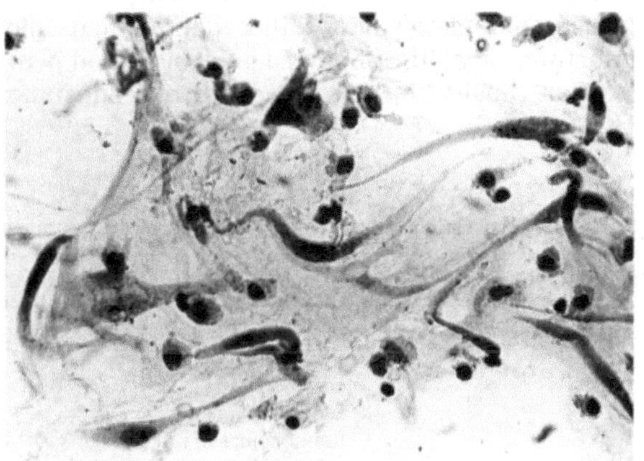

a

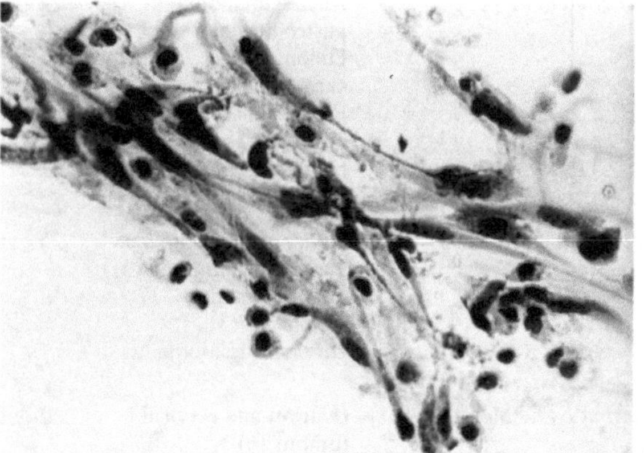

b

c

Fig. 38.14a–c. Aspirate from benign periureteral stenosis contains loose clusters of fibroblastic cells (**a**), sometimes with a corkscrew shape (**b**), and fine uniform nuclear texture (**c**). H & E, × 61)

Results

Of the 22 patients, 19 had malignant aspiration biopsy results. In two patients very enlarged pelvic nodes were the cause of obstruction of the distal ureter (Cases 11 and 12). A retroperitoneal mass of 4 × 5 cm in diameter involving the tail of the pancreas and the left pelviureteral junction, was disclosed in Case 1. A diagnosis of retroperitoneal fibrosis was made in three patients (Cases 9, 15 and 21), while a periureteral metastatic spread was found in the remaining cases.

Complications

Minimal discomfort for the patients was experienced in all cases and severe complications did not occur. In two cases a small periureteral extravasation of contrast medium occurred, which cleared completely in 24 h without other further consequences. Fever lasting 24–48 h appeared in three patients. It is possible that this might have been due to the visualisation manoeuvres rather than to the needle puncture. Given the improved guidability and penetration ability of the needle used, adequate material was obtained quickly and promptly in a high percentage of cases with a noteworthy reduction of the exposure of both the patient and physician to direct radiation.

Discussion

The occurrence of ureteral obstruction in patients with known cancer is generally considered a terminal event which dramatically influences the prognosis [7]. Nevertheless, in some cases stenosis of the ureters is secondary to radiation or surgical therapy and is caused by retroperitoneal fibrosis or other complications (tubo-ovarian abscess or lymphocysts), which are especially frequent in the man-

Table 38.2. Aspiration biopsy cytology in diagnosing ureteral strictures as reported in literature[a]

Reference	Primary lesion	Needle	Guidance	Cytology	Histology
Göthlin and Barbaric [18]	Prostatic carcinoma (1) Irradiation for bladder carcinoma (1) Seminoma (1) Renal failure of unknown origin (1)	Not specified	Fluoroscopy	Metastases Metastases Metastases Lymphocytic lymphoma	NP
Freiman et al. [19]	Adenocarcinoma of gastrointestinal tract (2) Endometrial carcinoma (1) Prostatic carcinoma (1) Transitional cell carcinoma of bladder (1)	23-gauge Chiba	Fluoroscopy		NP
Meier et al. [20]	Periureteral masses (3)	22-gauge Chiba	Fluoroscopy	3 metastases	NP
Wein et al. [15]	Bladder carcinoma (4) Prostatic carcinoma (3) Gastrointestinal neoplasms (3) Cervical carcinoma (2)	23-gauge Chiba	Fluoroscopy	6 positive 5 negative 1 false negative	Exploratory surgery performed in only 4 of the 6 negative cases
Barbaric and McIntosh [21]	Ovarian and cervical tumour (6) Prostatic tumour (4) Tumour of colon and appendix (3) Tumour of lung and breast (5) Lymphoma (2) Tumour of kidney (1) Tumour of testis (1) Transitional cell carcinoma (1) Stricture (4) Appendiceal mucocele (1) Unknown (3)	22-gauge Chiba	Fluoroscopy	18 true positive 7 true negative 4 false negative 3 inadequate	Exploratory surgery performed in negative cases and in half of those with positive aspiration biopsies

NP, not performed.
[a] Luciani et al. [6]; reproduced by kind permission of the Editor of *Cancer* and the publisher, J. B. Lippincott Co., Philadelphia.

agement of tumours of gynaecological interest [2, 8–14].

The exact knowledge of the nature of the obstruction is vital in determining the planning of an adequate therapeutic strategy. However, the true nature of a ureteral obstruction is rarely depicted by current imaging methods, especially when the lesion is not large enough to be promptly visualised on a CT scan, nor is urinary cytology very helpful, since malignant cells are an unusual finding in urinary sediment [15, 16]. Exploratory laparotomy, with direct biopsies of the lesion, seems to be an extreme approach, especially in patients with advanced pelvic cancer previously treated by irradiation or chemotherapy who develop metastatic ureteral obstruction, since they are the worst candidates for open surgical procedure because of the high morbidity [17].

Although few experiences with ABC have been reported in the literature (Table 38.2), it is our impression that this procedure can ensure adequate material for a prompt and correct diagnosis of the nature of such ureteral strictures. Our data demonstrate that the aspiration of the periureteral tissue in patients with ureteral stenosis complicating known cancer is a safe, reliable, minimally invasive method helpful in differentiating metastatic, recurrent or persistent disease from inflammatory and radiation changes.

The accuracy of the cytological diagnosis obviously depends on the presence of adequate cytological material in the aspirate, its proper preparation and the diagnostic ability of the cytopathologist. Unless aspirations are inadequate because of difficulties experienced during the procedure, the problem of a non-representative sampling is diminished with the use of a new needle with improved aspiration ability. On the other hand, the danger of a false positive diagnosis should never be underestimated, since most patients undergo irradiation, chemotherapy or surgery, which instigate inflammatory or reparative processes. Spindle, stellate, or fibroblastic cells may simulate malignant cells; however, the absence of nuclear chromatin abnormalities and the knowledge of the morphology of the primary cancer are the most important indications to differentiate inflammatory changes from malignancy.

In conclusion, we can say that ABC is an effective, safe, accurate procedure for the diagnosis of the nature of ureteral obstruction in patients with known malignancy. On the basis of ABC results clinicians can decide upon the best treatment for their patients, by evaluating the need for open surgical procedures in each case individually.

References

1. Brin EN, Shiff M Jr, Weis RM (1975) Palliative urinary diversion of pelvic malignancy. J Urol 113:619–622
2. Persky L, Kursh ED, Feldman S (1978) Extrinsic obstruction of the ureter. In: Harrison HJ, Gitter RF, Perlmutter AD, Stamey TA, Walsh PL (eds) Campbell's urology, 4th edn. Saunders, Philadelphia, pp 438–440
3. Ortlip SA, Fraley EE (1982) Indications for palliative urinary diversion in patients with cancer. Urol Clin North Am 9 (1):79–84
4. Dean RJ, Lytton B (1978) Urologic complications of pelvic irradiation. J Urol 119:64–67
5. Hillman BJ, Clark RL, Babbitt G (1984) Efficacy of the excretory urogram in the staging of gynecologic malignancies. AJR 143:997–999
6. Luciani L, Scappini P, Pusiol T, Piscioli F (1987) The role of aspiration biopsy cytology in the management of ureteral obstruction in patients with known cancer. Cancer 59:1936–1946
7. Zubler MA (1984) Genitourinary complications of cancer. In: Smith FE, Lane M (eds) Medical complications of malignancy. Wiley, New York, pp 65–72
8. Welch JS, Pratt JH, Symmonds RE (1961) The wertheim hysterectomy for squamous cell carcinoma of the uterine cervix. Am J Obstet Gynecol 81:978–987
9. Green TH, Meigs JV, Ulfelder H, Curtin RR (1962) Urologic complications of radical wertheim hysterectomy. Incidence, etiology, management and prevention. Obstet Gynecol 20:293–312
10. Christensen A, Zange P, Nielsen E (1964) Surgical and radiation treatment of invasive carcinoma of the uterine cervix. Acta Obstet Gynecol Scand 43:60–87
11. Decker DG, Smith RA (1968) Sequential radiation therapy and surgery for stage I and stage II cancer of the cervix. Am J Roentgenol Radium Ther Nucl Med 102:152–160
12. Knipper S (1972) Stènoses urétérales après irradiation à haute énergie pour cancer du col utérin. Acta Radiol 32:110–132
13. Kaplan AL (1977) Post-radiation ureteral obstruction. Obstet Gynecol Surv 32:108
14. Zerbib M, Teyssier P, Steg A (1983) Les stènoses urétérales après traitment des cancers du col utérin: fibrose post radiothérapique ou récidive neoplastique? J Chir (Paris) 120:503–513
15. Wein AJ, Ring EJ, Freiman AB, Oleaga JA, Carpiniello VL, Banner MP, Pollack HM (1979). Applications of thin needle aspiration biopsy in urology. J Urol 121:626–629
16. Luciani L, Aldovini D, Piscioli F, Pusiol T, Polla E, Menichelli E (1983) Cytologic detection of ureteral metastasis from breast carcinoma causing peripelvic extravasation. Urology 22:56–58
17. Jones CR, Woodhouse CRJ, Hendry WF (1984) Urological problems following treatment of carcinoma of the cervix. Br J Urol 56:609–613
18. Göthlin JH, Barbaric ZL (1978) Fluoroscopy-guided percutaneous transperitoneal fine-needle biopsy of renal masses. Urology 11:300–302
19. Freiman DB, Ring EJ, Oleaga JA, Carpiniello VL, Wein AJ (1978) Thin needle biopsy in the diagnosis of ureteral obstruction with malignancy. Cancer 42:714–716
20. Meier WL, Willscher MK, Novicki DE, Pischinger RJ (1979) Evaluation of perihilar and central renal masses using the Chiba needle. J Urol 121:414–416
21. Barbaric ZL, MacIntosh PK (1981) Periureteral thin-needle aspiration biopsy. Urol Radiol 2:181–185

Chapter 39

Percutaneous Bone Biopsy

J. Zornoza

Treatment of the various malignant diseases arising primarily from the genitourinary system has undergone considerable change in recent decades and, in many areas, remains in a state of flux or controversy. However, as increasing cooperative efforts have developed among many disciplines (urology, pathology, radiology, radiotherapy and oncology), the emergence of multimodal therapy has significantly improved the outlook for patients with many of these diseases.

There is no universally accepted method of clinical staging, but the TNM system is one of the most popular. The M stands for distant metastases which are absent (M0) or present (M1, M2, M3). The presence of bone metastases can be determined by nuclear bone scan or metastatic bone survey. Osseous metastases constitute the most common form of haematogenous metastases from prostatic carcinoma.

Bone lesions were among the first pathological abnormalities to be histologically diagnosed by percutaneous biopsy. Since the original reports three factors have contributed to the successful evolution of the bone biopsy procedure: the development of new instruments specifically designed for bone biopsy, the introduction of fluoroscopy guidance for accurate needle placement and the recognition by the medical community of the safety and accuracy of the biopsy procedure and its applicability to almost any part of the skeleton [1–3].

The need for histological diagnosis prior to the treatment of any pathological bone process is a well-recognised principle of therapy [4]. Although the value of percutaneous needle biopsy of bone has been supported many times in the literature [4–27], it has been practised in only a few centres in the USA. In most cases, tissue diagnosis of bone lesions is based on material obtained by surgical biopsy. This may be the result of a lack of information as to the true nature of percutaneous needle biopsy and a lack of confidence in the pathological evaluation of the material. Percutaneous bone biopsy is composed of three elements, each one a separate procedure in itself. First, an aspiration biopsy is performed for cytological evaluation, second, a core biopsy is performed for histopathological examination and third, material is obtained for culture and microbiological evaluation. The combination of the cytological features and those of the sections obtained from the core of tissue gained in percutaneous biopsy is usually comparable to tissue obtained by surgical biopsy.

As a rule, any bone lesion may be subjected to percutaneous aspiration biopsy and should be first approached by this route because the procedure involves minimal discomfort and complications and is less costly than surgery. Metastatic disease to bone has been and remains the most common indication for percutaneous biopsy.

The percutaneous needle biopsy procedure is especially preferable to that of open biopsy for areas that are difficult to reach surgically, such as the spine and certain regions located deep in the pelvis. The surgical technique is cumbersome and carries a significant morbidity. Percutaneous needle biopsy is simpler, faster and safer [28–31]. The clavicle and other bones with relatively poor vascularity are also well suited for needle biopsy. Open biopsy of these bones can lead to poor healing, infection and other complications. The use of percutaneous biopsy instead of open biopsy is even more important in cases in which the lesion itself comprises the mechanical stability of the bone involved. In such cases, the surgical removal of tissue may lead to a critical weakening of the bone and a pathological fracture, with all the attendant complications. Percutaneous needle biopsy is also an excellent technique for the evaluation of sudden changes of bone texture, such as osteoporosis and vertebral collapse, because the radiographical changes are non-specific and multiple biopsies may be required.

Traditionally, metastatic disease has been the primary indication for percutaneous needle bone biopsy. Even in a case of clear clinical evidence of a relationship between the presence of a neoplasm and the appearance of a destructive bone lesion, a histological diagnosis is necessary [4]. Percutaneous needle biopsy can provide a sample of tissue with which to establish a definite diagnosis.

Percutaneous needle bone biopsy is indicated for the following groups of patients:

1. Patients with known multiple primary tumours who present with a bone lesion. Percutaneous bone biopsy is helpful in determining the type of tumour that is causing the bone lesion, thereby allowing proper therapy to be established.

2. Patients for whom there is a need to determine whether viable tumour cells are present in a radiographically stable metastatic bone lesion. In patients with carcinoma of the prostate, tumorous cells have been found by percutaneous biopsy after several years of continuous chemotherapy and occasional radiation therapy of the lesion. The pres-

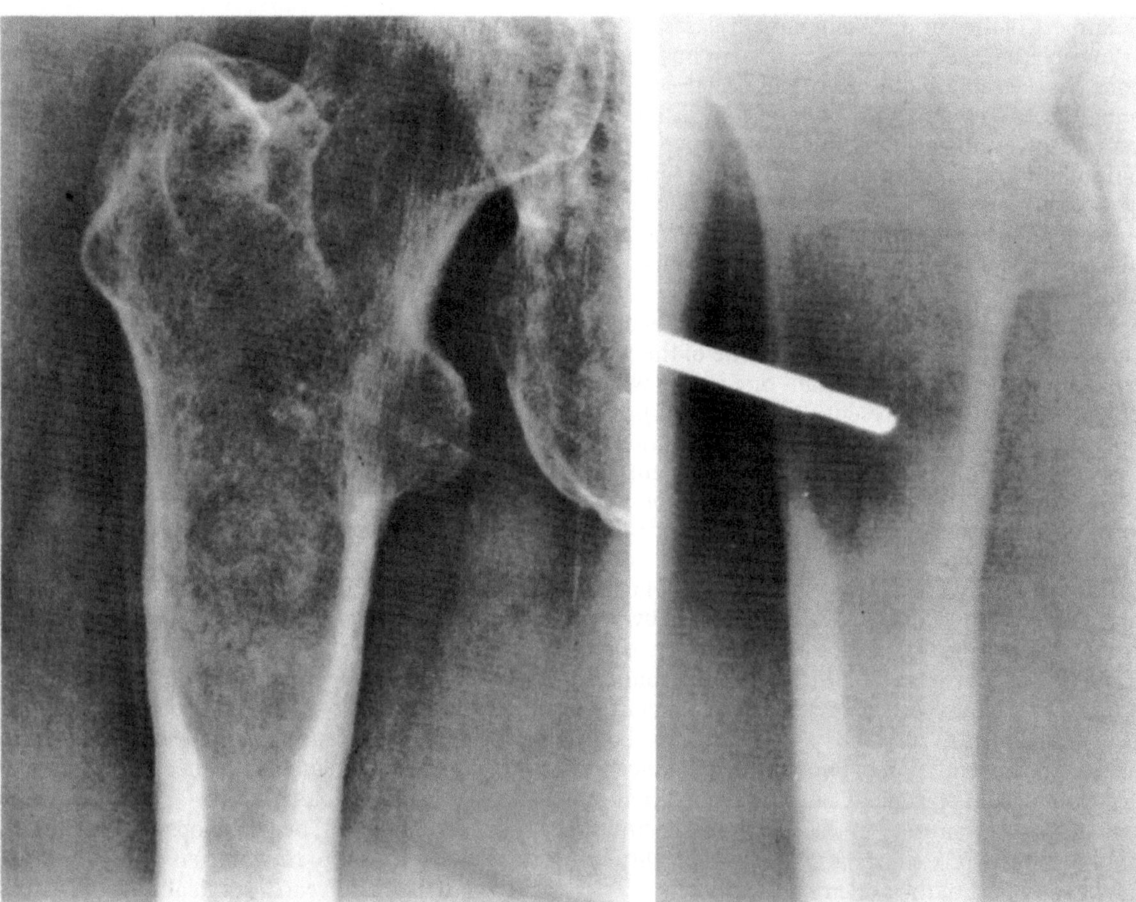

Fig. 39.1a, b. An elderly male with prostatic carcinoma. **a** Posteroanterior radiograph of the femur shows an osteolytic lesion involving the proximal diaphysis. **b** Percutaneous bone biopsy revealed metastatic carcinoma.

ence of dormant tumour tissue in such patients may be the cause of the recurrence of metastatic disease after many years of apparently complete remission.

3. Patients with a known primary tumour and a positive radionuclide bone scan. A positive radionuclide scan is not always an indication of metastatic disease. A normal congenital variation or the result of a previous injury can cause an increased uptake in the area in question. When it is difficult to determine by clinical or radiographic means whether the lesion is metastatic or benign, percutaneous needle biopsy is indicated.

4. Patients with a previously known primary tumour which has been in clinical remission for several years, who present with a new bone lesion. Percutaneous needle biopsy can determine whether a new bone lesion represents metastatic disease or a new primary.

5. Patients with a known primary tumour who develop a bone lesion which does not resemble the metastatic bone involvement usually seen for that particular primary tumour. Examples of this are the presence of a purely lytic bone lesion in a patient with known prostatic carcinoma (Fig. 39.1). In such cases percutaneous needle biopsy is an extremely useful method for determining the true nature of the metastatic deposit and either establishing or eliminating the possibility of a second primary tumour.

6. Patients with no known primary tumour who present with a bone lesion characteristic of metastatic disease. Traditionally such patients undergo multiple tests in an attempt to determine the site of origin. The yield of these studies is rather low and the primary tumour frequently remains undiscovered. At our institution, percutaneous needle biopsy is the primary tool used in the evaluation of such patients. At times the biopsy fails to indicate the site of origin. Nevertheless, the pathological diagnosis (adenocarcinoma or squamous carcinoma) limits the search for the primary tumour to certain organs and therefore shortens the diagnostic process.

7. Patients with Paget's disease and a lytic lesion. The sensation of pain accompanying the appearance of a lytic lesion in a region of Paget's disease usually indicates malignant degeneration, particularly osteosarcoma. Metastatic disease to an area of Paget's disease can be clinically and radiographically indistinguishable from malignant degeneration. In such cases, percutaneous needle biopsy can determine the correct diagnosis (Fig. 39.2).

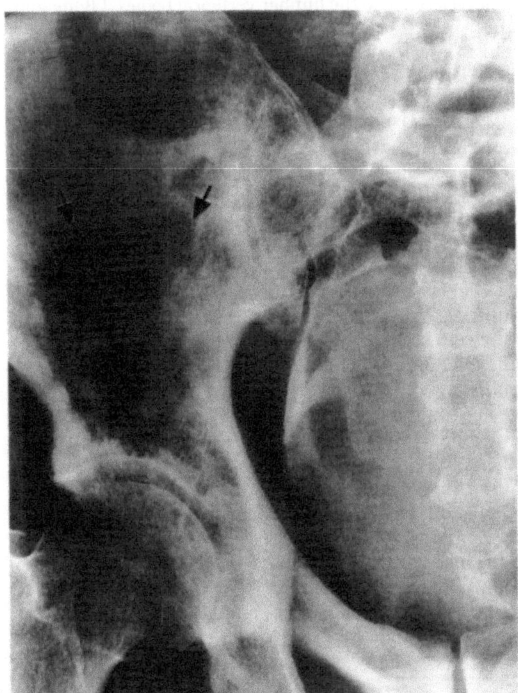

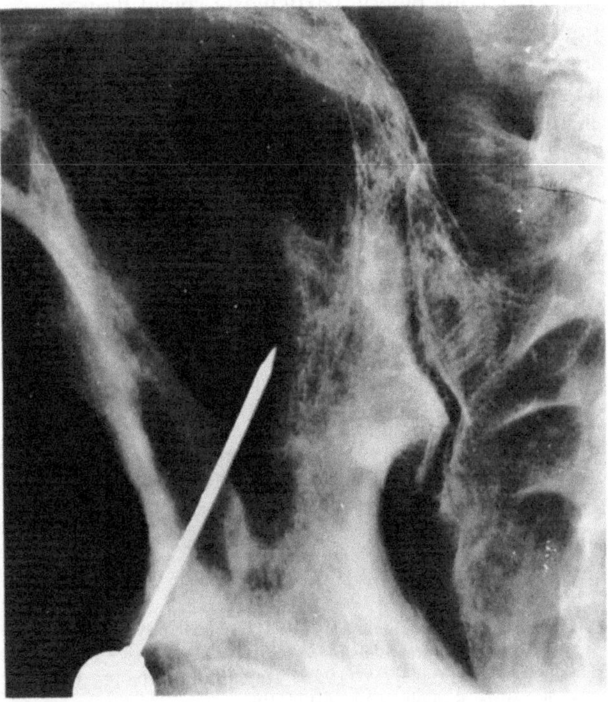

Fig. 39.2a, b. A 73-year-old male with Paget's disease and prostatic carcinoma. **a** Anteroposterior radiograph of the pelvis shows an osteolytic lesion of the right iliac bone (*arrows*). **b** Percutaneous bone biopsy revealed metastatic adenocarcinoma.

8. Patients who are at high surgical risk, but whose clinical situation calls for the evaluation of bone tissue.

There are few contraindications to the performance of percutaneous needle biopsy of bone. A percutaneous biopsy should not be performed on a patient with a high risk of haemorrhage. A minimal count of 40 000 platelets is necessary before the procedure may be safely undertaken. The risk of infection is not a deterrent, provided the technique is carefully followed.

Pain is the most common complication of percutaneous bone biopsy and is usually associated with the negative pressure created in the medullary cavity when suction is applied to retrieve the tissue sample. The introduction of the needle in spinal biopsies may also cause some pain.

Bleeding is another potential complication of percutaneous bone biopsy, particularly biopsy of areas containing large vessels. However, its occurrence is quite low.

Neurological damage is the most serious complication of percutaneous bone biopsy that has been reported [7, 8, 32–36]. Some instances of foot-drop, caused by damage to either the sciatic nerve or roots of the spine, have occurred after lower lumbar biopsies [37].

It should be stressed that percutaneous bone biopsy, although relatively uncomplicated, safe and valuable, is a procedure that requires expertise. Percutaneous bone biopsy, particularly spinal biopsy, should always be performed by an experienced radiologist, skilled in the techniques, knowledgeable of the factors and familiar with the potential complications and precautionary measures associated with this procedure.

References

1. Lalli AF (1970) Roentgen-guided aspiration biopsies of skeletal lesions. J Can Assoc Radiol 21:71–73
2. Lalli AF (1972) The direct fluoroscopically guided approach to renal, thoracic, and skeletal lesions. Curr Probl Radiol 2:30–41
3. Pepe RG, Lalli AF (1976) Percutaneous aspiration bone biopsy by fluoroscopic guidance. Cleve Clin Q 43:77–83
4. Adler O, Rosenberger A (1979) Fine needle aspiration biopsy of osteolytic metastatic lesions. AJR 133:15–18
5. Collins JD, Bassett L, Main GD, Kagan C (1979) Percutaneous biopsy following bone scans. Radiology 132:439–442
6. Cramer LE, Kuhn C, Stein AH (1964) Needle biopsy of bone. Surg Gynecol Obstet 118:1253–1256
7. Debnam JW, Staple TW (1975) Trephine bone biopsy by radiologists. Radiology 116:607–609
8. Debnam JW, Staple TW (1975) Needle biopsy of bone. Radiol Clin North Am 13:157–164
9. deSantos LA, Wallace S, Murray JA, Lukeman JM (1978) Percutaneous needle biopsy in the cancer patient. Am J Roentgenol Radium Ther Nucl Med 130:641–649
10. Evarts CM (1975) Diagnostic techniques: closed biopsy of bone. Clin Orthop 107:100–111
11. Fornasier VL, Cameron HU (1975) Techniques of closed bone biopsy. CRC Crit Rev Clin Lab Sci 6:145–155
12. Frankel CJ (1954) Aspiration biopsy of the spine. J Bone Joint Surg [Am] 36:69–74
13. Hanafee WN, Tobin PL (1969) Closed bone biopsy by a radiologist. Radiology 92:605-606
14. Hartman JT, Sombeck JB (1962) Use of the Craig needle for biopsy of bone. Cleve Clin Q 29:200–203
15. Hartman JT, Phalen GS (1967) Needle biopsy of bone. Report of three representative cases. JAMA 200:113–115
16. Legge D, Ennis JT, Dempsey J (1978) Percutaneous needle biopsy in the management of solitary lesion of bone. Clin Radiol 29:497–500
17. McCollister EC (1975) Diagnostic techniques: closed biopsy of bone. Clin Orthop 107:100–111
18. Martin HE, Ellis EB (1934) Aspiration biopsy. Surg Gynecol Obstet 59:578–589
19. Mazet R, Cozen L (1952) The diagnostic value of vertebral body needle biopsy. Ann Surg 135:245–252
20. Ray RD (1953) Needle biopsy of the lumbar vertebral bodies. A modification of the Valls technique. J Bone Joint Surg [Am] 35:760–762
21. Rix RR, Brooks SM (1952) Needle biopsy in bone lesions. N Engl J Med 246:373–375
22. Robertson RC, Ball RP (1935) Destructive spine lesions. Diagnosis by needle biopsy. J Bone Joint Surg 17:749–758
23. Schajowicz F (1955) Aspiration biopsy in bone lesions. J Bone Joint Surg [Am] 37:465–477
24. Schajowicz F, Hokama J (1976) Aspiration (puncture of needle) biopsy of bone lesions. Recent Results Cancer Res 54:439–444
25. Siffert RS, Arkin AM (1949) Trephine biopsy of bone with special reference to the lumbar vertebral bodies. J Bone Joint Surg [Am] 31:146–149
26. Stahl DC, Jacobs B (1967) Diagnosis of obscure lesions of the skeleton: evaluation of biopsy methods. JAMA 201:83–85
27. Stewart FW (1933) The diagnosis of tumors by aspiration. Am J Pathol 9:801–811
28. Crenshaw AH (1971) In: Campbell's operative orthopedics, 5th edn, vol 1. Mosby, St Louis
29. Fang HSY, Ong GB (1962) Direct anterior approach to the upper cervical spine. J Bone Joint Surg [Am] 44:1588–1604
30. Michele AA, Krueger FJ (1949) Surgical approach to the vertebral body. J Bone Joint Surg [Am] 31:837–878
31. Southwick WO, Robinson RD (1957) Surgical approaches to the vertebral bodies in the cervical and lumbar regions. J Bone Joint Surg [Am] 39:631–644
32. Armstrong P, Chalmers AH, Green G, Irving JD (1978) Needle aspiration biopsy of the spine in suspected disc space infection. Br J Radiol 51:333–337
33. Fisher WB (1971) Hazard in bone-marrow biopsy. N. Engl J Med 285:804
34. McLaughlin RE, Miller WR, Miller CW (1976) Quadriparesis after needle aspiration of the cervical spine. Report of a case. J Bone Joint Surg [Am] 58:1167–1168
35. McNutt DR, Fudenberg HH (1972) Bone marrow biopsy and osteoporosis. N Engl J Med 286:46
36. Nagel DA, Albright JA, Keggi KJ (1965) Closer look at spinal lesions: open biopsy of vertebral lesions. JAMA 191:975–978
37. Moore TM, Meyers MH, Patzakis MJ, Terry R, Harvey JP (1979) Closed biopsy of musculoskeletal lesions. J Bone Joint Surg [Am] 61:375–380

Subject Index